SOCIETY FOR EXPERIMENTAL BIOLOGY

SEMINAR SERIES: 33

NEUROHORMONES IN INVERTEBRATES

SOCIETY FOR EXPERIMENTAL BIOLOGY SEMINAR SERIES

A series of multi-author volumes developed from seminars held by the Society for Experimental Biology. Each volume serves not only as an introductory review of a specific topic, but also introduces the reader to experimental evidence to support the theories and principles discussed, and points the way to new research.

1. Effects of air pollution on plants. *Edited by T.A. Mansfield*
2. Effects of pollutants on aquatic organisms. *Edited by A.P.M. Lockwood*
3. Analytical and quantitative methods. *Edited by J.A. Meek & H.Y. Elder*
4. Isolation of plant growth substances. *Edited by J.R. Hillman*
5. Aspects of animal movement. *Edited by H.Y. Elder & E.R. Truman*
6. Neurones without impulses: their significance for vertebrate and invertebrate systems. *Edited by A. Roberts & B.M.H. Bush*
7. Development and specialisation of skeletal muscle. *Edited by D.F. Goldspink*
8. Stomatal physiology. *Edited by P.G. Jarvis & T.A. Mansfield*
9. Brain mechanisms of behaviour in lower vertebrates. *Edited by P.R. Laming*
10. The cell cycle. *Edited by P.C.L. John*
11. Effects of disease on the physiology of the growing plant. *Edited by P.G. Ayres*
12. Biology of the chemotactic response. *Edited by J.M. Lackie & P.C. Williamson*
13. Animal migration. *Edited by D.J. Aidley*
14. Biological timekeeping. *Edited by J. Brady*
15. The nucleolus. *Edited by E.G. Jordan & C.A. Cullis*
16. Gills. *Edited by D.F. Houlihan, J.C. Rankin & T.J. Shuttleworth*
17. Cellular acclimatisation to environmental change. *Edited by A.R.Cossins & P. Sheterline*
18. Plant biotechnology. *Edited by S.H. Mantell & H. Smith*
19. Storage carbohydrates in vascular plants. Edited by *D.H. Lewis*
20. The physiology and biochemistry of plant respiration. *Edited by J.M. Palmer*
21. Chloroplast biogenesis. *Edited by R.J. Ellis*
22. Instrumentation for environmental physiology. *Edited by B. Marshall & F.I. Woodward*
23. The biosynthesis and metabolism of plant hormones. *Edited by A. Crozier & J.R.Hillman*
24. Coordination of motor behaviour. *Edited by B.M.H. Bush & F. Clarac*
25. Cell ageing and cell death. *Edited by I. Davies & D.C. Sigee*
26. The cell division cycle in plants. Edited by *J.A. Bryant & D. Francis*
27. Control of leaf growth. *Edited by N.R. Baker, W.J. Davies & C. Ong*
28. Biochemistry of plant cell walls. *Edited by C.T. Brett & J.R. Hillman*
29. Immunology in plant science. *Edited by T.L. Wang*
30. Root development and function. *Edited by P.J. Gregory, J.V. Lake & D.A. Rose*
31. Plant canopies: their growth, form and function. *Edited by G. Russell, B. Marshall & P.G. Jarvis*
32. Developmental mutants in higher plants. *Edited by H. Thomas & D. Grierson*
33. Neurohormones in invertebrates. *Edited by M.C. Thorndyke & G.J. Goldsworthy*
34. Acid toxicity and aquatic animals. *Edited by R. Morris, E.W. Taylor, D.J.A. Brown & J.A. Brown*
35. The division and segregation of organelles. *Edited by S.A. Boffey & D. Lloyd*

NEUROHORMONES IN INVERTEBRATES

Edited by

M.C. Thorndyke

Department of Biology, Royal Holloway and Bedford New College University of London, Egham, Surrey.

G.J. Goldsworthy

Department of Biology, Birkbeck College University of London

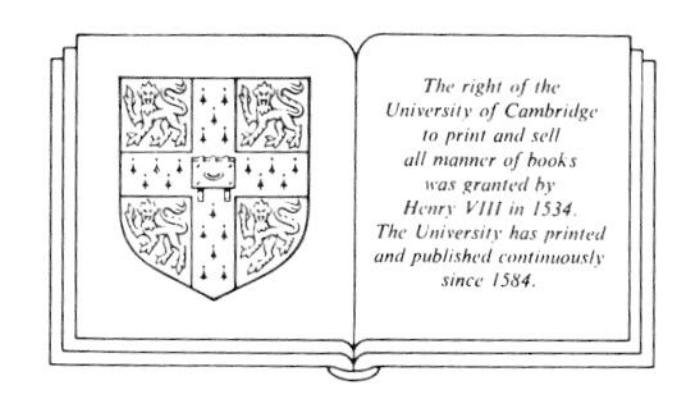

CAMBRIDGE UNIVERSITY PRESS

Cambridge

New York New Rochelle Melbourne Sydney

Published by the Press Syndicate of the University of Cambridge
The Pitt Building, Trumpington Street, Cambridge CB2 1RP
32 East 57th Street, New York, NY 10022, USA
10 Stamford Road, Oakleigh, Melbourne 3166, Australia

First published 1988

Printed in Great Britain at the University Press, Cambridge

British Library cataloguing in publication data

Neurohormones in invertebrates – (Society for Experimental Biology seminar series; 33)
1. Nervous system – Invertebrates
2. Neuropeptides 3. Neurohormones
I. Thorndyke, M.C. II. Goldsworthy,
Graham J. III. Series
592'.0188 QL364

Library of Congress cataloguing in publication data

Neurohormones in invertebrates.
(Seminar series/Society for Experimental Biology; 33)
Includes index.
1. Invertebrates–Physiology. 2. Neuroendocrinology.
I. Thorndyke, M.C. II. Goldsworthy, G.J. III. Series:
Seminar series (Society for Experimental Biology
(Great Britain)); 33
QL364.N49 1987 592'.0188 87–15100

ISBN 0 521 32843 8

DQ

CONTENTS

CONTRIBUTORS

Boer, H.H.
Biologisch Laboratorium, Vrije Universiteit, De Boelelaan 1087, 1007MC Amsterdam, Netherlands.

Cottrell, G.A.
Department of Physiology & Pharmacology, St. Andrew's University, Fife KY16 9TS.

Davies, N.W.
Department of Physiology & Pharmacology, St. Andrew's University, Fife KY16 9TS.

Dhainaut-Courtois, N.
Lab. D'Endocrinologie des Invertebres, Universitè des Sciences et Techniques, Biologie Animale de Lille, 59655 Villeneuve D'Ascq Cedex, France.

Duve, H.
School of Biological Sciences, Queen Mary College, University of London, London E1 4NS.

Ebberink, R.H.M.
Department of Biology, Free University, De Boelelaan 1087, 1081 HV Amsterdam, Netherlands.

Gäde, G.
Institut fur Zoologie IV fur Universitat Dusseldorf, Universitatstr. 1, Federal Republic of Germany.

Georges, G.
Laboratoire de Zoologies et Biologies Animales, Universite Scientifique et Medicale de Grenoble, BP 68 38402 Saint-Martin d'Heres Cedex, France.

Geraerts, W.P.M.
Department of Biology, Free University, De Boelelaan 1087, 1081 HV Amsterdam, Netherlands.

Golding, D.W.
Dove Marine Laboratory, University of Newcastle upon Tyne, Cullercoats, Tyne & Wear.

Goldsworthy, G.J.
Department of Biology, Birkbeck College, University of London, Malet Street, London WC1E 7HX, UK.

Graff, D.
Zoological Institute, University of Heidelberg, Im Neumenheimer Feld 230, D-6900 Heidelberg 1, Federal Republic of Germany.

Grimmelikhuijzen, C.J.P.
Zoologisches Institut, Der Universitat Heidelberg, Fachrichtung Physiologie, Im Neumenheimer Feld 230, D-6900 Heidelberg 1, Federal Republic Germany.

Hekimi, S.
Laboratoire de Neurobiologie, Universite de Geneve, 20 rue de l'Ecole-de-Medicine, CH-1211 Geneve 4, Switzerland.

Joosse, J.
Department of Biology, Free University, De Boelelaan 1087, 1981 HV Amsterdam.

Keller, R.
Institut fur Zoophysiologie der Universitat Bonn, Endenicher Allee 11—13, 10 Bonn, Germany.

Martin, G.
Physiologie et Genetique des Crustaces, Laboratoire de Biologie Animale, Universite de Poitiers, 40 Avenue du Recteur Pineau, 86022 Poitiers Cedex, France.

van Minnen, J.
Department of Biology, Free University, De Boelelaan 1087, 1081 HV Amsterdam, Netherlands.

Mordue, W.
Department of Zoology, University of Aberdeen, Tillydrone Avenue, Aberdeen AB9 2TN.

Nambu, J.R.
Department of Biological Sciences, Stanford University, Stanford, CA 94305, USA.

Oates, A.
Department of Physiology & Pharmacology, St. Andrew's University, Fife KY16 9TS.

O'Shea, M.
Lab. de Neurobiologie, Dept. de Biologie Animale, Pavilion des Isotopes, 20 Boulevarde D'Yvoy, CH-1211, Geneve 4, Switzerland.

Porchet, M.
Lab. D'Endocrinologie des Invertebres, Universite des Sciences et Techniques, Biologie Animale de Lille, 59655 Villeneuve D'Ascq Cedex, France.

Pow, D.V.
Department of Zoology, University of Newcastle upon Tyne, UK.

Remy, C.
Lab. de Neuroendocrinologie, Universite de Bordeaux 1, Avenue des Facultes, 33405 Talence Cedex, France.

Scheller, R.H.
Department of Biological Sciences, Stanford University, Stanford, California 94305, USA.

Siegert, K.J.
Department of Zoology, University of Aberdeen, Tillydrone Avenue, Aberdeen AB9 2TN.

Spencer, A.N.
Department of Zoology, University of Alberta, Edmonton, Alberta, Canada T6G 2E9.

Taussig, R.
Department of Biological Sciences, Stanford University, Stanford, CA 94305, USA.

Thorndyke, M.C.
School of Life Sciences, RHBNC, Egham, Surrey TW20 0EX.

Thorpe, A.
School of Biological Sciences, Queen Mary College, University of London, London E1 4NS.

Turner, J.
Department of Physiology & Pharmacology, St. Andrew's University, Fife KY16 9TS.

Vieillemaringe, J.
Lab. de Neuroendocrinologie, Universite de Bordeaux 1, Avenue des Facultes, 33405 Talence Cedex, France.

Vreugdenhil, E.
Department of Biology, Free University, De Boelelaan 1087, 1981 HV Amsterdam, Netherlands.

Webster, S.G.
Institut fur Zoophysiologie der Universitat Bonn, Endenicher Allee 11—13, 10 Bonn, Germany.

Wheeler, C.H.
Department of Biology, Birkbeck College, University of London, Malet Street, London WC1E 7HX, UK.

Witten, J.
Laboratoire de Neurobiologie, Universite de Geneve, 20 rue de l'Ecole-de-Medecin, CH1211 Geneve 4, Switzerland.

Worden, M.K.
Laboratoire de Neurobiologie, Universite de Geneve, 20 rue de l'Ecole-de-Medecin, CH1211 Geneve 4, Switzerland.

PREFACE

This volume arose from an International Congress held in Bordeaux during 1986 and organised jointly by the Comparative Endocrinology Group of the Society for Experimental Biology, Laboratoire de Neurobiologie, Universite de Bordeaux I and Centre National de la Recherche Scientifique (CNRS).

The chapters which follow have been prepared by the invited seminar series speakers attending that meeting, and are designed as broad overviews of their particular specialities.

For the original meeting in Bordeaux we particularly extend warm and grateful thanks to our friend and colleague Professor Adrien Giradie and his collaborators in the Neurobiology Laboratory, Bordeaux, without whom the symposium could not have taken place, and this volume would not have been produced.

The symposium also benefited from the support of the following organisations: Society for Experimental Biology, UK; Centre National de la Recherche Scientifique, France; Direction de la Cooperation et des Relations Internationales du Minitere de l'Education Nationale, France; Universite de Bordeaux 1; Beckman; Bioblock Scientific; Bordeaux Chimie-Cofralab; Etablissements Laurent; Imperial Chemical Industries plc. (Plant Protection, Jealot's Hill, UK); Laboratory Data Control; Mairie de Bordeaux; Mairie de Gradignan; Office du Tourisme de Bordeaux; Peninsula Laboratories Europe; Pfizer Research Ltd.; Poly-Labo; Rohm Haas Chemical Co; Shell Research Ltd.; Sofranie-Mettler; Wild Leitz France.

MCT & GJG
February 1988

J. JOOSSE

What is special about peptides as neuronal messengers?

Nerve cells use various chemical messengers, including acetylcholine, monoamines, amino acids and neuropeptides. In the past 10 years it has become clear that there is a great molecular diversity of such messengers particularly amongst the amino acids. However, the molecular diversity of neuropeptides exceeds by far that of the amino acids. There is now evidence that in one animal species as many as 100-200 different peptide molecules may serve as neuronal messengers (Kandel 1983; Nieuwenhuys 1985; Joosse 1986).

Neuropeptides occur in nervous systems of all animals, even in the most simple such as coelenterates (Schaller *et al.*, 1984). For many years neuronal peptides were thought to have only a neuroendocrine role. The present view however is that, in addition to their role as neurohormones, they may act as typical synaptic neurotransmitters and as paracrine or neurocrine regulators at non-synaptic sites (Buma & Roubos 1985; Nieuwenhuys 1985).

Immunocytochemical and biochemical studies provide substantial evidence that structurally related peptides are found in many different phyla. Well-known examples are insulin-like peptides and FMRFamide. An important consequence of these findings is that some of the peptides found so far may have arisen from a smaller number of ancestral molecules, possibly at the prokaryote stage (Joosse 1987).

A large part of the present volume concerns the molecular diversity and functions of neuropeptides in invertebrates. In view of the great diversity in structure and function of neuropeptides and their presence throughout the animal kingdom, the question arises as to the adaptive value of peptide variety.

This question is discussed in the literature only occasionally, and emphasis is placed on the fact that, at least until now, unlike classical transmitters no special enzymatic machinery has been identified which effects the rapid breakdown of peptides at the sites of their release. Therefore, their actions last longer than those of classical transmitters. In the nervous system, peptides may diffuse to more widely distributed receptors (Kandel 1983) and this is likely to be important, for example, when peptides are co-released with classical transmitters. Here, peptides may recruit numerous neurones in a nucleus (Jan & Jan 1983). Others have stressed that some neuropeptides are able to bind to several different receptors, which means that these molecules contain more information than classical transmitters (Schwyzer 1980).

Previously, little attention has been given to the eco-physiological aspects of neuronal messengers. For example, what are the energy costs involved in the production of various messengers? In this respect, studies should be made of the energy budget of peptide transport from the site of synthesis (cell body), to the release point. Classical transmitters are often synthesized near their site of release, which reduces transport costs. Recycling of materials is also important. Little is known about reprocessing of peptides and their amino acid components. (Probably the single amino acid transmitters are the most economic signal molecules.) Acetylcholine reabsorption from the synaptic cleft is a well-known recycling phenomenon: after exocytosis, the membranes of the elementary granules are reprocessed locally. In peptidergic neurones, the multilamellar bodies which contain similar membrane material need to be transported retrogressively to the cell bodies at considerable energy cost (see also Golding & Pow, Chapter 1).

Any consideration of the selective value of neurotransmitter variety must include their genetics. Classical neurotransmitters are synthesized by one or a series of enzymes, from precursors in the blood or in cells, as components of typical metabolic pathways. Thus, it is not the structures of the neurotransmitters, but that of the synthesizing enzymes that are encoded by the genome. On the other hand, with neuropeptides the primary molecular structures of the signal molecules themselves are encoded by the genome; triplets of nucleotides code for each amino acid of the peptide molecules. Here we perceive a basic difference between categories of neuronal messenger (cf. Joosse 1987). Classical transmitters have an indirect relationship with the genome, whereas for peptides this relationship is direct. The mechanisms which operate in changing the genome are not restricted to point mutations, but show great diversity. All such changes may affect the nucleotide sequence in genes coding for peptides, and consequently the composition and sequence of the amino acids comprising the peptide messengers. Clearly, genomic changes will also affect genes coding for enzymes involved in the synthesis of classical transmitters. The effects of the latter changes are difficult to predict, but it may be that the quantity of messenger produced changes, or that a particular enzyme in the synthetic pathway of the messenger is unable to function. However, the chances are extremely low that the structure of the messenger concerned is changed. Moreover, classical transmitters are small molecules, and a change in their structure will often be lethal, since their functioning is not only dependent on receptor binding, but also on subsequent enzymatic breakdown and/or reabsorption. On the other hand, changes in the amino acid composition of peptides have the advantage of being gradual: a change from one hydrophobic amino acid to another may alter the characteristics of the complete molecule only slightly, whereas a change from any amino acid into proline could alter the shape of the entire molecule with severe consequences for receptor binding. This graduality of structural alterations makes peptide messengers suitable candidates for a crucial role in the evolutionary development of regulatory systems. Small changes in molecular structure may alter the rate or duration of the process controlled and therefore may be more easily successful when exposed to selection pressure. Another positive aspect of the direct genomic coding of neuropeptide structure is the occurrence of families of genes coding for the same neuropeptide precursor molecule, but with slightly different molecular structures for the final product (Hakanson & Thorell 1985). From the viewpoint of population genetics this is a highly advantageous situation for the successful control of adaptive changes.

In conclusion it appears that peptides are a special category of messenger. These molecules are suitable candidates for the control of the great number of processes related to animal adaption. This may explain why peptides are involved in the control of great diversity of physiological processes and why these molecules have persisted as messengers throughout animal evolution, despite their apparently higher energy budget compared with classical transmitters.

References

Buma, P. & Roubos, E.W. (1985). Ultrastructural demonstration of nonsynaptic release sites in the central nervous system of the snail Lymnaea stagnalis, the insect Periplaneta americana, and the rat. *Neuroscience* **17**, 867-79.

Hakanson, R. & Thorell, J. (eds.) (1985). Biogenetics of Neurohormonal Peptides. Academic Press, pp. 303.

Jan, Y.N. & Jan, L.Y. (1983). A LHRH-like peptidergic neurotransmitter capable of 'action at a distance' in autonomic ganglia. TINS 6, 320-25.

Joosse, J. (1986). Neuropeptides: peripheral and central messengers of the brain. In: Comparative endocrinology, developments and directions, ed. C.L. Ralph, pp. 12-32. *Progr. Clin. biol. Res.*, Vol. 205. New York: Allan Liss.

Joosse, J. (1987). Functional and evolutionary perspectives of neuropeptides and their precursors. *Proc. Firstt Internat. Congress Neuroend.* (In press).

Kandel, E.R. (1983). Neurobiology and molecular biology: the second encounter, *Cold Spring Harb. Symp. quant. Biol.* **48**, 891-908.

Nieuwenhuys, R. (1985). *Chemoarchitecture of the brain,* pp.1-246. Berlin: Springer.

Schaller, H.C., Hoffmeister, S. & Bodenmuller, H. (1984). Hormonal control of regeneration in Hydra. In: *Biosynthesis, metabolism and mode of action of invertebrate hormones*, ed. J.A. Hoffman & M. Porchet, pp. 5-9. Berlin: Springer.

Schwyzer, R. (1980). Organization and transduction of peptide information. *Trends in Pharmacol. Sc.* **3**, 327-31.

PART I

Immunocytochemistry and Ultrastructure

D.W. GOLDING & D.V. POW

The new neurobiology – ultrastructural aspects of peptide release as revealed by studies of invertebrate nervous systems

Ultrastructure of invertebrate nervous systems

Neurosecretory and other neurones

Examination of invertebrate nervous systems reveals that many are richly endowed with neurones resembling classical neurosecretory cells in cytology and ultrastructure. Such cells are clearly specialized for peptide secretion. They contain an abundance of rough endoplasmic reticulum (RER), and secretory granules (variously known as elementary granules, large dense-cored vesicles, etc.) generated by Golgi bodies, accumulate in large numbers within the perikarya. Although many are doubtless endocrine cells, others (Figs. 1–3) have axons which extend not to blood cavities, but into the central neuropile where the secretory material is discharged.

Furthermore, some secretory granules are evident in virtually all neurones (Golding & Whittle 1977), and this is consistent with the finding that many and perhaps all neurones, including those with conventional transmitters, also secrete peptides (review by Hokfelt, Johansson & Goldstein 1984).

Nerve terminals: vesicles and granules

In most nerve terminals, whether in the central or peripheral nervous systems, large numbers of synaptic vesicles are encountered (Figs. 4 & 5). Measuring 20–50nm in diameter, the vesicles in all but a small minority of terminals (Fig. 6) have lucent contents, except following exposure to a mixture of Zinc iodide and Osmium tetroxide (the ZIO reagent) (Fig. 7) which deposits extremely electron-dense material within them (May & Golding 1982). The vesicles cluster densely adjacent to sites of specialized contact with other cells. Pre- and postsynaptic thickenings are present and the synaptic clefts are wider and often more regular in form than other intercellular spaces, and contain moderately dense material.

Secretory granules also accumulate in synaptic terminals (Figs. 4–7). In contrast to synaptic vesicles, granules show great variety in size (80–200nm in diameter), form, electron density of the core, presence of a clear halo, etc. (Fig. 8). The majority 'stand back' from the synaptic junction and occupy more peripheral regions of the terminal. Unlike synaptic vesicles, they show little or no affinity for the ZIO reacent.

Figs. 1–3. Non-endocrine peptidergic cells. g. secretory granules; m, multivesicular body; n, nucleus; r, rough endoplasmic reticulum; arrows, Golgi apparatus giving rise to granules. Fig. 1. *Lumbricus terrestris* (Annelida), cerebral ganglion; bar 1000 nm. Fig. 2. *L. terrestris*, cerebral ganglion; bar 500 nm. Fig. 3. *Dendrocoelum lacteum* (Platyhelminthes), cerebral ganglion; bar 500 nm.

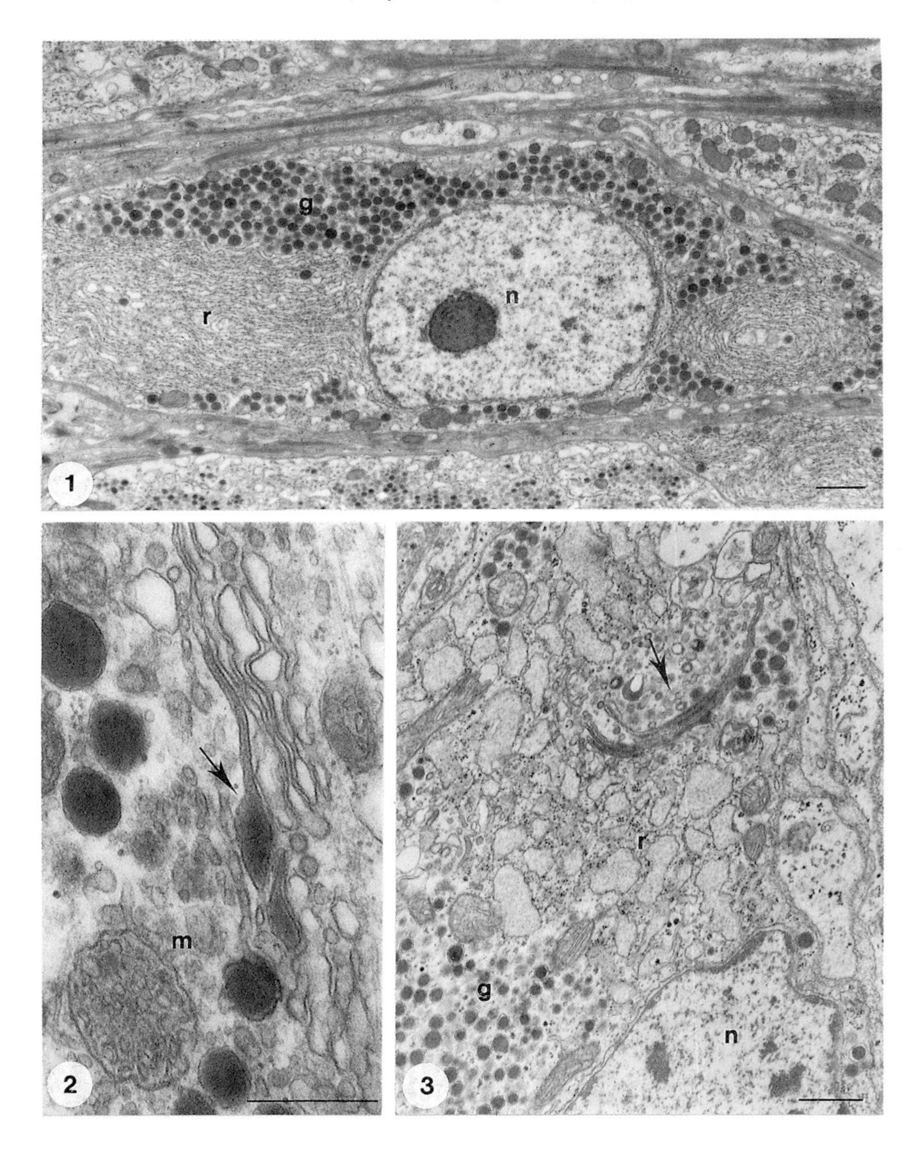

Immunocytochemical and other studies have established that neuropeptides are stored in the granules and not within synaptic vesicles (review by Hokfelt *et al.* 1984). Their apparent presence in the cytoplasm outside the granules is generally regarded as an artefact. Comparable findings relate to endocrine peptides in neurosecretory cells (review by Morris, Nordmann & Dyball 1978).

Secretory discharge by exocytosis

Synaptic vesicles engaged in exocytosis are encountered in extremely rare cases, the bounding membrane of the vesicle having fused with the presynaptic membrane, presumably allowing discharge into the synaptic cleft.

In contrast, exocytosis of secretory granules occurs mainly from within apparently unspecialized regions of the phasmalemma of non-endocrine nerve terminals. Demonstrated clearly during studies of annelid central nervous systems (Golding & May, 1982), the process is now recognized as a phenomenon of fundamental neurobiological mportance and has been reported with respect to a wide range of invertebrates (Figs. 8–11) (Shkolnik & Schwartz, 1980; Golding & Bayraktaroglu, 1984; Buma, Roubos & Buijs, 1984). Similar findings relate to vertebrates (Buma & Roubos, 1986), including the classic cholinergic synapses innervating the adrenal medulla and its homologues (Golding & Pow, 1987). Several lines of evidence indicate that the phenomenon is not a fixation artefact, but a correlate of secretory release resulting from neural activity. Sites of discharge are marked by the formation of omega profiles, with material of varying density being present within the indentations, and several sites of discharge may be present in a single fibre profile. Furthermore, compound exocytosis, in which one or more granules have apparently fused with another already engaged in discharge, is sometimes encountered (Fig. 9). Dissolution of the material does not always keep pace with release and pools of material then accumulate in the extracellular space.

Although granule exocytosis is associated mainly with areas of undifferentiated membrane, discharge into synaptic clefts does occur, albeit more rarely. It seems to be encountered with unusual frequency within regions of membrane immediately adjacent to areas occupied by membrane thickenings (Fig. 12). Exocytosis also occurs from within areas of the terminal in contact with glia as well as those adjacent to neuronal elements. Indeed, in the corpus cardiacum of the locust, quantitative studies show that exocytosis is not targeted towards the postsynaptic gland cells but is equally likely to be associated with regions of the plasmalemma adjacent to other fibres, glia, etc. (Pow & Golding, 1987). Last, granule discharge in the corpus cardiacum is associated with varicosities and does not apparently occur from within the narrow regions of fibres which connect them (Pow & Golding, 1987).

In most cases, sites of granule exocytosis and synaptic contact, respectively, are not sharply segregated from each other within separate varicosities. However, in

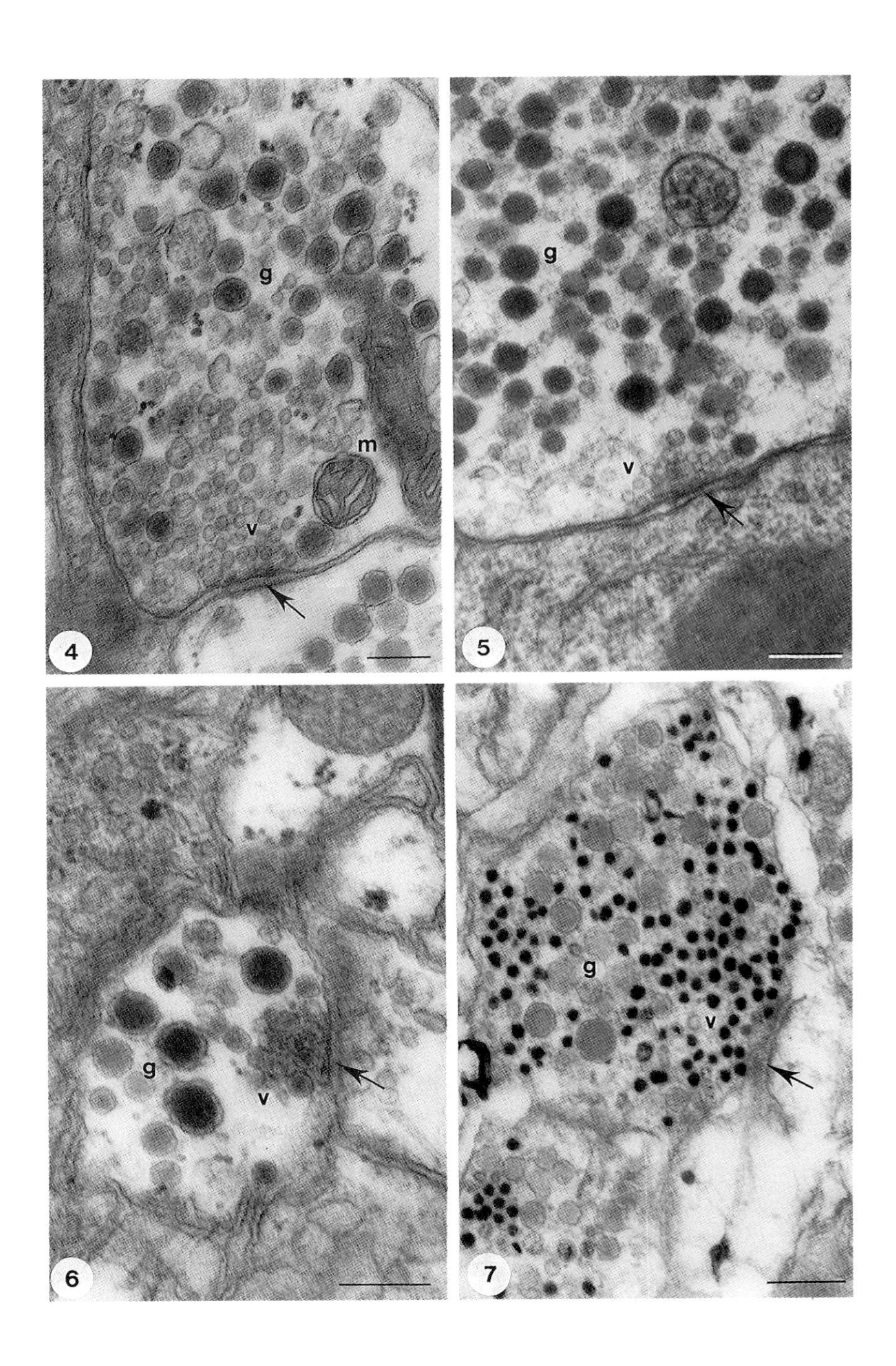
g
m
v
4
g
v
5
g
v
6
g
v
7

some types of neurone in annelids, in which granule exocytosis is abundant, synapses and synaptic vesicles are at most extremely rare, and similar findings apparently apply to *Lymnaea* (Buma & Roubos 1986).

Exoycytotic profiles captured by tannic acid

The study of glandular discharge has been greatly facilitated by the use of tannic acid (Fig.10), a complex mixture of glucosides whose orthophenol radicles are negatively charged at neutral pH. Infused into living ganglia maintained in saline solutions *in vitro* (the TARI method, Buma, Roubos & Buijs, 1984), or injected *in vivo* (the TAIV method, Pow & Golding, 1987), tannic acid prevents the dissolution of granule cores as they are discharged and enhances their eventual electron density. Exocytotic profiles, 'frozen' *in situ*, are thus progressively accumulated as secretion continues (Fig. 11). However, the two techniques should be applied with caution, since in at least some organs, cytolysis and aberrant granule discharge results from exposure to tannic acid for extended periods of time (Pow, unpublished).

Membrane retrieval

Omega profiles resulting from granule exocytosis are sometimes etched by one or more coated pits (Fig. 13), apparently giving rise to vesicles 30–50 nm in diameter (Golding & May 1982). In other cases, omega profiles may be coated *in toto* (Golding & Bayraktaroglu 1984; Buma & Roubos 1985) probably indicating that granule membrane is being retrieved by endocytosis of 'vacuoles' similar in size to the granules.

A dual vesicle hypothesis

Correlation of ultrastructural observations with information drawn more widely, led Golding & May (1982) to conclude that neurones typically possess two, dichotomous, secretory mechanisms for the elaboration and discharge of neurochemical mediators (Fig. 14).

Origins

Secretory inclusions may well differ, first, in their respective origins within the neurone (Fig. 14a). Granules are undoubtedly generated by the Golgi apparatus in

Figs. 4–7. Synaptic terminals containing secretory granules (g), and synaptic vesicles (v), focussed on membrane thickenings with differentiated clefts (arrows). Fig. 4. *L. terrestris; m,* mitochondria; bar 200 nm. Fig. 5. *Schistocerca gregaria* (Insecta); bar 200 nm. Fig. 6. *Nereis diversicolor* (Annelida), showing dense-cored synaptic vesicles; bar 200 nm. Fig. 7. *N. diversicolor*, showing synaptic vesicles 'stained' with ZIO; bar 200 nm. Courtesy, Dr. Barbara A. May, University of Newcastle upon Tyne.

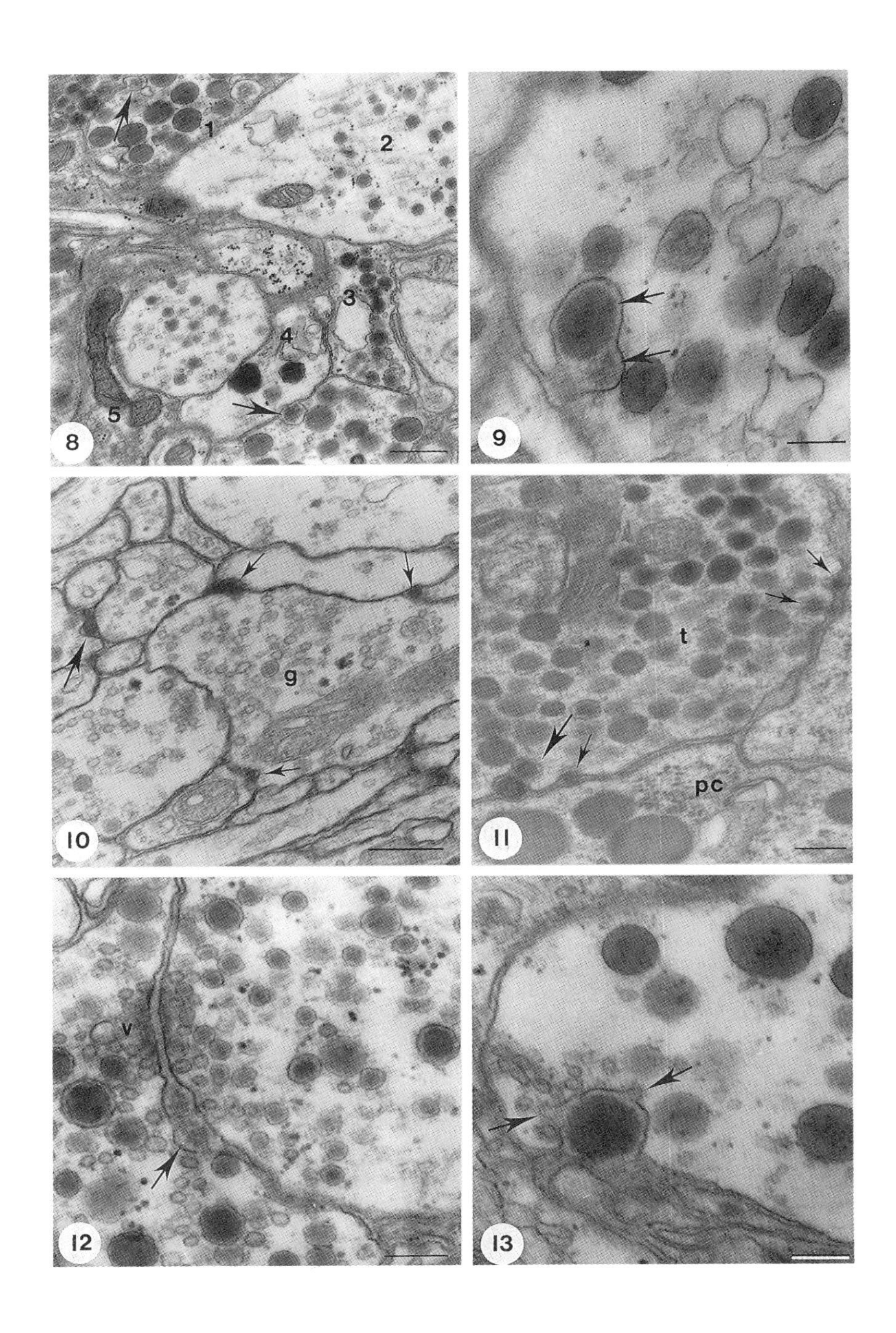

1
2
3
4
5
8
9
t
pc
10
g
11
v
12
13

the perikaryon, whereas synaptic vesicles may well originate from smooth endoplasmic reticulum (Hernandez–Nicaise 1973), in some cases locally, within the nerve terminals (Droz, Rambourg & Koenig 1975).

The respective chemical mediators associated with the two categories of inclusions show a more firmly established disparity in origins. Peptides are synthesized by the RER, and condensed and packaged within secretory granules by the Golgi apparatus, all within the cell body. Peptide processing, but no further synthesis, takes place peripherally. In contrast, conventional transmitters are synthesized throughout the neurone. Transmitter stores within granules in, for example, a sympathetic neurone, are thus enhanced during axonal transport, and further incorporation into these elements, and into synaptic vesicles, occurs within the nerve terminals (review by Hokfelt *et al.* 1984).

Release

Our ultrastructural studies suggest that neurones typically exhibit a three-fold differentiation of their surfaces with respect to adaptations for secretory release. Presynaptive thickenings are specialized for the fusion of synaptic vesicles, although some granule discharge may also take place here. Other expansive regions of the terminal surface are, though *apparently* unspecialized morphologically, adapted as sites of granule exocytosis (which may be compound)(Fig. 14a). Last, the ultra-structural evidence that preterminal axons etc., show, at most, low levels of secretory discharge is indicative of their further functional differentiation.

Nerve terminals thus show a variety of modes of secretory release. Conventional *synaptic* discharge is associated with specialized junctions with the postsynaptic cell. In contrast, exocytosis may occur at unspecialized sites within the terminal membrane, being untargeted with respect to postsynaptic elements, and constitute *non-synaptic* (Dismukes 1979) release. Various patterns of secretory release may result from the combination of these modes of discharge. For example, synaptic specializations may be absent and both classes of inclusions then show exocytosis at

Figs. 8–13. Secretory granule exocytosis. Fig. 8. *L. terrestris*, showing abundance and diversity (numbers 1–5) of secretory inclusions within the neuropile. arrows, exocytosis; bar 500 nm. Fig. 9. *L. terrestris*, showing compound exocytosis in which a single profile contains the remains of two granule cores (arrows); bar 100 nm. Fig. 10. *D. lacteum*, tannic acid. Discharged granule cores at site (arrow) and possible sites (small arrows) of exocytosis show enhanced density when compared with undischarged granules (g); bar 500 nm. Fig. 11. *S. gregaria*, TAIV method. Simple (arrows) and compound exocytosis (large arrow) from within nerve terminal (*t*) in the corpus cardiacum, resulting from neural activation. *pc*, Postsynaptic cell; bar 200 nm. Fig. 12. *L. terrestris*. Site of granule exocytosis (arrows) immediately adjacent to synaptic contact with clustered vesicles (v); bar 200 nm. Fig. 13. *L. terrestris*. Exocytotic profile has two coated pits (arrows) and associated microvesicles; bar 200 nm.

Fig. 14(a). Diverse origins, and patterns of discharge, of synaptic vesicles and secretory granules, respectively. (b) Diverse retrieval patterns and fates of vesicles and granules, respectively.

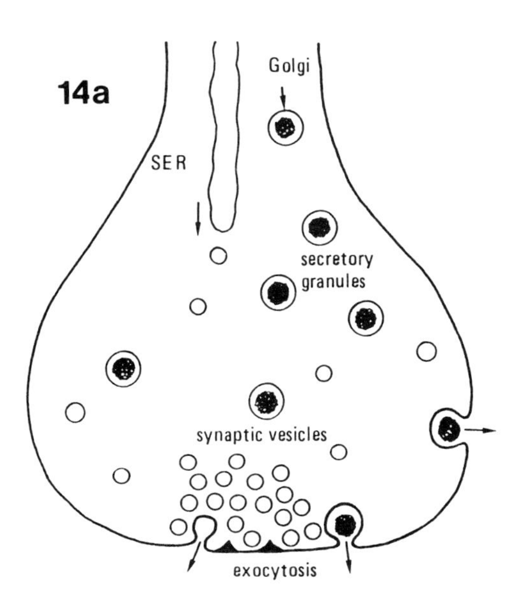

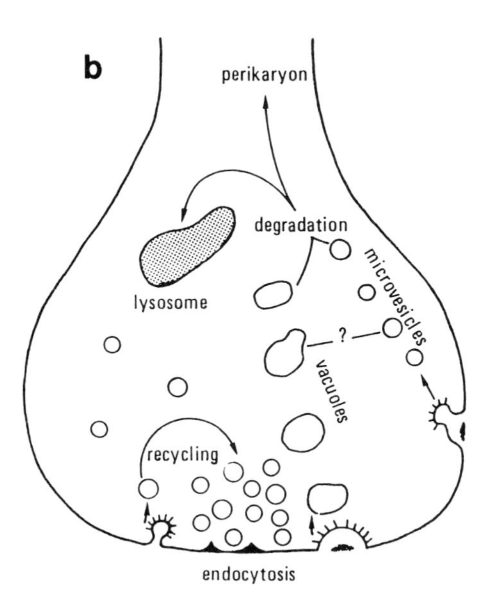

morphologically unspecialized sites (Thureson–Klein 1983). Nevertheless, our observations suggest that the majority of nerve terminals are characterized by a combination of vesicle exocytosis at specialized junctions and granule discharge from within morphologically undifferentiated sites, and this would be consistent with the wider sphere of influence of peptide mediators (e.g., see Jan & Jan 1982).

Retrieval

Secretory release by exocytosis is invariably followed by membrane retrieval, first, because of the increase in surface area of the plasmalemma that would otherwise result, and second, because the bounding membranes of secretory inclusions, and plasmalemma, respectively, have their own distinctive morphological and functional specializations (Zinder *et al.* 1978).

It is clear that the pattern of retrieval of secretory granule membrane is markedly different from that of synaptic vesicles (Fig. 14b), whichever interpretation of the latter is adopted (*cf.* reviews by Heuser 1978; Ceccarelli & Hurlbut 1980). Membrane is internalized from the exact site of exocytosis, takes place at least in part whilst dissolution of the extruded core is in progress, and involves the formation of coated pits and vesicles.

In some cases, membranes of individual granules are probably retrieved intact as 'vacuoles', as in the neurohypophysis (review by Morris *et al.* 1978). However, in other systems containing large secretory granules, endocytosis is apparently accompanied by fragmentation of granule membranes to form 'microvesicles' – a process common among both endocrine (e.g., Nagasawa & Douglas 1972) and exocrine glands (review by Herzog 1981).

Although internalized microvesicles are usually identical in appearance to synaptic vesicles *individually,* the respective ultrastructural configurations to which these

Fig. 15. Sites of granule exocytosis (a) and synaptic contact (b) are readily distinguishable.

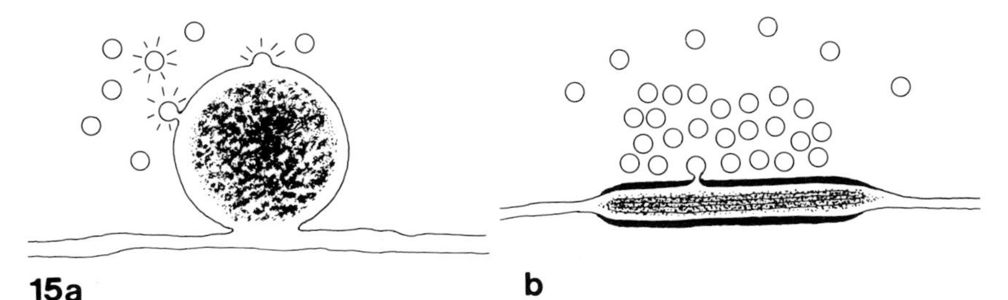

inclusions contribute are readily distinguishable (Fig. 15). Furthermore, the clathrin coats of endocytotic vesicles are sometimes prominent, whereas synaptic vesicles are uncoated. Some synapses feature dense-cored vesicles of a type never associated with sites of granule exocytosis. Last, retrieval of granule membrane does not lead to the accumulation of the large numbers of vesicles associated with many synapses.

Recycling

Membrane retrieval is the first part of the process by which synaptic vesicles are recycled, being followed by shedding, if formed, of clathrin coats (Heuser 1978) and repletion with neurotransmitter. In contrast, such 'short-loop' recycling (Fig. 14b) is not to be expected in regard to secretory granules, since their peptide and protein contents cannot be replenished within the terminals. Consequently, chronically stimulated terminals contain reduced numbers of granules, whereas the population of synaptic vesicles is maintained (Lynch 1980). Two possible fates await retrieved granule membranes. As with the membranes of neurohypophysial granules, they may be subject to lysosomal degradation (Theodosis 1982), either peripherally or after transport to the perikaryon by retrograde axoplasmic flow, so that their biochemical constituents become available for re-use. Alternatively, the internalized membranes may enter a 'long-loop' recycling process by being returned to the perikaryon and re-used by the Golgi apparatus in granule production, as in many non-nervous gland cells (review by Herzog 1981).

Conclusion

The uniqueness of neuronal function may be regarded as deriving from the capacity for *spatially-specific glandular activity*, by which secretory products are discharged at selected, highly localized sites. The morphology of the neurone with its slender, elongate processes is fundamental to this function. However, it could be said to pose great 'problems' for glandular activity, since the cell body with its nucleus, RER and Golgi complex is far distant from many of the sites of discharge. By virtue of the process of axoplasmic flow, secretory granules can be transported to distant nerve terminals, but both they and their peptide stores can be replenished only from the perikaryon – a process that may take hours. In contrast, synaptic vesicles can be recycled locally and their contents replenished with locally synthesized transmitter. Taken together with the rapidity of vesicle discharge and the involvement of specialized sites of release, the synaptic apparatus may justifiably be regarded as the most highly advanced glandular mechanism in the living world. In contrast, the secretory granules of neurones are conventional in almost every aspect of their cytophysiology – including their mode of discharge – and it is ironic that granule exocytosis from within synaptic terminals should have eluded detection for so long

Acknowledgements

The authors gratefully acknowledge the assistance of Mr. R.M. Hewit. Support by research grant GR/A88743 from the SERC to D.W. Golding, and a research studentship from the SERC to D.V. Pow.

References

Buma, P. & Roubos, E.W. (1986). Ultrastructural demonstration of nonsynaptic release sites in the central nervous system of the snail (*Lymnaea stagnalis*), the insect (*Periplaneta americana*), and the rat. *Neuroscience*, **17**, 867–78.

Buma, P., Roubos, E.W. & Buijs, R.M. (1984). Ultrastructural demonstration of exocytosis of neural, neuroendocrine and endocrine secretions with an *in vitro* tannic acid (TARI-) method. *Histochemistry*, **80**, 247–56.

Ceccarelli, B. & Hurlbut, W.P. (1980). Vesicle hypothesis of the release of quanta of acetylcholine. *Physiological Reviews*, **60**, 396–441.

Couteaux, R. & Pecot–Dechavassine, M. (1970). Vésicules synaptiques et poches an niveau des 'zone actives' de la jonction neuromusculaire. *Comptes rendues, Academie de Sciences, Paris*, **271D**, 2346–9.

Dismukes, R.K. (1979). New concepts of molecular communication among neurons. *Behavioural Brain Science*, **2**, 409–48.

Droz, B., Rambourg, A. & Koenig, H.L. (1975). The smooth endoplasmic reticulum: structure and role in the renewal of axonal membrane and synaptic vesicles by fast axonal transport. *Brain Research*, **93**, 1–13

Golding, D.W. & Bayraktaroglu, E. (1984). Exocytosis of secretory granules – a probable mechanism for the release of neuromodulators in invertebrate neuropiles. *Experientia*, **40**, 1277–80.

Golding, D.W. & May, B.A. (1982). Duality of secretory inclusions in neurones – ultrastructure of the corresponding sites of release in invertebrate nervous systems. *Acta Zoologica (Stockholm)*, **63**, 229–38.

Golding, D.W. & Whittle, A.C. (1977). Neurosecretion and related phenomena in annelids. *International Review of Cytology*, Suppl. 5, 189–302.

Golding, D.W. & Pow, D.V.(1987). 'Neurosecretion' by a classic cholinergic innervation apparatus. A comparative study of adrenal chromaffin glands in four vertebrate species (Teleosts, anurans, mammals). *Cell and Tissue Research*, **249**, 421–5.

Hernandez–Nicaise, M-L. (1973). The nervous system of ctenophores III. Ultrastructure of synapses. *Journal of Neurocytology*, **2**, 249–63.

Herzog, V. (1981). Endocytosis in secretory cells. *Philosophical Transactions of the Royal Society of London*, B, **296**, 67–72.

Heuser, J.E. (1978). Synaptic vesicle exocytosis and recycling during transmitter discharge from the neuromuscular junction. In *Transport of macromolecules in cellular systems*, ed. S.C.Silverstein, pp. 445–464. Dahlem Konferenzen, Berlin.

Hökfelt, T., Johansson, O. & Goldstein, M. (1984). Chemical anatomy of the brain. *Science*, **225**, 1326–34.

Jan, L.Y. & Jan, Y.N. (1982). Peptidergic transmission in sympathetic ganglia of the frog. *Journal of Physiology*, **327**, 219–46.

Lynch, R. (1980). Stimulation-induced reduction of large dense core vesicle number in cholinergic motor nerve endings. *Brain Research*, **194**, 249–54.

May, B.A. & Golding, D.W. (1982). Synaptic and synaptoid vesicles constitute a single category of inclusions. New evidence from ZI0 impregnation. *Acta Zoologica (Stockholm)*, **63**, 171–6.

Morris, J.F., Nordmann, J.J. & Dyball, R.E.J. (1978). Structure–function correlation in mammalian neurosecretion. *International Reviews in Experimental Pathology*, **18**, 1–97.

Nagasawa, J. & Douglas, W.W. (1972). Thorium dioxide uptake into adrenal medullary cells and the problem of recapture of granule membrane following exocytosis. *Brain Research*, **37**, 141–5.

Pow, D.V. & Golding, D.W. (1987). 'Neurosecretion' by aminergic synaptic terminals *in vivo* – a study of secretory granule exocytosis in the corpus cardiacum of the flying locust. *Neuroscience* (in press).

Shkolnik, L.J. & Schwartz, J.H. (1980). Genesis and maturation of serotonergic vescicles in identified giant cerebral neuron of *Aplysia*. *Journal of Neurophysiology*, **43**, 945–67.

Theodosis, D.T. (1982). Secretion-related accumulation of horseradish peroxidase in magnocellular cell bodies of the rat supraoptic nucleus. *Brain Research*, **233**, 3–16.

Thureson–Klein, A. (1983). Exocytosis from large and small dense-cored vesicles in noradrenergic nerve terminals. *Neuroscience*, **10**, 245–52.

Zinder, O., Hoffman, P.G., Bonner, W.M. & Pollard, H.B. (1978). Comparison of chemical properties of purified plasma membranes and secretory vesicle membranes from the bovine adrenal medulla. *Cell and Tissue Research*, **188**, 153–70.

H.H. BOER & J. VAN MINNEN

Immunocytochemistry of hormonal peptides in molluscs: optical and electron microscopy and the use of monoclonal antibodies

Introduction

Study of model-systems has played a major role in the developments which have taken place in neurobiology during the past decades. Studies of opisthobranch and pulmonate molluscs have made important contributions in this area. These animals make excellent models, because they possess a relatively small central nervous system (CNS), which contains a limited number of large (polyploid, e.g. Boer *et al.*, 1977) and readily accessible neurons. Two species have been studied in particular; the opisthobranch *Aplysia californica* (e.g. Strumwasser *et al.* 1980) and the pulmonate freshwater snail *Lymnaea stagnalis* (e.g. Joosse & Geraerts 1983; Roubos 1984).

Immunocytochemistry has become an important tool in neurobiology. In our laboratory the CNS of *L. stagnalis* has been investigated extensively using immunocytochemical methods (Boer *et al.*, 1979, 1980, 1984a,b, 1986; Schot & Boer 1982; Schot *et al.* 1981, 1983, 1984, see also Chapter 13, this volume). Recently monoclonal antibodies were raised to homogenates of whole CNS from *L. stagnalis* (Boer & Van Minnen 1985). These investigations, in conjunction with functional and electrophysiological studies can illustrate the central position of the peptidergic neuron in neurotransmission and in neuro-endocrine control processes (Joosse 1986).

Neural and hormonal communication

Animals possess two major systems for the regulation and coordination of body functions: the nervous system and the endocrine system. In the classic view these systems differ in a number of aspects. The chemical messengers of the nervous system (neurotransmitters) are small molecules (acetylcholine, biogenic amines, amino acids), which are released at sites of direct contact (synapse) between neurons and their targets. Transmission of information is very rapid and the number of transmitters is relatively low (about a dozen). On the other hand, the messengers of the endocrine system (hormones) are released into and transported by the body fluids. Signal transmission usually takes longer. Hormones are more numerous and more diverse than neurotransmitters. They belong essentially to one of the following three categories of compounds: glycoproteins, proteins or peptides; steroid or steroid-like molecules; and other, less common molecules (e.g. thyroid hormones). The study of the two messenger systems has largely been conducted in different domains of biology: for the nervous system, neuroanatomy, neurophysiology and pharmacology; for the endocrine system, surgical manipulation, chemical physiology and histology. In the classic view an important principle common to both systems is that a signal transmitting cell produces only one chemical messenger.

Unifying concepts

Since the discovery of the phenomenon of neurosecretion by Ernst Scharrer in 1928, but particularly during the past decade, due to the introduction of new techniques, it has become increasingly clear that the nervous system and that part of the endocrine system that produces proteinaceous hormones are closely related.

Neurosecretion

Neurosecretory cells are neurons that possess the properties of endocrine cells. Their products are peptides, which are released into and transported by the

Fig. 1. Aorta of *Helix aspersa*. (a.) Light micrograph of cross section stained with anti–FMRFamide. Immunoreactive axons are present at the border of the outer longitudinal (arrows) and inner circular muscular layer (im). ×95. (b) Electron micrograph of immunocytochemically labelled (PAP) secretion vesicles in one of these axons. ×85,000. The figure superimposed on a and b shows that the circular muscles contract *in vitro* upon addition of FMRFamide to the bathing medium (concentrations at left arrow 3×10^{-7} M, middle arrow 4×10^{-7} M, right arrow 8×10^{7} M). Time scale: distance between first and second arrow, 40 min (Griffond *et al.* 1986).

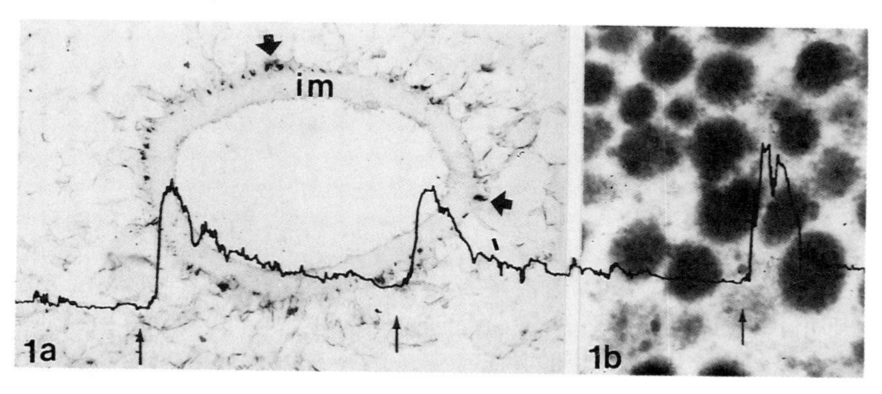

body fluids. Morphological observations have further narrowed the distinction between the two control systems. Axons of some neurosecretory systems appeared to terminate not at blood vessels or sinuses, but at short distances from target cells, where they release their product (periperhal neurosecretion). In *Lymnaea stagnalis* this phenomenon has been observed in various organs, e.g. in the kidney (Wendelaar Bonga 1972) and in muscles (Plesch 1977). Experimental studies on the neurosecretory system innervating the kidney indicated that here the messenger released by the system acts as a neurohormone. On the other hand, the presence of neurosecretory axon terminals on muscle cells suggests that here the messenger acts as a neuro-muscular transmitter. In this way it is possible that neurosecretory substances can act either as hormones or as neurotransmitters, or both.

Fig. 2. Electron micrograph of a cross section of the cerebral commissure of *L. stagnalis* showing numerous axon profiles with microtubules (thin arrows). Some profiles contain secretion vesicles (sv), the contents of which may be released non-synaptically by exocytosis (arrows). Tannic Acid Ringer Incubation (TARI method). ×50,000. (Courtesy Dr. E.D. Schmidt.)

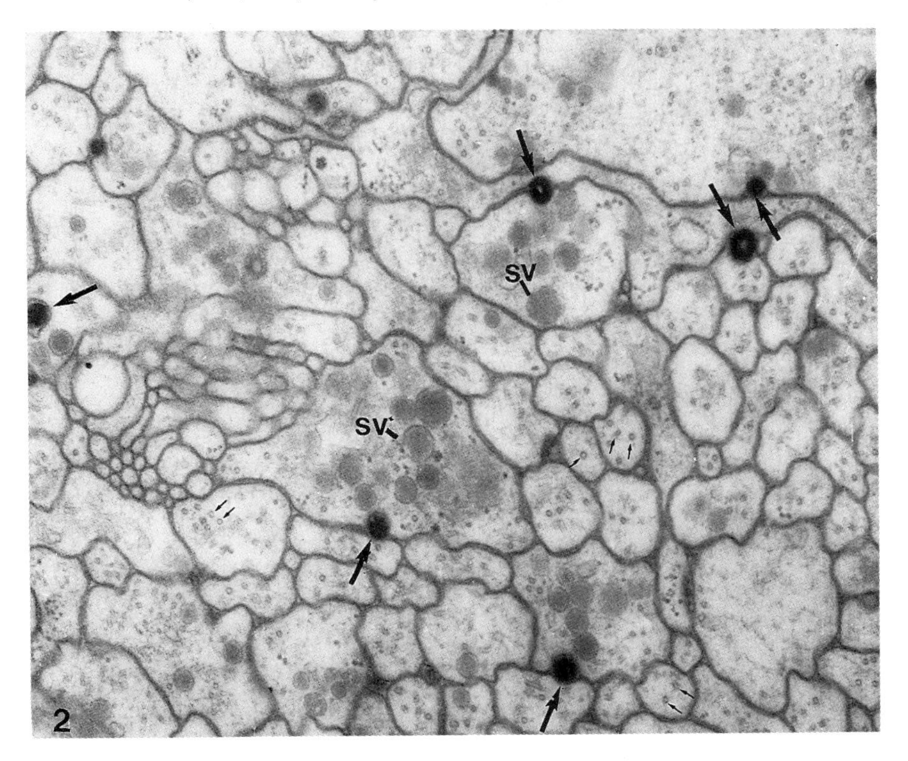

Peptidergic synapses

The suggestion that peptides can act as neurotransmitters has been supported by numerous investigations, especially by immunocytochemical and pharmacological studies. Thus, peptidergic synapses have been observed in the rat brain at the electron microscope level (Buys & Swaab 1979). Similarly, in *L. stagnalis* axons immunoreactive to antisera raised against various neuropeptides have been observed to form synapse-like structures with neurons or with muscle cells (Boer *et al.* 1980, 1984a; Schot *et al.* 1981, 1983). Furthermore, electrophysiological and pharmacological evidence showing that neuropeptides act as neurotransmitters, is accumulating rapidly. This evidence has been obtained from various species, including *L. stagnalis,* (Fig.1), *Helix aspersa, Mytilus edulis* and *Aplysia californica* (e.g. Cottrell *et al.* 1983; Stefano & Leung 1984; Ter Maat *et al.* 1986; Griffond *et al.* 1986; see also Chapter 14, this volume).

Non-synaptic neurotransmission

Electrophysiological studies on both the central and the peripheral nervous system show that follower cells of a neuron (e.g. other neurons) may act upon electrical stimulation of this neuron with a delayed response (e.g. Beaudet & Descarrier 1978; Mayeri & Rothman 1981; Jan & Jan 1983; Schmitt 1984; Vizi 1984). It has been suggested that the stimulated neuron does not form regular synapses with its followers, but that it releases its messenger in a "non-synaptic" fashion into the intercellular space between neuron and target (see Chapter 1, this volume). The messenger will have to travel over some distance before it can exert its action, which explains the delay in the response of the target. The similarity between the concepts of non-synaptic communication and that of peripheral neurosecretion is evident (cf. also the paracrine system, where endocrine cells regulate adjacent cells rather than distant targets).

Recently the concept of non-synaptic communication has been supported in ultrastructural studies of the CNS of *Lymnaea stagnalis* (Roubos *et al.* 1983; Buma & Roubos 1986). In these studies using the TARI method (Tannic Acid Ringer Incubation) it was shown that neurons may release their secretory product by exocytosis at sites (e.g. along the axons) showing no morphological specializations (Fig. 2). In a detailed study of the Caudo-Dorsal Cells (CDC), which produce the ovulation hormone, Schmidt & Roubos (1987) showed that some of these peptidergic neurons (the ventral CDC) release their products in two ways: non-synaptically from a system of collaterals inside the cerebral commissure; and from neurohaemal axon terminals in the periphery of the commissure (cf. Roubos 1984; Fig. 3). The authors suggest that the non-synaptically released messenger(s) act upon neuronal systems in the commissure. A possible target is the Ring Neuron, as this neuron reacts with a delayed response upon electrical stimulation of the CDC (Jansen & Bos 1983; Ter Maat & Jansen 1984).

Co-transmission

There are also neurons that produce and release more than one chemical messenger (Burnstock 1976, 1981; Hökfelt *et al.* 1978; Lundberg & Hökfelt 1983). Many combinations of messengers have been found: several combinations of two classical transmitters (Cottrell 1977; Burnstock & Hökfelt 1979; Cournil *et al.* 1984), of a classical transmitter and a neuropeptide (Lundberg & Hökfelt 1983) and of two or more neuropeptides (Larson 1980; Guillemin 1981; Geraerts *et al.* 1983; Viveros *et al.* 1983). The studies on co-transmission have been carried out on various species, including molluscs (Cottrell 1977; Osborne 1977; Osborne & Dockray 1982). In *L. stagnalis* immunocytochemistry has shown that neurons exist which contain two classical transmitters (serotonin, dopamine) and at least one (vasotocin-like) neuropeptide (Boer *et al.* 1984b).

The functional significance of co-transmission may be diverse. One of the messengers may act as an autotransmitter (Moed *et al.* 1987), i.e. inhibit or stimulate release of messengers from the releasing axon terminal itself. The other messengers may act upon the same or on different targets. In the peripheral nervous system (e.g. rat salivary gland), the classical transmitter causes a rapid response of short duration, whereas the co-released peptide is responsible for a more long-lasting effect (Lundberg & Hökfelt 1983; see also Chapter 15, this volume). In insects octopamine and the peptide proctolin are involved in co-transmission (see Chapter 8 this volume). It has also been considered that a neuron might form two types of synapses, each containing a particular transmitter (Dismukes 1979).

Fig. 3. Morphology of a ventral CDC located in the left cerebral ganglion. The CDC has two axons: one directly entering the neurohaemal area (nh) in the periphery of the cerebral commissure; the other running to the contralateral ganglion before entering the nh. In the commissure this latter axon gives rise to a system of collaterals (cs). Here non-synaptic release of secretory products takes place. RN, Ring Neuron. (Courtesy Drs. E.D. Schmidt and E.W. Roubos.)

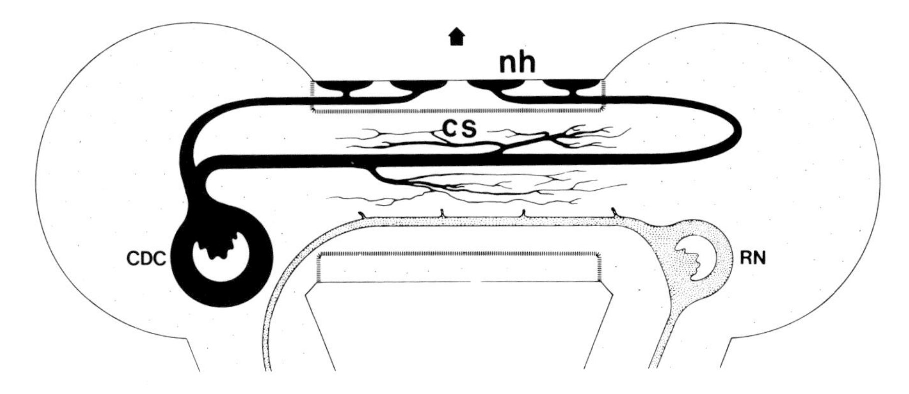

Multiple peptides

It is well established that neuropeptides are derived from large precursor molecules. In this volume Taussig *et al.* (Chapter 15) and Geraerts *et al.* (Chapter 13), report on the study of the peptidergic systems in *Aplysia* (Bag Cell system) and *Lymnaea* (Caudo-Dorsal Cells), respectively. These cells produce a precursor which is cleaved into several smaller peptides which apparently integrate the complex processes of ovulation and egg laying behaviour. Some of them act as neurohormones, exerting effects on distant targets, such as the gonad and the accessory sex organs, whereas others act as neurotransmitters, acting upon neurons located in the CNS.

With respect to the production and release of multiple peptides, another phenomenon should be mentioned briefly. It has now been established that certain peptidergic systems contain peptides derived from unrelated precursors, e.g. from

Fig. 4. The Caudo-Dorsal Cell (CDC) of *Lymnaea*, a multifunctional neuron. See text. N, neuron; MN, motor neuron; RN, Ring Neuron; CDCA, autotransmitter; CDCH, ovulation hormone; CaFl, calfluxin. (Courtesy Dr. E.W. Roubos.)

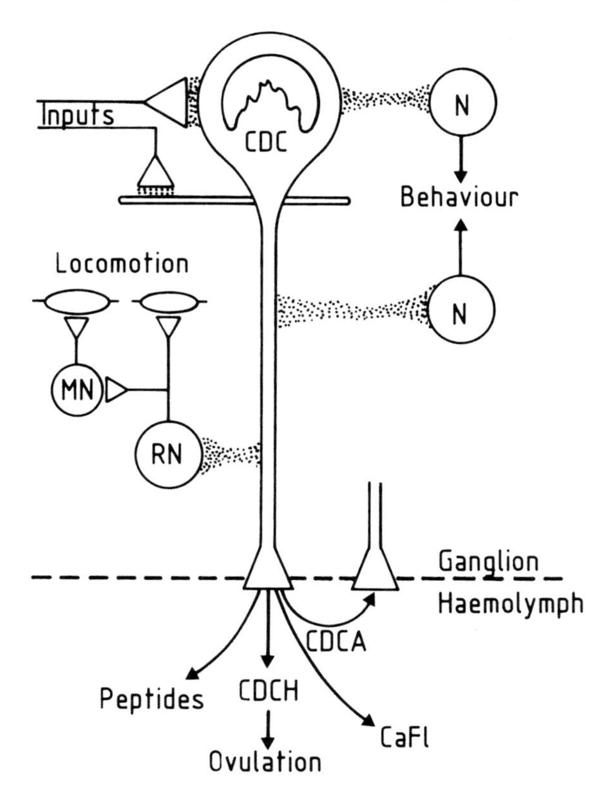

pro-opiomelanocortin and from the oxytocin or the vasopressin precursor (Martin *et al.* 1983). This suggests that the complexity of these systems is even greater than hitherto thought.

The peptidergic neuron, a multifunctional cell

From the evidence available, our image of the peptidergic neuron emerges as one of a multifunctional cell, which takes a central position in the regulation ofboth nervous and endocrine processes. In this respect it is interesting to note that, in contrast to vertebrates, most invertebrates possess few real endocrine organs. For example, in basommatophoran pulmonates the only endocrine organs recognised so far are the Dorsal Bodies. These structures regulate oocyte growth and the development and secretory activity of female accessory sex organs (Joosse & Geraerts 1983). Most endocrine functions in these and other invertebrates are carried out by peptidergic neurons, which suggests that the peptidergic neuron as a regulator of diverse body functions is a phylogenetically ancient phenomenon. The fact that in primitive animals like *Hydra* numerous peptidergic cell types occur, whereas cells containing classical transmitters seem to be absent sustains this hypothesis (see Chapter 10, this volume).

The CDC of *L. stagnalis* is an example of a multifunctional peptidergic cell (Fig. 4). Upon stimulation, the CDC release their peptidergic products into the haemolymph and also (non-synaptically) into the CNS (Fig. 3). Among these peptides are the CDCA (CDC autotransmitter) which stimulates other CDC, and the CDCH (CDC Hormone) which induces ovulation (Geraerts & Bohliken 1976). Recently a third peptide was identified, calfluxin (CaFl). This peptide, which is also produced by the CDC causes an influx of Ca^{++} into the mitochondria of the cells of one of the accessory sex organs, the albumen gland (Dictus *et al.*, 1987). In addition, other peptides with still unknown functions are released. Probably the non-synaptically released products are involved in evoking particular behaviours related to the egg laying process by influencing other neurons in the CNS, including the Ring Neuron.

Universality of neuropeptides

There are several indications that not only the classical neurotransmitters, but also neuropeptides have a wide distribution in the animal kingdom. For example, the (vertebrate) opioids met– and leu–enkephalin and met–enkephalin–Arg–Phe have been isolated from CNS of the mussel *Mytilus edulis* (Leung & Stefano 1984). Recently met–enkephalin and a leu–enkephalin-like substance have also been extracted from *Lymnaea* CNS (Boer *et al.* 1987). In both species opioid receptors were identified. Moreover, insulin-like molecules have been found in insects and molluscs (Duve & Thorpe 1980; Hemminga 1984; see also Chapter 6, this volume). On the other hand, the head activator from *Hydra* has also been isolated from the rat and from the human hypothalamus (Bodenmüller *et al.* 1980). Furthermore, vertebrate neuropeptides exert electrophysiological effects in invertebrate neuronal systems (Ter Maat *et al.* 1986), while the molluscan cardio-active tetrapeptide FMRFamide has effects on arterial blood pressure in the rat (Mues *et al.* 1982; Barnard & Dockray 1984). Although these observations suggest universality of neuropeptides, there are also reports indicating that neuropeptides can be species- or genus-specific in their activity (Dogterom & Van Loenhout 1983).

Immunocytochemistry

The idea that vertebrate neuropeptides occur in invertebrates has received particular support from immunocytochemical studies. Numerous investigations have been carried out on diverse invertebrate groups using antisera against vertebrate neuropeptides; e.g. coelenterates (Grimmelikhuizen 1984), insects (Rémy *et al.* 1979; Duve & Thorpe 1980; Veenstra *et al.* 1984), crustaceans (Van Deynen 1985), molluscs (Grimm–Jørgensen 1978; Schot *et al.* 1981; Marchand *et al.* 1982; Voigt *et al.* 1983) and tunicates (Fritsch *et al.* 1979; Thorndyke & Probert 1979). (See also Chapters 3, 6, 9 and 12, this volume,)

Specificity

Results obtained with immunocytochemistry cannot be used to confirm the absolute identification of a particular peptide, unless carried out in conjunction with chemical analyses or bioassay experiments (Schot *et al.* 1984). A positive immunological reaction does not necessarily indicate that the molecule to which the antiserum was raised is indeed present in the immunoreactive cell. Antibodies (predominantly IgG molecules) recognise a particular antigenic determinant and this may be a short sequence of amino acids. Molecules sharing this determinant will react with an antibody raised to this determinant; the reacting molecules may well differ in other structural aspects. Moreover, a polyclonal antiserum may contain populations of IgG molecules with different specificities (i.e. directed to different parts of the molecule used as antigen). One way around this latter problem is the use of monoclonal antibodies. Another difficulty with immunocytochemistry is that fixation may affect the antigenicity of a molecule. Despite all this, immunocytochemistry is of great value for the identification of cell types, for morphological studies and as a basis for further biochemical and recombinant-DNA studies. (See Chapter 3, this volume.)

Lymnaea stagnalis

Among the molluscs *Lymnaea* probably is immunocytochemically the most extensively investigated species (Boer *et al.* 1979, 1980, 1984a,b, 1986; Schot *et al.* 1981, 1983, 1984; Schot & Boer 1982; Boer & Van Minnen 1985; Van Minnen & Vreugdenhil 1987). Using antisera to vertebrate peptides and biogenic amines numerous immunoreactive neurons have been observed (Table I) and maps prepared

Table I. *Antisera to which neurons of* L. stagnalis *are immunoreactive*

vasopressin	somatostatin	GIP
vasotocin	substance P	PP
oxytocin	glucagon	FMRFamide
CRF	insulin	AKH
ACTH	secretin	dopamine
αMSH	gastrin	serotonin
met–enkephalin	calcitonin	histamine
leu–enkephalin	VIP	octopamine
dynorphin		

of the localisation of the immunoreactive cells in the CNS. With some antisera immunoreactive neurons are found throughout the CNS, with others only in particular ganglia . With only a small number of antisera (anti-CCK, β–endorphin, neurophysin I and II, α–neoendorphin) are negative results obtained. Because fixation may affect immunoreactivity, these negative results do not necessarily imply the absence of peptides reactive to these antisera, as only a limited number of fixatives has been used so far. In addition to antisera to vertebrate peptides and to classical transmitters, antisera to the molluscan cardio-active peptide FMRFamide and to the insect adipokinetic hormone (AKH) gave positive results (Table I).

Studying neurons in serial paraffin sections shows that some immunoreactive axons terminate on other neurons or on muscle fibres in a synapse-like fashion (Fig. 1), whereas others end in neurohaemal areas in the peripheries of nerves or commissures (Boer *et al.* 1980; Schot *et al.* 1981). These observations, which are confirmed at the electron microscope level (Schot *et al.* 1983; Boer *et al.* 1984a), suggest that the immunoreactive substances act either as neurotransmitters or as neurohormones.

Neurons of *Lymnaea* are large (diameter 20–150 μm), and this makes it possible to study consecutive sections of the same neuron with different antisera. This alternate section method showed that many neurons are immunoreactive to two or more antisera (Table II). In most cases this co-immunoreactivity is not due to lack of specificity of the antiserum, because in addition to neurons stained with two antisera, others stain with only one of the pair (Boer & Schot 1983). However, it is not clear whether the reactive antigenic determinants are present on the same or on different molecules. Since in a number of cases the antisera used were raised to essentially unrelated peptides, there might be more than one peptide and hence more than one precursor in the double and triple immunoreactive cells. It should be mentioned, however, that peptides considered unrelated, like CRF and FMRFamide have been shown to occur on the same precursor (see Chapter 15, this volume). Co-existence of

Table II. *Combinations of antisera to which neurons of* L. stagnalis *show co-existent immunoreactivity*

vasotocin – FMRFamide
vasotocin – vasopressin
vasotocin – FMRFamide – gastrin
vasotocin–serotonin–dopamine
αMSH – met – enkephalin – leu–enkephalin
αMSH – serotonin – dopamine
ACTH – met–enkephalin
met–enkephalin – leu–enkephalin
FMRFamide – gastrin
gastrin – PP
gastrin – serotonin – dopamine
gastrin – substance P
serotonin – dopamine

immunoreactivity might indicate separate packaging, transport and release of different messengers (Roubos 1984).

The results suggest that many peptides, first described in vertebrates, are also present in *Lymnaea* and that probably co-existence of peptides derived from different precursors does occur in this species. However, as we have said, speculations on the structure of immunoreactive substances solely on the basis of immunocytochemical results, are precarious.

A family of FMRF-like peptides

The cardio-active tetrapeptide FMRFamide (Phe–Met–Arg–Phe–NH_2) was isolated originally from the CNS of the clam *Macrocallista nimbosa* (Price & Greenberg 1977). It exerts pharmacological effects on a variety of molluscan muscles and neurons (e.g. Price & Greenberg 1980; Geraerts *et al.* 1981; Lloyd 1982; Austin *et al.* 1983; Cottrell *et al* . 1983; Boyd *et al.* 1984; Griffond *et al.* 1986; Ter Maat *et al.* 1986; see also Chapter 14, this volume). Furthermore, with antisera to FMRFamide, neurons have been identified in representatives of all major phyla, including the vertebrates (Boer *et al.* 1980; Greenberg *et al.* 1985).

Recently the FMRFamide precursor of *Aplysia* has been studied with recombinant-DNA techniques. It contains 28 copies of the tetrapeptide (Schaefer *et al.* 1985). It is not certain whether the tetrapeptide occurs also in other species, but peptides closely related to FMRFamide have been isolated from various molluscan species (Table III; Greenberg *et al.* 1985). There are also indications that FMRFamide-like substances occur in other phyla (Dockray *et al.* 1983). These peptides are all immunoreactive

Table III. *FMRFamide-like peptides of molluscs*

Octopus	Tyr–Gly–Gly–Phe–Met–Arg–Phe–NH_2
Macrocallista	Phe–Met–Arg–Phe–NH_2
Aplysia	Phe–Met–Arg–Phe–NH_2
Pomacea	Phe–Leu–Arg–Phe–NH_2
Helix	
	pGlu–Asp–Pro–Phe–Leu–Arg–Phe–NH_2
Lymnaea	Ser–Asp–Pro–Phe–Leu–Arg–Phe–NH_2
	Gly–Asp–Pro–Phe–Leu–Arg–Phe–NH_2

with anti-FMRFamide (aFM). The antigenicity of the FMRFamide-like substances seems largely dependent on the amidated C-terminal of the molecules (Price & Greenberg 1980). They evoke different reactions in electrophysiological test-systems; four different receptors have been distinguished for FMRFamide and related peptides (Cottrell *et al.* 1987; see also Chapter 14, this volume). The data indicate that there exists a "family" of FMRFamide-like peptides in the animal kingdom.

Lymnaea stagnalis

Immunocytochemical and biochemical results indicate that several members of the FMRFamide-family occur in *Lymnaea*. With aFM numerous immunoreactive neurons and fibre tracts were observed in the CNS (Schot & Boer 1982; Schot *et al.* 1984). As many as 12 histochemically different aFM-immunoreactive cell types can be recognised. Distinction among cell types is made on the basis of effects of fixation on the immunoreactivity of the cells, the staining results obtained with other antisera (co-existence of immunoreactivities), and immuno-adsorption results (antisera are adsorbed with homologous and heterologous antigens). Analysis of the results shows that among the 12 aFM positive cell types, 7 probably contain an antigenic determinant that is identical or closely related to the Phe–Met–Arg–Phe–NH_2 sequence (Fig. 5, Table IV); the remaining 5 aFM positive cell types apparently stain with less specific (cross reacting) IgG molecules present in the polyclonal aFM (Schot *et al.* 1984). Subsequent ultrastructural studies indicated that a histochemically defined population of aFM immunoreactive cells may comprise more than one cell type: the mean diameter of the secretory granules in cells of the same histochemical type may differ (Fig. 5; Boer & Van Minnen 1985). All this evidence suggests that there are at least 7, but probably more members of the FMRFamide peptide family present in *Lymnaea*. This notion has since been confirmed by peptide-isolation and -characterization; at least 8 peptides from *Lymnaea* react like FMRFamide in bioassay and RIA-systems. Two of these peptides have been sequenced and synthesized. They

Table IV. *Histochemical characteristics of anti-FMRFamide immunoreactive neurons in the CNS of* L. stagnalis . *Three fixatives were used; PF (paraformaldehyde), G (glutaraldehyde) or GPA (glutaraldehyde–picric acid–acetic acid). Adsorptions were carried out with homologous (aFM+FM: aFM adsorbed with FMRF–NH$_2$) or with heterologous antigens (aVT+FM: anti-vasotocin adsorbed with FMRFamide). For location of cell types see Fig. 5*

cell		antisera			
type	fixation	aFM	aFM+FM	aVT	aVT+FM
1	PF				
2	PF,GPA		–		
3	PF,G		–		
4	PF,G		–		
5	PF,G,GPA		–		
6	PF,G,GPA		–		
7	PF,G,GPA		–		

are heptapeptides of which only the N-terminal amino acid is different (Table III; Ebberink & Joosse 1985).

The axons of certain aFM immunoreactive systems form synapse-like structures with other neurons or with muscle cells (Schot *et al.* 1983; Boer *et al.* 1984a). These observations support the hypothesis that the FMRFamide-like substances contained in these systems act as neurotransmitters (see Chapter 14, this volume). In some systems the axons end in neurohaemal areas, which suggests that the messenger is released into the haemolymph as a neurohormone (system 3, Fig. 5). Because FMRFamide and related peptides obviously occur in various systems (tentacle, digestive tract, genital tract) unrelated to circulatory functions, the term "cardio-active" would seem inappropriate for these peptides (cf. Griffond *et al.* 1986).

Monoclonal antibodies

The evidence presented above indicates that a large percentage of neuronal systems are peptidergic, that many systems may contain still unknown peptides, that several members of the same peptide family may occur in one species and that peptides may occur in an identical or closely related form in different species. With this in mind, it seems worthwhile to try to identify new peptidergic systems and to further characterise established systems. This view was recently approached in *Lymnaea* with the monoclonal antibody technique (Boer & Van Minnen 1985; Van

Minnen & Vreugdenhil 1987). In invertebrates a similar approach has been followed in studies on leeches (Zipser & McKay 1981; Zipser 1982) and on *Drosophila* (Miller & Benzer 1983).

Whole CNS of *Lymnaea* were homogenized. After centrifugation, the supernatant was injected intraperitoneally into BALB/c mice (15 CNS per mouse). The mice were boosted with the same amount of antigen. Five days after the booster the spleens were removed from the mice. Spleen plasmablasts were fused with Sp 2/0–ag 14

Fig. 5. Localisation of histochemically different anti-FMRFamide immunoreactive elements in the CNS of *L. stagnalis*. Numbers in boxes indicate types (cf. Table IV). Other numbers indicate mean size (in nm) of secretory vesicles in the cells. BU, buccal; CE, cerebral; PE, pedal; PL, pleural; LP and RP, left and right parietal; V, visceral ganglion; DB, Dorsal Body, LL, lateral lobe.

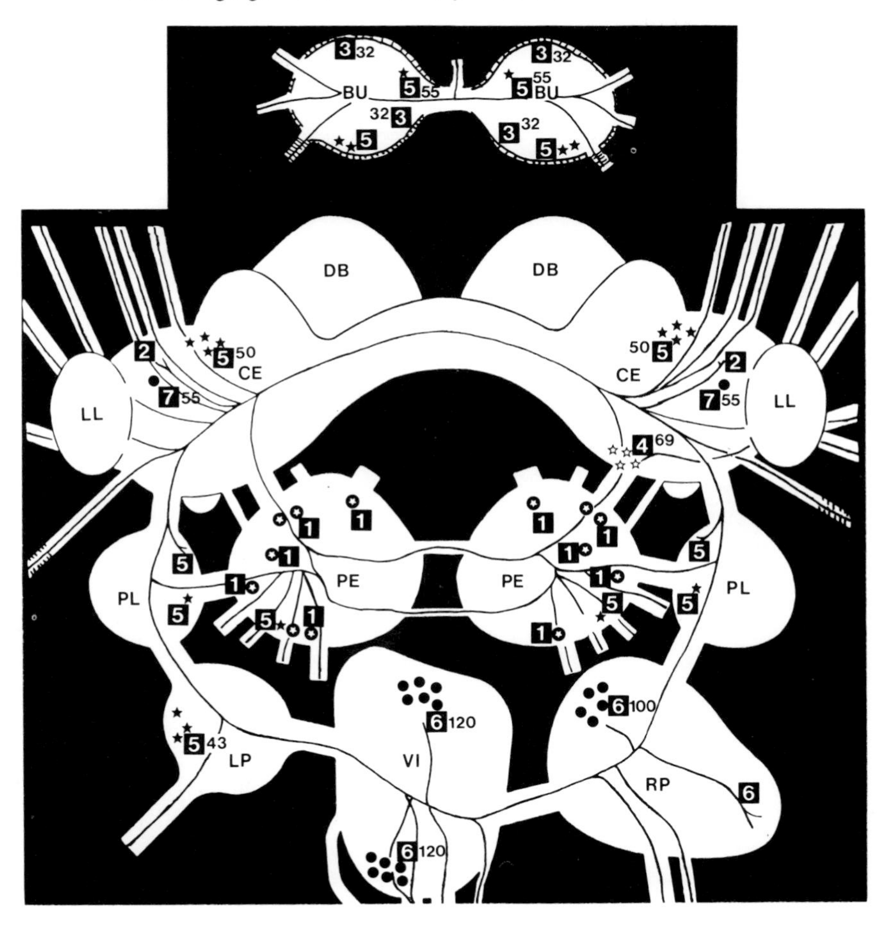

myeloma cells (Köhler & Milstein 1975). The supernatants of the hybridomas were tested for the presence of antibodies (ALMA: Anti–*Lymnaea*–Monoclonal–Antibodies) with immunocytochemistry. CNS of *Lymnaea* were quickly frozen in liquid N_2–cooled freon 22, lyophilized, fixed in paraformaldehyde vapour (1 h at 60°C) and embedded in paraffin. From these CNS, 8 μm sections were made. One slide contained one CNS. The sections were incubated for 1–2 h at room temperature with the culture medium of the clones. As a second antibody, rabbit anti-mouse IgG conjugated with horseradish peroxidase, was used. Cells of cultures producing

Figs. 6–10. ALMA (Anti *Lymnaea* Monoclonal Antibody) stain the Dorsal Bodies (Fig. 6, ×100), the LGC (Fig. 7, ×150), the CDC (Fig. 8, DB Dorsal Body, ×150), the LYC (Fig. 9, ×115) and anti-FMRFamide (aFM) positive neurons (Fig. 10a and b adjacent sections stained with aFM and ALMA, respectively). At arrows, cell positive with aFM, negative with ALMA. Counterstained with haemalum. ×150.

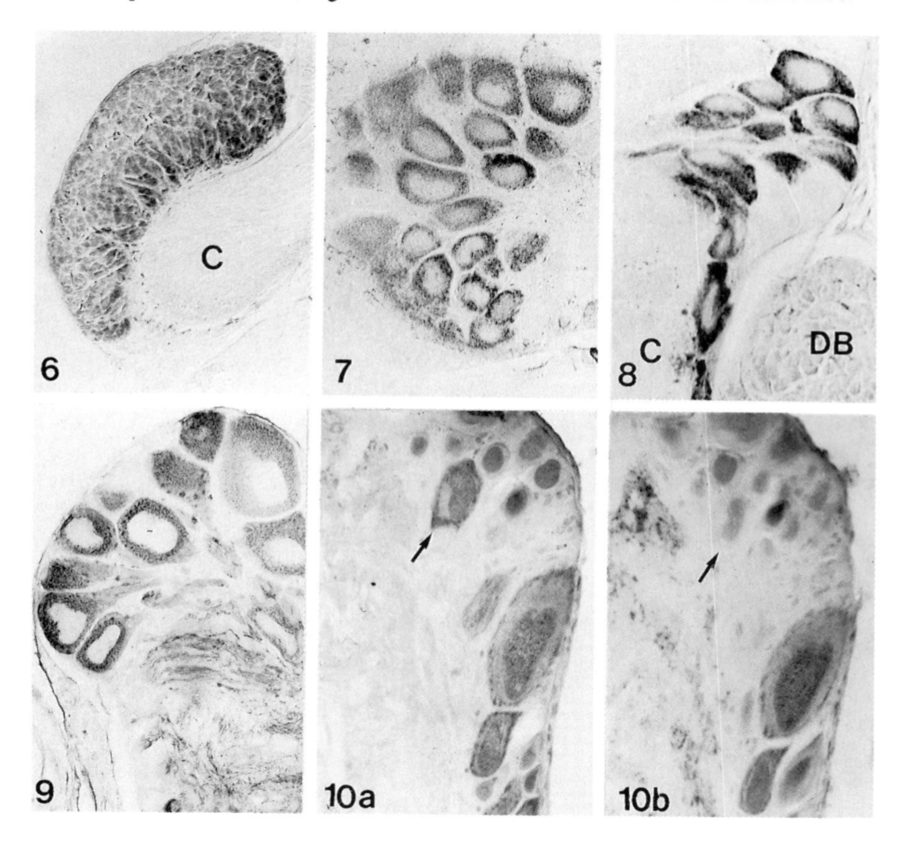

monoclonal antibodies to particular neuron types were used for the production of ascites in mice.

Antibodies present in the culture medium of many of the wells were directed against non-neural elements (muscle cells, connective tissue cells), to all neurons or to all axons. These hybridomas were discarded, as it was considered likely that the antibodies were directed to non-neural peptides, or to common organelles (e.g. microtubules) or substances (e.g. enzymes), rather than to specific neuropeptides. The attention was focused on hybridomas that produced antibodies to previously characterised peptidergic systems, to the endocrine Dorsal Bodies and to well defined, hitherto unidentified groups of neurons.

Figs. 6–10 give some examples. In these figures we see selective staining of the Dorsal Bodies (Fig. 6), of the cerebral LGC (Fig. 7, Light Green Cells, which produce a growth promoting factor (Geraerts 1976; see also Chapter 13, this volume), the CDC (Fig. 8), the LYC (Fig. 9, Light Yellow Cells, right parietal ganglion, function unknown) and of some of the aFM immunoreactive cells (Fig. 10). Here, not only the cell bodies appeared to be stained, but also the axons and axon terminals. With the ALMA to the LGC and LYC, neurons and axon tracts were also stained in other ganglia.

It seems plausible that the ALMA reacting with the previously characterised (neuro)endocrine cells are directed to peptidergic secretion products. By analogy this may also be assumed for the hitherto unidentified neurons. It cannot be decided on the basis of the evidence presently available whether the antibodies were raised against single peptides or to precursor molecules or to intermediates. Some of the peptides will be too small (<1000 D; see also Chapter 13, this volume) to evoke an immune response (Roitt 1984). However, if an ALMA is formed against an epitope on a precursor molecule, the end product containing this epitope will also recognise the antibody.

Studies with ALMA

Preliminary experiments with the anti-LGC antibody show that ALMA can be used for the purification of peptides: an extract of the CNS prepared with an affinity column loaded with anti-LGC appeared to possess a high level of bioactivity in an assay for growth hormone (Fig. 11, R. Kerkhoven & A. Doderer, personal communication).

One of the selected ALMA appears to react with two giant neurons. Previously these neurons were identified using anti-ACTH (Boer *et al.* 1979). Further studies using the alternate section method showed that this ALMA also reacts with neurons staining with anti-αMSH, with anti-met–enkephalin or with anti-αMSH and with anti-met–enkephalin (Figs. 12–14, Table V). These cell types are related and contain a substance possessing the same antigenic determinant. Because the cells are all

Table V. *The four cell types in* L. stagnalis *which are immunoreactive to a particular monoclonal antibody (ALMA) and to antisera raised against ACTH (a–ACTH), αMSH (a–αMSH) or/and met–enkephalin (a–MetEn).*

Neuron type	antisera			
	ALMA	a–αMSH	a–ACTH	a–MetEn
1	+	–	+	–
2	+	+	–	–
3	+	+	–	+
4	+	–	–	+

Fig. 11 (a.) Anti-LGC (see Fig. 6) ALMA was used for affinity chromatography (Affinity Chromatography, principles and methods. Pharmacia Fine Chemicals, ed., 1983) of crude extracts of CNS of *Lymnaea.* The figure shows HPLC runs of fraction 4 and of fractions 5 and 6 (pooled); fraction 4 has a peak (arrow) in the same position as observed in the procedure for isolation of the growth hormone. (b.)Incorporation of ^{45}Ca (cpm) into the shell edge of *Lymnaea* upon injection of snails with fraction 4 (n=5) is significantly higher than upon injection of distilled water (n=7) or fractions 5 and 6 (n=7).

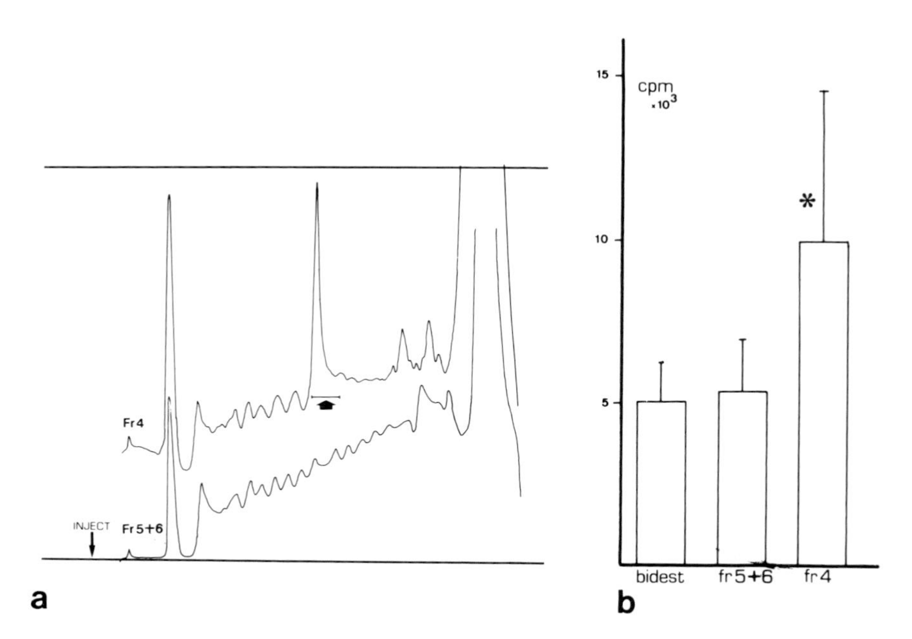

reactive to antibodies raised to peptides derived from vertebrate proopiomelanocortin (POMC), it is tempting to speculate that in *Lymnaea* a similar precursor exists (Boer *et al.* 1987).

Based on the idea of the universality of various biologically active peptides (Scharrer 1978), it may be that certain ALMA will also react with neurons in other, unrelated species. Preliminary observations sustain this assumption. For example, the anti-LYC monoclonal antibody reacts with particular neurons in the Colorado potato beetle and with neurons in the guppy (Fig. 15, 16) and the ALMA staining neurons also reacting with antibodies against POMC-derived peptides, reacts with neuron populations in the hypothalamus of the frog (R. Kerkhoven, personal communication). There are also instances where peptides are species- or genus-specific. Here it would seem possible to use monoclonal antibodies to study the relatedness of species. Thus, preliminary observations indicate that the ALMA against the DB and the CDC react only with neurons of *L. stagnalis*, whereas the ALMA against the LGC also react with neurons of other Lymnaeidae (*L. ovata, L. palustris*). None of these three ALMA react with neurons of species of other pulmonate families (*Biomphalaria glabrata, Bulinus truncatus, Helix aspersa, Limax maximus*).

Figs. 12–14. Alternate sections of neurons immunoreactive to a particular ALMA (Figs. 12a–14a) stain either with anti-ACTH (Fig. 12b, × 250), anti-αMSH (Fig. 13b, × 150) or with anti-met–enkephalin (Fig. 14b, × 250). At arrows in Fig. 13a,b cells that are only anti-αMSH positive.

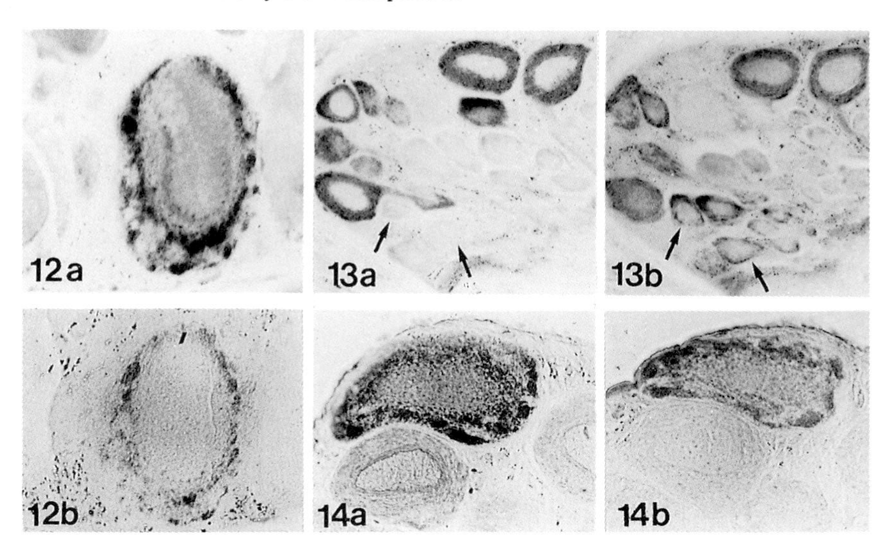

Figs. 15, 16. With anti-LYC ALMA, neurons are stained in the hypothalamic area of the guppy (Fig. 15, ×50, V ventricle; 15a high magnification of some positive neurons. Note beaded fibres at arrows. ×625) and in the brain of the Colorado potato beetle (Fig. 16, ×400).

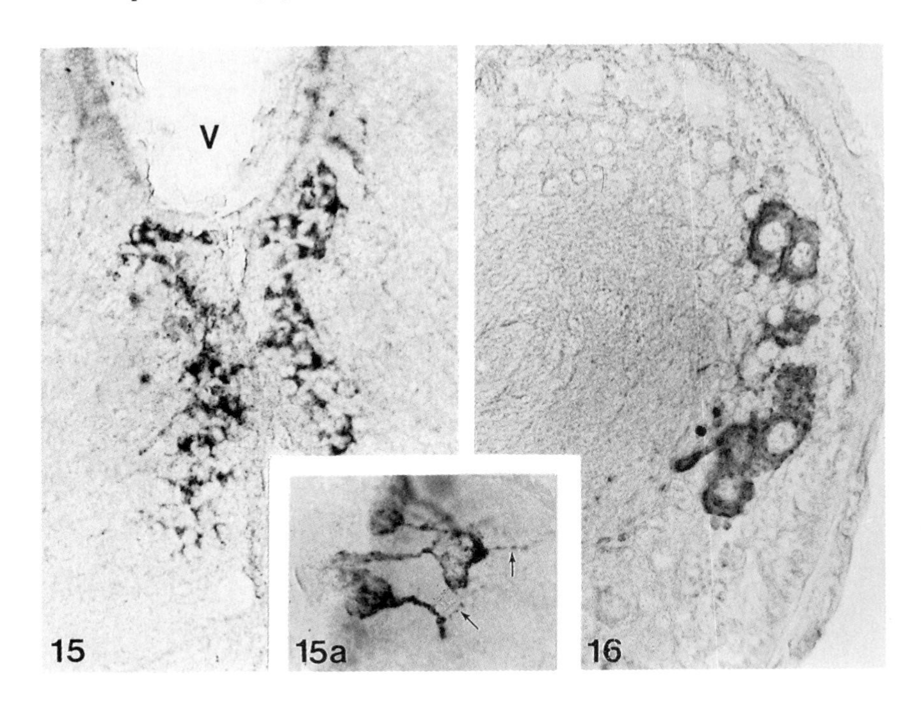

References

Austin, T., Weiss, S. & Lukowiak, K. (1983). FMRFamide effects on spontaneous and induced contractions of the anterior gizzard in *Aplysia*. *Can. J. Physiol. Pharmcol.*, **61**, 949–953.

Barnard, C.S. & Dockray, G.J. (1984). Increases in arterial blood pressure in the rat in response to a new vertebrate neuropeptide, LPLRFamide, and a related molluscan peptide, FMRF–amide. *Regulatory Peptides*, **8**, 209–215.

Beaudet, A. & Descarries, L. (1978). The monoamine innervation of rat cerebral cortex: synaptic and nonsynaptic axon terminals. *Neuroscience*, **3**, 851–860.

Bodenmüller, H., Schaller, H.C. & Darai, G. (1980). Human hypothalamus and intestine contain a *Hydra* neuropeptide. *Neurosci. Lett.*, **16**, 71.

Boer, H.H. & Minnen, J. van (1985). Immunocytochemistry of peptidergic systems in the pond snail *Lymnaea stagnalis*. *Peptides*, **6**, Suppl. 3, 459–463.

Boer, H.H. & Schot, L.P.C. (1983). Phylogenetic aspects of peptidergic systems. In: *Molluscan Neuro-Endocrinology* (ed. Lever, J. & Boer, H.H.). North Holland Publ. Co., Amsterdam, Oxford, New York, pp. 9–13.

Boer, H.H., Groot, C., Jong–Brink, M.de & Cornelisse, C.J. (1977). Polyploidy in the freshwater snail *Lymnaea stagnalis* (Gastropoda, Pulmonata). A cytophotometric analysis of the DNA in neurons and some other cell types. *Neth. J. Zool.*, 245–252.

Boer, H.H., Geraerts, W.P.M., Schot, L.P.C. & Ebberink, R.H.M. (1986). Immunocytochemical and physiological studies on (neuroendocrine) neurons of the pond snail *Lymnaea stagnalis*. In: *Handbook of comparative aspects of opioid and related neuropeptide mechanisms* (ed. Stefano, G.B.) Vol. 1, pp. 223–241. CRC Press, Boca Raton Fl., USA.

Boer, H.H., Schot, L.P.C., Veenstra, J.A. & Reichelt, D. (1980). Immunocytochemical identification of neural elements in the central nervous system of a snail, some insects, a fish and a mammal with an antiserum to the molluscan cardio-excitatory tetrapeptide FMRFamide. *Cell Tissue Res.*, **213**, 21–27.

Boer, H.H., Schot, L.P.C., Reichelt, D., Brand, H. & Maat, A. ter, (1984a). Ultrastructural immunocytochemical evidence for peptidergic neurotransmission in the pond snail *Lymnaea stagnalis*. *Cell Tissue Res.* **238**, 197–201.

Boer, H.H., Schot, L.P.C., Roubos, E.W., Maat, A. ter, Lodder, J.C., Reichelt, D. & Swaab, D.F. (1979). ACTH-like immunoreactivity in two electrotonically coupled giant neurons in the pond snail *Lymnaea stagnalis*. *Cell Tissue Res.*, **202**, 231–240.

Boer, H.H., Schot, L.P.C., Steinbusch, H.W.M., Montagne, C. & Reichelt, D. (1984b). Co-existence of immunoreactivity to anti-dopamine, anti-serotonin and anti-vasotocin in the cerebral giant neuron of the pond snail *Lymnaea stagnalis*. *Cell Tissue Res.*, **238**, 411–412.

Boer, H.H., Minnen, J. van, Stefano, G.B. & Leung, M.K. (1987). Opioid peptides in the central nervous sytem (CNS) of *Lymnaea stagnalis*. In: *Neurobiology, Molluscan Models* (ed. Boer, H.H., Geraerts, W.P.M., & Joosse, J.) North Holland Publ. Co., Amsterdam, Oxford, New York, pp. 68–72.

Boyd, P.J., Osborne, N.N. & Walker, R.J. (1984). The pharmacological actions of 5–hydroxytryptamine, FMRF-amide and substance P and their possible occurrence in the heart of the snail *Helix aspersa* L. *Neurochem. Int.*, **6**, 633–640.

Buma, P. & Roubos, E.W. (1986). Ultrastructural demonstration of nonsynaptic release sites in the brain of the snail *Lymnaea stagnalis*, the insect *Periplaneta americana*, and the rat. *Neuroscience*, **17**, 867–870.

Burnstock, G. (1976). Do some nerve cells release more than one transmitter? *Neuroscience*, **1**, 239–248.

Burnstock, G. (1981). Neurotransmitters and trophic factors in the autonomic nervous system. *J. Physiol.*, **313**, 11–25.

Burnstock, G. & Höfkelt, T. (1979). Non-adrenergic, non-cholinergic neuroeffector mechanisms in the nervous system. *Neurosciences Res. Prog. Bulll.*, **17**, 383–487.

Buijs, R.M. & Swaab, D.F. (1979). Immunoelectronmicroscopical demonstration of vasopressin and oxytocin synapses in the rat limbic system. *Cell Tissue Res.*, **204**, 355–367.

Cottrell, G.A. (1977). Identified amine-containing neurones and their synaptic connexions. *Neuroscience*, **2**, 1–18.

Cottrell, G.A., Schot, L.P.C. & Dockray, G.J. (1983). Identification and probable role of a single neurone containing the neuropeptide *Helix*–FMRF–amide. *Nature*, **304**, 638–640.

Cottrell, G.A., Bewick, G.S. & Davies, N.W. (1987). Multiple receptors of the "FMRFamide-series" of intercellular messengers. In: *Neurobiology, Molluscan Models* (ed. Boer. H.H., Geraerts, W.P.M. & Joosse, J.), North Holland Publ. Co., Amsterdam, Oxford, New York, pp. 115–123.

Cournil, I., Geffard, M., Moulins, M. & Le Moal, M. (1984). Coexistence of dopamine and serotonin in an identified neuron of the lobster nervous system. *Brain Res.*, **310**, 397–400.

Deijnen, J.E. van (1985). An immunocytochemical study of the optic and supraoesophageal ganglia of the crayfish *Astacus leptodactylus* (Nordmann 1840) with antisera against biologically active peptides of vertebrates and invertebrates. *Cell Tissue Res.*, **240**, 175–183.

Dictus, W.J.A.G., Jong–Brink, M. de & Boer, H.H. (1987). A neuropeptide (Calfluxin) is involved in the regulation of calcium influx into mitochondria of the albumen gland of the freshwater snail *Lymnaea stagnalis*. *Gen. Comp. Endocrinol.* **65**, 439–450.

Dismukes, R.K. (1979). New concepts of molecular communication among neurons. *Behav. Br. Sciences*, **2**, 409–448.

Dockray, G.J., Reese, J.R., Shively, J., Gayton, R.J. & Barnard, C.S. (1983). A novel active pentapeptide from chicken brain identified by antibodies to FMRFamide. *Nature*, **305**, 328–330.

Dogterom, G. & Loenhout, H. van (1983). Specificity of ovulation hormones of some basommatophoran species studied by means of iso- and heterospecific injections. *Gen. Comp. Endocrinol.*, **52**, 121–125.

Duve, H. & Thorpe, A. (1980). Isolation and localization of an insect insulin-like material: immunological, biological and physical characteristics. *Gen. Comp. Endocrinol.*, **40**, 363–364.

Ebberink, R.H.M. & Joosse, J. (1985). Molecular properties of various snail peptides from brain and gut. *Peptides*, **6**, Suppl. 3:451–457.

Fritsch, H.A.R., Noorden, S. van & Pearse, A.G.E. (1979). Localization of somatostatin-, Substance P- and calcitonin-like immunoreactivity in the neural ganglion of *Ciona intestinalis* L. (Ascidiaceae). *Cell Tissue Res.*, **202**, 263–274.

Geraerts, W.P.M. (1976). Control of growth by the neurosecretory hormone of the Light Green Cells in the freshwater snail *Lymnaea stagnalis*. *Gen. Comp. Endocrinol.*, **29**, 61–71.

Geraerts, W.P.M. & Bohliken, S. (1976). The control of ovulation in the hermaphroditic freshwater snail *Lymnaea stagnalis* by the neurohormone of the caudodorsal cells. *Gen. Comp. Endocrinol.*, **28**, 350–357.

Geraerts, W.P.M., Leeuwen, J.P.Th.M. van, Nuyt, K. & With, N.D. de (1981). Cardioactive peptides of the CNS of the pulmonate snail *Lymnaea stagnalis*. *Experientia*, **37**, 1168–1169.

Geraerts, W.P.M., Tensen, C.P. & Hogenes, Th.M. (1983). Multiple release of peptides by electrically active neurosecretory Caudo-Dorsal Cells of *Lymnaea stagnalis*. *Neurosci. Lett.*, **41**, 151–155.

Greenberg, M.J., Price, D.A. & Lehman, H.K. (1985). The FMRFamide-like peptides of molluscs and vertebrates: distribution and evidences of function. In: *Proc. Int. Symp. Neurosecretion*, Tokyo, Sept. 1984.

Griffond, B., Boer, H.H. & Wijdenes, J. (1986). Localization and function of an FMRFamide-like substance in the aorta of *Helix aspersa*. *Cell Tissue Res.*, **246**, 303–307.

Grimmelikhuijzen, C.J.P. (1984). Peptides in the nervous system of coelenterates. In: Falkmer, S. Håkanson, R. & Sundler, F. (ed.) *Evolution and tumor pathology of the neuroendocrine system*. Elsevier, Amsterdam, pp. 39–58.

Grimm–Jørgensen, Y. (1978). Immunoreactive thyrotropin-releasing factor in a gastropod. Distribution in the central nervous system and haemolymph of *Lymnaea stagnalis*. *Gen. Comp. Endocrinol.*, **35**, 387–390.

Guillemin, R. (1981). Hypothalamic hypophysiotropic peptides, known and unknown. *J. Endocr.*, **90**, 3P–10P.

Hemminga, M.A. (1984). Regulation of glycogen metabolism in the freshwater snail *Lymnaea stagnalis*. Thesis Vrije Universiteit Amsterdam, pp. 1–149.

Hökfelt, T., Ljungdahl, A., Steinbusch, H., Verhofstad, A., Nilsson, G., Brodin, E., Pernow, B. & Goldstein, M. (1978). Immunohistochemical evidence of substance P-like immunoreactivity in some 5–hydroxytryptamine-containing neurons in the rat central nervous system. *Neuroscience*, **3**, 517–538.

Jan, Y.N. & Jan, L.Y. (1983). A LHRH-like peptidergic neurotransmitter capable of "action at a distance" in autonomic ganglia. *TINS*, **6**, 320–325.

Jansen, R.F. & Bos, N.P.A. (1983). Neurophysiology of an identified neuron modulating ovulation hormone producing caudo-dorsal cells of *Lymnaea stagnalis*. In: *Molluscan Neuro-Endocrinology* (eds. Lever, J. & Boer, H.H.), North Holland Publ. Co., Amsterdam, Oxford, New York, pp. 78–81.

Joosse, J. (1986). Neuropeptides: peripheral and central messengers of the brain. In: *Comparative Endocrinology: Developments and Directions*, (ed. Ralph, Ch.). Alan R. Liss Inc., pp. 13–22.

Joosse, J., Geraerts, W.P.M. (1983). Endocrinology. In: *The Mollusca 4* (ed. Wilbur, K.M. & Saleuddin, A.S.M.). Academic Press, New York, pp. 317–406.

Köhler, G. & Millstein, C. (1975). Continuous cultures of fused cells secreting antibody of defined specificity. *Nature*, **256**, 495.

Larsson, L–I. (1980). On the possible existence of multiple endocrine, paracrine and neurocrine messengers in secretory cell systems. *Invest. Cell Path.*, **3**, 73–85.

Leung, M.K. & Stefano, G.B. (1984). Isolation and identification of enkephalins in pedal ganglia of *Mytilus edulus* (Mollusca). *Proc. Natl. Acad. Sci. USA*, **81**, 955–956.

Lloyd, P.E. (1982). Cardioactive neuropeptides in gastropods. *Fed. Proc.*, **41**, 2948–2952.

Lundberg, J.M. & Hökfelt, T. (1983). Coexistence of peptides and classical neurotransmitters. *TINS*, **6**, 325–333.

Maat, A. ter & Jansen, R.F. (1984). The egg-laying behaviour of the pond snail: electrophysiological aspects. In: *Biosynthesis, Metabolism and Mode of Action of Invertebrate Hormones* (ed. Hoffmann, J. & Porchet, M.), Springer, Berlin, pp. 57–62.

Maat, A. Ter, Jansen, R.F. & Bos, N.P.A. (1986). The effect of met–enkephalin and FMRFamide on the excitability of the neurosecretory Caudo-Dorsal Cells of the pond snail. In: *Handbook of comparative aspects of opioid and related neuropeptide mechanisms* (ed. Stefano, G.B.) , CRC Press, Boca Raton, Fl., USA, Vol. IIpp. 145–157.

Marchand, C.R., Wijdenes, J. & Schot, L.P.C. (1982). Localisation par la technique cytoimmunoenzymologique d'un neuropeptide cardio-excitateur (le FMRF-amide) dans le collier nerveux péri-oesophagien d'*Helix aspersa* Müller (Gastéropode, pulmoné, stylommmatophore) *C.R. Acad. Sci. Paris*, **294**, 39–44.

Martin, R., Geis, R., Holl, R., Schäfer, M. & Voigt, K.H. (1983). Co-existence of unrelated peptides in ocytocin and vasopressin terminals of rat neurohypophyses: immunoreactive methionine–enkephalin, leucine–enkephalin, and cholecystokinin-like substances. *Neuroscience*, **8**, 213–227.

Mayeri, E. & Rothman, B.S. (1981). Nonsynaptic peptidergic neurotransmission in the abdominal ganglion of *Aplysia*. In: *Neurosecretion – Molecules, Cells, Systems* (ed. Farner, D.S. & Lederis, K.), pp. 305–316. Plenum Press, New York.

Miller, C.A. & Benzer, S. (1983). Monoclonal antibodies cross-reacting between *Drosophila* and human brain. *Proc. Natl. Acad. Sci. USA*, **80**, 7641–7645.

Minnen, J. van & Vreugdenhil, E. (1987). The occurrence of gonadotropic hormones in the central nervous system and reproductive tract of *Lymnaea stagnalis*. An immunocytochemical and *in situ* hybridization study. In: *Neurobiology, Molluscan Models* (ed. Boer, H.H., Geraerts, W.P.M. & Joosse, J.) North Holland Publ. Co., Amsterdam, Oxford, New York, pp. 62–67.

Moed, P.J., Bos, N.P.A., Kits, K.S. & Maat, A. ter (1987). The role of release products and second messengers in the regulation of electrical activity of the neuroendocrine Caudo-Dorsal Cells of *Lymnaea stagnalis*. In: *Neurobiology, Mollusan Models* (ed. Boer, H.H., Geraerts, W.P.M. & Joosse, J.) North Holland Publ. Co., Amsterdam, Oxford, New York, 11. 194–199.

Mues, G., Fuchs, I., Wei, E.T., Weber, E., Evans, C.J., Barchas, J.D. & Chang, J.K. (1982). Blood pressure elevation in rats by peripheral administration of Tyr–Gly–Gly–Phe–Met–Arg–Phe and the invertebrate neuropeptide, Phe–Met–Arg–Phe–NH_2. *Life Sciences*, **31**, 2555–2561.

Osborne, N.N. (1977). Do snail neurones contain more than one neurotransmitter? *Nature*, **270**, 622–623.

Osborne, N.N. & Dockray, G.J. (1982). Bombesin-like immunoreactivity in specific neurones of the snail *Helix aspersa* and an example of the coexistence of substance P and serotonin in an invertebrate neurone. *Neurochem. Intern.*, **4**, 175–180.

Plesch, B. (1977). An ultrastructural study of the musculature of the pond snail *Lymnaea stagnalis* (L.), *Cell Tissue Res.*, **180**, 317–340.

Price, D.A. & Greenberg, M.J. (1977). Purification and characterization of a cardioexcitotatory neuropeptide from the central ganglia of a bivalve mollusc. *Prep. Biochem.*, **7**, 261–281.

Price, D.A. & Greenberg, M.J. (1980). Pharmacology of the molluscan cardioexcitatory neuropeptide FMRFamide. *Gen. Pharmac.*, **11**, 237–241.

Roitt, I. (1984). *Essential Immunology*, 5th ed. Blackwell Scientific Publications, Oxford.

Roubos, E.W. (1984). Cytobiology of the ovulation neurohormone producing neuroendocrine Caudo-Dorsal Cells of *Lymnaea stagnalis*. *Int. Rev. Cytology*, **89**, 295–347.

Roubos, E.W., Buma, P. & Roos, W.F. de (1983). Ultrastructural correlates of electrotonic and neurochemical communication in *Lymnaea stagnalis* with particular reference to nonsynaptic transmission and neuroendocrine cells. In: *Molluscan Neuro-Endocrinology* (ed. Lever, J. & Boer, H.H.). North Holland Publ. Co., Amsterdam, New York, Oxford, pp. 68–74.

Schaefer, M., Piciotto, M.R., Kreiner, T., Kaldany, R.R., Taussig, R. & Scheller, R.H. (1985). *Aplysia* neurons express a gene encoding multiple FMRFamide neuropeptides. *Cell*, **41**, 457–467.

Scharrer, B. (1978). Peptidergic neurons: facts and trends. *Gen. Comp. Endocrinol.*, **34**, 50–62.

Schmidt, E.D. & Roubos, E.W. (1987). Morphological basis for nonsynaptic communication within the central nervous system by exocytotic release of secretory material from the egg-laying stimulating neuroendocrine caudodorsal cells of *Lymnaea stagnalis*. *Neuroscience* , **20**, 247–257.

Schmitt, F.O. (1984). Molecular regulators of brain function: A new view. *Neuroscience*, **13**, 991–1001.

Schot, L.P.C. & Boer, H.H. (1982). Immunocytochemical demonstration of peptidergic cells in the pond snail *Lymnaea stagnalis* with an antiserum to the molluscan cardioactive tetrapeptide FMRFamide. *Cell Tissue Res.*, **225**, 347–254.

Schot, L.P.C., Boer, H.H. & Montagne–Wajer C. (1984). Characterisation of multiple immunoreactive neurons in the central nervous system of the pond snail *Lymnaea stagnalis*. *Histochem.*, **81**, 373–378.

Schot, L.P.C., Boer, H.H. & Wijdenes, J. (1983). Localization of neurons innervating the heart of *Lymnaea stagnalis* studied immunocytochemically with anti-FMRFamide and anti-vasotocin. In: *Molluscan Neuro-Endocrinology* (ed. Lever, J. & Boer, H.H.), pp. 203–208. North Holland Publ. Co., Amsterdam, Oxford, New York.

Schot, L.P.C., Boer, H.H., Swaab, D.F. & Noorden, S. van (1981). Immunocytochemical demonstration of peptidergic neurons in the central nervous system of the pond snail *Lymnae stagnalis* with antisera raised to biologically active peptides of vertebrates. *Cell Tissue Res.*, **216**, 272–291.

Stefano, G.B. & Leung, M.K. (1984). Presence of met–enkephalin–Arg–Phe in molluscan neural tissues. *Brain Res.*, **298**, 362–365.

Strumwasser, F., Kaczmarek, L.K., Chiu, A.Y., Keller, E., Jennings, K.R. & Viele, D.P. (1980). *Peptides: integrators of cell and tissue function* (ed. Bloom, F.E.), pp. 197–218. Raven Press, New York.

Thorndyke, M.C. & Probert, L. (1979). Calcitonin-like cells in the pharynx of the ascidian *Styela clava*. *Cell Tissue Res.*, **203**, 301–309.

Veenstra, J.A., Romberg–Privee, H.M., Schooneveld, H. & Polak, J.M. (1984). Immunocytochemical localization of peptidergic neurons and neurosecretory cells in the neuro-endocrine system of the Colorado potato beetle with antisera to vertebrate regulatory peptides. *Histochemistry*, **82**, 9–18.

Viveros, O.H., Diliberto, E.J. & Daniels, A.J. (1983). Biochemical and functional evidence for the cosecretion of multiple messengers from single and multiple compartments. *Federation Proc.*, **42**, 2923–2928.

Vizi, E.S. (ed.) (1984). *Non-synaptic Interactions between Neurons: Modulation of Neurochemical Transmission, Pharmacological and Clinical Aspects*. Wiley, Chichester.

Voigt, K.H., Kiehling, C., Frösch, D., Bickel, U., Geis, R. & Martin R. (1983). Identity and function of neuropeptides in the vena cava neuropil of *Octopus*. In: *Molluscan Neuro-Endocrinology* (ed. Lever, J. & Boer, H.H.), pp. 228–235. North Holland Publ. Co., Amsterdam, Oxford, New York.

Wendelaar Bonga, S.E. (1972). Neuroendocrine involvement in osmoregulation in a freshwater mollusc, *Lymnaea stagnalis*. *Gen. Comp. Endocrinol. Suppl.***3**, 308–316.

Zipser, B. (1982). Complete distribution patterns of neurons with characteristic antigens in the leech central nervous system. *J. Neurosci.*, **2**, 1453–1464.

Zipser, B. & McKay, R. (1981). Monoclonal antibodies distinguish identifiable neurons in the leech. *Nature*, **289**, 549–554.

CHRISTIAN RÉMY & JEAN VIEILLEMARINGE

Immunocytology of insect peptides and amines

In vertebrate endocrinology, significant progress has been made over the past 30 years due to the availability of immunochemical techniques. Marshall (1951) obtained the first immunohistological result in the adenohypophysis of several mammalian species using an anti-ACTH antiserum. These immunohistological investigations were made possible because of the early isolation of vertebrate hormones and neurohormones.

Invertebrate neuropeptides, however, were not purified and sequenced until more recently because of the small size of central nervous systems (CNS) or neurohaemal organs. The first invertebrate neuropeptide to be isolated was proctolin from the cockroach, *Periplaneta americana* (Brown & Starratt 1975). Invertebrate neurosecretory cells have however been visualized by histochemical staining methods for about forty years.

Most antibodies used in vertebrate immunocytochemistry are raised against mammalian hormones or neurohormones. They nevertheless give positive immunoreactions at the level of lower vertebrate hypophyseal and hypothalamic cells. It was tempting therefore to discover if such antisera could also generate positive results in invertebrates. The first evidence of the existence of an immunochemical relationship between vertebrate and invertebrate neuropeptides was produced in 1975 by Grimm–Jorgensen & MacKelvy. Using radioimmunoassays, they found an immunoreactive thyrotropin releasing hormone (TRH)-like substance in gastropod ganglia.

In 1977, some invertebrate neurosecretory cells synthesizing neurosecretory products related to vertebrate neuropeptides or vertebrate gastro-entero-pancreatic peptides, were visualized by immunocytochemical techniques, in earthworm ganglia (Sundler *et al.* 1977) and insect suboesophageal ganglia (Rémy *et al.* 1977). These early findings caused surprise and even scepticism among several scientists. There are now over two hundred different immunocytochemical results in this area, about seventy five of which concern insects.

In insects, two kinds of amines (catecholamines and indolylalkylamines) have been localized in the nervous system by classical histofluorescence techniques (for reviews see Klemm 1976; Evans 1980). However these formaldehyde-induced fluorescence

Table 1

Orders	Species	Localization	Authors
Neurophysin antiserum			
Dic.	*Periplaneta americana*	Brian, CC	Verhaert *et al.*, 1984
Neurophysin, vasopressin antisera			
Dic.	*Periplaneta americana*	Brain	Verhaert *et al.*, 1984
Or.	*Locusta migratoria*	SOG	Rémy *et al.*, 1979
Neurophysin, vasopressin, vasotocin antisera			
P.	*Clitumnus extradentatus*	SOG	Rémy *et al.*, 1977
Vasopressin antiserum			
Dic.	*Periplaneta americana*	Brain, CC, CA	Verhaert *et al.*, 1984
Or.	*Acheta domesticus*	SOG	Strambi *et al.*, 1979
Or.	*Locusta migratoria*	SOG	Veenstra, 1984.
C.	*Leptinotarsa decemlineata*	SOG	Veenstra, 1984
Oxytocin antiserum			
Dic.	*Leucophaea maderae*	CC	Hansen *et al.*, 1982
Dic.	*Periplaneta americana*	Brain, CC, CA	Verhaert *et al.*, 1984
Vasopressin, vasotocin, oxytocin antisera			
C.	*Leptinotarsa decemlineata*	Brain, SOG	Veenstra *et al.*, 1984

CA. Corpora allata, CC.: Ca cardiaca, SOG.: suboesophageal ganglion. C.: Coleoptera, Dic.: Dictyoptera, Or.: Orthoptera, P.: Phasmida

methods, even when a microspectrofluorometric analysis is carried out, permit only tentative conclusions about the precise identity of the transmitters. To overcome this difficulty, antisera raised against the different amines have been used since 1983.

A. Insect neuropeptides related to vertebrate or invertebrate brain-gut neuropeptides

I. Immunocytochemistry

Immunocytochemical investigations of neuropeptides have been conducted in about thirty-five different species of insects belonging to eight distinct orders. Immunoreactive cells have been located in the entire CNS: brain, suboesophageal ganglion, neurohaemal organs such as corpora cardiaca, thoracic and abdominal ganglia. They were also located in the midgut. In most cases perikarya and only some of their processes were observed. Descriptions concerning whole immunoreactive neurosecretory systems are more scarce.

1. *Relationships with vertebrate brain–gut neuropeptides*

Relationships between insect neurosecretory products and number of vertebrate brain-gut neuropeptides have been found: neurohormones, opioids, and gastro-entero-pancreatic (GEP) neurohormonal polypeptides.

a. Vertebrate neurohormones

Except for thyroliberin (TRH) antiserum, antisera raised against all the vertebrate neurohormones have been successfully used: neurohypophyseal polypeptides such as vasopressin (the first to be reported), vasotocin, oxytocin, neurophysin and releasing hormones or release-inhibiting hormones such as gonadoliberin (LH–RH), somatostatin (SRIF), corticoliberin (CRF), somatoliberin (GRF).

These results are summarized in Table 1. Vasopressin or vasotocin-like neuropeptides were mostly located in suboesophageal ganglia. In *Clitumnus extradentatus* and *Locusta migratoria* (Rémy *et al.* 1977, 1979) for example, they were found exclusively at the level of two paraldehyde fuchsin-positive perikarya (Fig. 1). Some differences were noted between these two insects: in *Clitumnus*, vasopressin and vasotocin antisera give positive immunoreactions, but in *Locusta* only vasopressin antiserum crossreacts with the suboesophageal neuropeptide, thereby showing that these two insect neuropeptides are not identical.

In these immunocytochemical studies, the specificity of the antisera used is of primary importance. In *Locusta migratoria* we observed identical results with two distinct vasopressin antisera, whereas by using two vasopressin antisera from other laboratories, Veenstra (1984) obtained crossreactions in the two large paraldehyde fuchsin-positive perikarya and at the level of two clusters of small suboesophageal cells. These small cells are paraldehyde fuchsin-negative and it is likely that the neuropeptide synthesized here is different from that produced by the two large perikarya.

In vertebrates, neurohypophyseal hormones are linked to protein carriers: the neurophysins. In insects, a similar association can be inferred with some neuropeptides. In *Clitumnus* or *Locusta* suboesophageal ganglia for example, by using neurophysin antisera, positive immunoreactions are obtained in the two large cell bodies synthesizing the vasopressin-like substance (Fig. 2). However, there is as yet no found evidence to suggest that insect neurohormones also have carriers.

Neurosecretory products related to vertebrate releasing or release inhibiting

Fig. 1. *Locusta migratoria* (Vasopressin antiserum). Transverse section of the two immunoreactive suboesophageal cells. ×320.

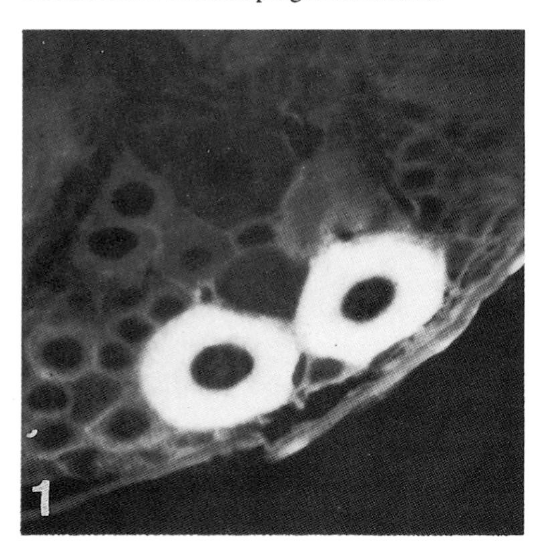

Fig. 2. *Locusta migratoria* (Neurophysin antiserum). Longitudinal section of the two suboesophageal cells synthesizing the vasopressin-like substance. × 320.

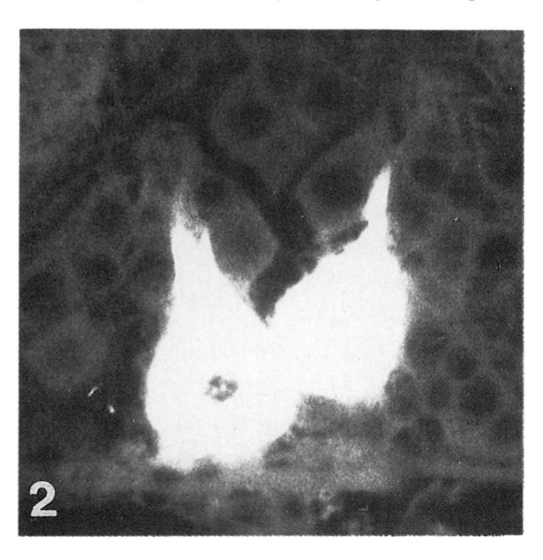

Table 2

Orders	Species	Localization	Authors
Gondoliberin			
LH–RH) antiserum			
Od.	*Aeshna cyanea*	Midgut	Andriès & Tramu, 1984
Dic.	*Blaberus craniifer*	Midgut	Andriès & Tramu, 1985
Somatostatin (SRIF) antiserum			
Dic.	*Periplaneta americana*	Midgut	Iwanaga *et al.*, 1981
Dic.	*Leucophaea maderae*	CC	Hansen *et al.*, 1982
Or.	*Locusta migratoria*	Brain, CC	Doerr–Schott *et al.* 1978
L.	*Manduca sexta*	Brain, CC	El–Salhy *et al.*, 1983
Dip.	*Eristalis aeneus*	Brain, ventral G.	El–Salhy *et al.*, 1980
Hy.	*Apis mellifica*	Midgut	Bounias & Dubois, 1982
Corticoliberin (CRF) antiserum			
Dic.	*Periplaneta americana*	Brain, SOG, CC	Verhaert *et al.*, 1984
Somatoliberin (GRF) antiserum			
Od.	*Aeshna cyanea*	Brain, SOG	Andriès *et al.*, 1984
Od.	*Aeshna cyanea*	Midgut	Andriès & Tramu, 1984
Dic.	*Blaberus craniifer*	Midgut	Andriès & Tramu, 1985

CC. Corpora cardiaca, SOG.: suboesophageal ganglion, Ventral G.: ventral ganglia; Dic.: Dictyoptera, Dip. Diptera, Hy.: Hymenoptera, L.: Lepidoptera, Od.: Odonata

hormones have been found in a wide range of insect orders from Odonata to Hymenoptera (Table 2). CRF-like and SRIF-like neuropeptides are found in different parts of cerebral neurosecretory systems of *Periplaneta americana* (Verhaert *et al.* 1984), *Locusta migratoria* (Doerr–Schott *et al.* 1978) and *Manduca sexta* (El–Salhy *et al.* 1983). In *Locusta*, the SRIF-like substance is synthesized in numerous paraldehyde fuchsin-positive A type cells of the pars intercerebralis (Fig. 3). Immunocytochemical techniques have allowed a better insight into the true nature of the neurosecretory product just described by histochemical methods in locust median type A cells by Girardie (Girardie & Girardie 1967). SRIF-like material could be observed in fibres running through the brain to nervus corporis cardiaci (Fig. 4) and ultimately in terminal arborizations in the releasing area, the corpora cardiaca (Fig. 5).

In mammals, several neuropeptides occurring in the brain have also been found in the endocrine cells of the GEP system and, conversely several biologically active peptides isolated from the GEP region have been demonstrated in the brain (see Buve & Thorpe 1980). In *Aeshna cyanea* (Andriès *et al.* 1984; Andriès & Tramu 1984).

Figs. 3–5. *Locusta migratoria* (Somatotropin Release Inhibiting Factor SRIF antiserum). Fig. 3. Immunoreactive A1 type cells in the pars intercerebralis. ×250.

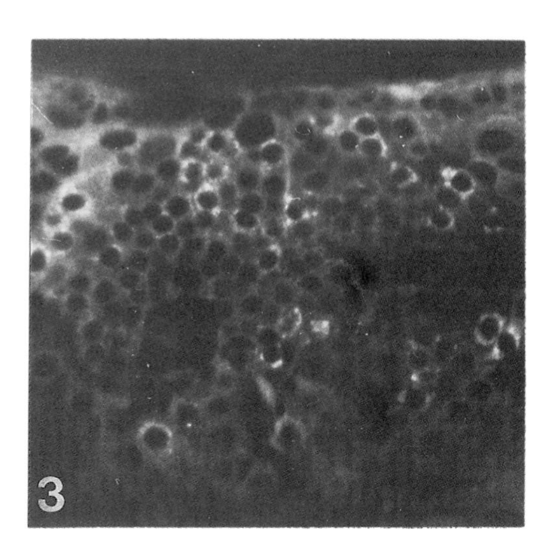

Fig. 4. Immunoreactive material in the tractus to the corpora cardiaca. pi, pars intercerebralis. ×250.

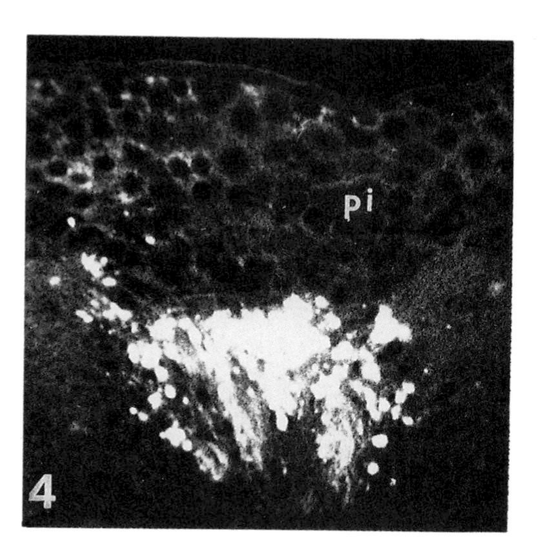

GRF-like substances were found both in cephalic neurons and midgut cells. In some cases therefore the peptidergic brain and midgut systems of insects can be said to be analogous to the brain-GEP system of vertebrates.

b. Opioid peptides

The existence of opioids in the vertebrate hypothalamo–hypophyseal complex was demonstrated in the 1970s (Guillemin *et al.* 1976). The immunochemical results obtained over the past few years show that similar peptides may also be present in invertebrate central nervous systems and, in particular, in those of insects. In vertebrates, endorphins (α and β) enkephalins (met– and leu–) are known to be derived from lipotropin (LPH) which, itself, is derived from a larger prohormone, the pro-opiomelanocortin which is also considered as the precursor of corticotropin (ACTH) and melanotropin (MSH). Though a precursor comparable to pro-opiomelanocortin has not been observed in insects, despite the assumptions of Simantov *et al.* (1976) that opioids could be limited to vertebrates, neuropeptides related to endophins, enkephalins, ACTH and MSH have been found in many parts of the CNS (Table 3).

The existence of neuropeptides related to α-endorphin was demonstrated in 2 clusters of suboesophageal cells in larvae of *Thaumetopoea pityocampa* and*Bombyx mori* (Rémy *et al.* 1978, 1979)(Fig. 6). These results have been confirmed in various insects belonging to 6 different orders. In addition to the suboesophageal ganglion,

Fig. 5. SRIF-like neuropeptide in the nervous corpora cardiaca. gcc, glandular corpora cardiaca. × 200.

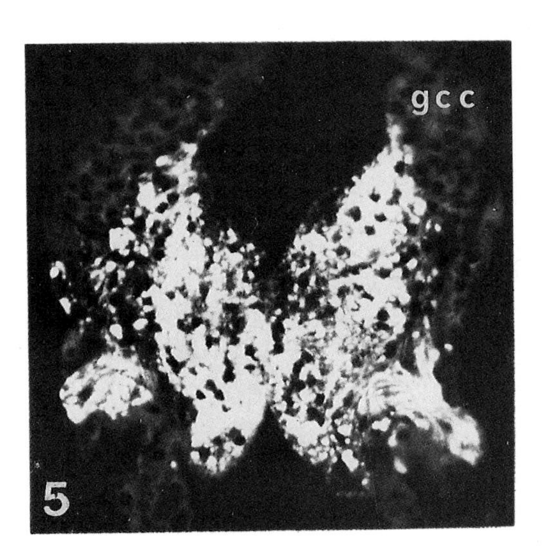

Table 3

Orders	Species	Localization	Authors
Endorphin antisera			
L.	*Thaumetopoea pityocampa*	SOG	Rémy *et al.*, 1978
L.	*Bombyx mori*	SOG	Rémy *et al.*, 1979
L.	*Manduca sexta*	Brain, CC	El–Salhy *et al.*, 1983
Dip.	*Eristalis aeneus*	Brain	El–Salhy *et al.*, 1980
Dip.	*Calliphora vomitoria*	Brain, CC	Duve & Thorpe, 1983
Enkephalin antisera			
Od.	*Aeshna cyanea*	Midgut	Andriès & Tramu, 1984
Dic.	*Leucophaea maderae*	CC	Hansen *et al.*, 1982
Dic.	*Periplaneta americana*	Brain, CC, CA	Verhaert & De Loof, 1985
Or.	*Locusta migratoria*	Brain	Rémy & Dubois, 1981
L.	*Manduca sexta*	Brain, CC, CA	El–Salhy *et al.*, 1985
Dip.	*Eristalis aeneus* Ventral G	Brian,	El–Salhy *et al.*, 1980
Dip.	*Drosophila melanogaster*	SOG, Ventral G.	Pagès *et al.*, 1983
Corticotropin (ACTH) antiserum			
Dic.	*Periplaneta americana*	Brain, SOG	Verhaert *et al.*, 1984
Melanotropin (MSH) antiserum			
Or.	*Locusta migratoria*	SOG	Veenstra, 1984
C.	*Leptinotarsa decemlineata*	SOG	Veenstra, 1984

CA.: Corpora allata, CC.: Corpora cardiaca, SOG.: Suboesophageal ganglion, Ventral G.: Ventral ganglia.
C.: Coleoptera, Dic.: Dictyoptera, Dip.: Diptera, L.: Lepidoptera, Od.: Odonata, Or.: Orthoptera.

opioids have been visualized in the brain of *Locusta migratoria* (Rémy & Dubois 1981), in ventral nerve cord and endocrine cells of the midgut (Andriès & Tramu 1984). These immunocytochemical observations were also confirmed by investigations carried out in *Leucophaea maderae* (Stefano & Scharrer 1981; Stefano *et al.* 1982). In this cockroach, specific high affinity binding sites for a synthetic enkephalin analog were demonstrated in the cerebral ganglia and in the midgut. These results provide strong evidence for the existence in insects of opioid receptors comparable to those in mammalian systems.

c. Gastro-entero-pancreatic (GEP) neurohormonal peptides

Substance P is widely distributed in the vertebrate peripheral CNS and acts as a sensory transmitter. Bombesin, which has been isolated from frog skin, influences the mammalian gastrointestinal tract and is a peptide neurotransmitter candidate in the vertebrate central nervous system (Snyder 1980). Immunochemical techniques have suggested that related neuropeptides may be present in the brains, suboesophageal ganglia and midguts of various insects (Table 4).

Several peptides are common to the brain and to the gastrointestinal tract of vertebrates (Straus & Yalow 1979) including gastrin/cholecystokinin (CCK), vasoactive intestinal polypeptide (VIP), pancreatic polypeptide (PP), insulin and glucagon.

As in vertebrates, GEP-like substances have been found in the midgut endocrine cells and CNS of insects (Tables 5 and 6). It is, however, worth mentioning that there seem to be very few insects presenting GEP-like neuropeptides simultaneously in the CNS and in the midgut, but this may be due to the fact that very few immunohistological investigations have been performed on the midgut, and studies have been focused mostly on the nervous system. The phenomenon has been observed in *Periplaneta americana* (PP-like and glucagon-like substances) and *Calliphora vomitoria* (PP-like material).

A gastrin/CCK-like material has been found in many cells of the brain and ventral ganglia of various insects (Fig. 7). This insect immunoreactive substance seems to possess properties more closely related to CCK than to gastrin. Duve & Thorpe (1981, 1984) and El–Salhy *et al.* (1983) showed that in *Calliphora erythrocephala,*

Fig. 6. *Thaumetopoea pityocampa* (α–endorphin antiserum). Primary cluster of immunoreactive perikarya in the anterior part of suboesophageal ganglion. n, neuropile. × 200.

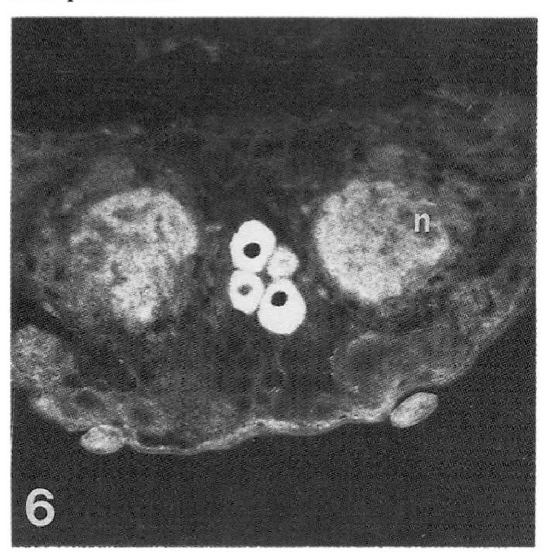

Table 4

Orders	Species	Localization	Authors
Substance P antiserum			
Od.	*Aeshna cyanea*	Midgut	Andriès & Tramu, 1984
Dic.	*Leucophaea maderae*	CC	Hansen *et al.*, 1982
Dic.	*Blaberus craniifer*	Midgut	Andriès & Tramu, 1985
Dic.	*Periplaneta americana*	Brain, SOG	Verhaert & De Loof, 1985
Or.	*Locusta migratoria*	Brain	Benedeczky *et al.*, 1982
L.	*Manduca sexta*	Brain, CC, CA	El-Salhy *et al.*, 1983
Dip.	*Eristalis aeneus*	Brain	El-Salhy *et al.*, 1980
Bombesin antiserum			
Dic.	*Leucophaea maderae*	CC	Haneen *et al.*, 1982
	Seven different species	Brain, SOG	Veenstra & Yanaihara, 1984

CA.: Corpora allata, CC.: Corpora cardiaca, SOG.: Suboesophageal ganglion.
Dic.: Dictyoptera, Dip.: Diptera, L.: Lepidoptera, Od.: Odonata, Or.: Orthoptera.

Fig. 7. *Locusta migratoria* (CHH antiserum). Immunoreactive perikarya in the pars intercerebralis (pi) and their axons into the tractus to the corpora cardiaca (arrow). × 350.

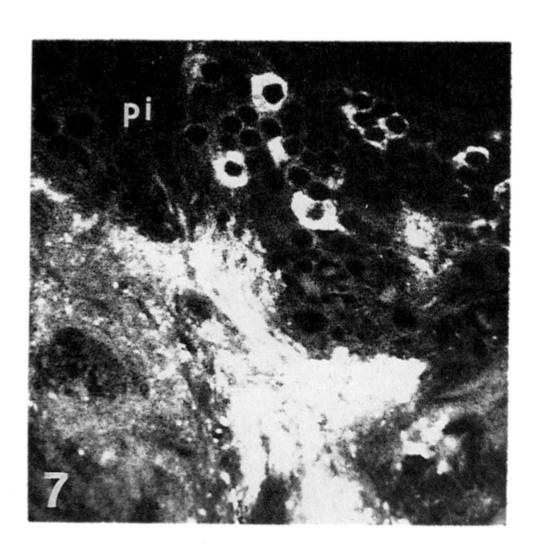

Table 5

Orders	Species	Localization	Authors
Gastrin/Cholecystokinin (CCK) antisera			
Od.	*Aeshna cyanea*	Midgut	Andriès & Tramu, 1984
L.	*Bombyx mori*	Brain, Ventral G.	Yui *et al.*, 1980
L.	*Manduca sexta*	Brain, CC	El–Salhy *et al.*, 1983
Dip.	*Eristalis aenus*	Brain, Ventral G.	El–Salhy *et al.*, 1980
Dip	*Calliphora erythrocephala*	Brain	Duve & Thorpe, 1981
			Duve *et al.*, 1983
Dip.	*Calliphora vomitoria*	Brain, Ventral G.	Duve & Thorpe, 1984
Vasoactive intestinal polypeptide (VIP) antiserum			
Od.	*Aeshna cyanea*	Midgut	Andriès & Tramu, 1984
Dic.	*Blaberus craniffer*	Midgut	Andriès & Tramu, 1985
L.	*Manduca sexta*	Brain, CC	El–Salhy *et al.*, 1983
Pancreatic polypeptide (PP) antiserum			
Od.	*Aeshna cyanea*	Midgut	Andriès & Tramu, 1984
Dic.	*Periplaneta americana*	Brain, SOG	Iwanaga *et al.*, 1981
		Ventral G, Midgut	Endo *et al.*, 1982
Or.	*Schistocerca gregaria*	Ventral G.	Myers & Evans, 1985
L.	*Bombyx mori*	Brain	Yui *et al.*, 1980
L.	*Manduca sexta*	Brain, CC	El–Salhy *et al.*, 1983
Dip.	*Calliphora erythrocephala*	Brain, SOG	Duve & Thorpe, 1980
Dip.	*Eristalis aeneus*	Brain, Ventral G.	El–Salhy *et al.*, 1980
Dip.	*Calliphora vomitoria*	Brain, SOG	Duve & Thorpe, 1982
		Ventral G, Midgut	

CC.: Corpus cardiaca, SOG.: Suboesophageal ganglion, Ventral G.: Ventral ganglia.
Dic.: Dictyoptera, Dip.: Diptera, L.: Lepidoptera, Od.: Odonata, Or.: Orthoptera.

Calliphora vomitoria and *Manduca sexta*, COOH–terminal specific antisera gave positive results while an antiserum directed against the NH_2–terminus was unreactive.

Invertebrate PP-like material was one of the first related neuropeptides detected by immunocytochemical methods (Sundler *et al.* 1977). It was subsequently described in the nervous tissues and midguts of several insects. A PP-like neuropeptide was extracted from *Calliphora vomitoria* heads, purified and subjected to amino acid analysis (Duve *et al.* 1982). The results obtained suggest that the molecular weight and the primary structure of this insect neuropeptide are very similar to that of mammalian PP. More recently other workers have pointed out a possible relationship between PP and the molluscan peptide FMRF-amide (see Chapter 14, this volume;

Table 6

Orders	Species	Localization	Authors
Insulin antiserum			
Dic.	*Periplaneta americana*	CC	Raabe, 1985
P.	*Carausius morosus*	CC	Raabe, 1985
Or.	*Locusta migratoria*	CC	Raabe, 1985
L.	*Bombyx mori*	Brain, CC	Yui *et al.*, 1980
L.	*Manduca sexta*	Brain, CC, CA	El–Salhy *et al.*, 1984
		Frontal G.	El–Salhy *et al.*, 1984
L.	*Ostrinia nbilalis*	Brain, CC	Lavenseau *et al.*, 1984
Dip.	*Calliphora vomitoria*	Brain	Duve & Thorpe, 1979
Dip.	*Eristalis aeneus*	Brain	El–Salhy *et al.*, 1980
Glucagon antiserum			
Dic.	*Periplaneta americana*	Brain, CC	Iwanaga *et al.*, 1981
		Midgut	Raabe, 1985
P.	*Carausius morosus*	Brain, CC	Raabe, 1985
Or.	*Locusta migratoria*	Brain, CC	Raabe, 1985
L.	*Manduca sexta*	Brain, CC	El–Salhy *et al.*,1983
Dip.	*Eristalis aeneus*	Brain	El–Salhy *et al.*, 1980

CA.: Corpora allata, CC.: Corpora cardiaca, Frontal G.: Frontal ganglion.
Dic.: Dictyoptera, Dip.: Diptera, L.: Lepidoptera, Or.: Orthoptera, P.: Phasmida

Veenstra & Schooneveld 1984; Myers & Evans 1985a,b; Verhaert *et al.* 1985). Using antisera raised against FMRF-amide immunoreactivity was observed in many neurons which also showed PP-like immunoreactivity. PP contains 36 amino acids and has an amidated tyrosine residue at its carboxyl terminus. FMRF-amide is a tetrapeptide with a sequence quite similar to that of the carboxyl terminus of PP. This insect neuropeptide would therefore seem to be more FMRF-amide-like than PP-like. Yet, the biochemical results mentioned above (Duve *et al.* 1982) argue against this conclusion. FMRF antiserum may just recognised C-terminus of the PP-like peptide and/or the PP-like peptide may be the 'precursor' to FMRF-amide.

Insect neuropeptides related to insulin and glucagon have been located exclusively in brain neurons and endocrine midgut cells (Fig. 8) and investigations directed at the purification and characterization of these peptides are more numerous than for the other biologically active insect peptides (see review in Thorpe & Duve 1984 and Chapter 6 of this volume). Although the ultimate proof, i.e. the elucidation of amino acid sequence is lacking, the results obtained corroborate the assumption that the insulin-like and glucagon-like substances of insects are very similar to vertebrate insulin and glucagon.

2. *Invertebrate neuropeptides and biologically active polypeptides*

Around fifteen neuropeptides or hormonal products related to neuropeptides have been isolated and sequenced from invertebrates over the past ten years (see Greenberg & Price 1983). Antibodies raised against four of them have allowed the detection of related neuropeptides in various insect neurons (Tables 7 and 8).

a. Proctolin

Proctolin is an insect pentapeptide first extracted from the cockroach *Periplaneta americana* (Brown & Starratt 1975) which acts as an excitatory neuromuscular transmitter in the hindgut. Proctolin is synthesized by a number of neurons in the terminal ganglion, about 185 neurons (*Periplaneta americana*; Bishop & O'Shea 1982). Proctolin-immunoreactive axon terminals are also found on the musculature of the fore- and hindgut, and on some segmental muscles.

b . Adipokinetic hormone (AKH)

AKH is a hormone originally extracted from the glandular lobe of the corpora cardiaca of locusts (Stone *et al.* 1976). Antisera raised against AKH have indicated a related peptide in the dictyopteran and phasmopteran glandular parts of the corpora cardiaca, and a related neuropeptide was detected in the brains of nineteen insect species belonging to nine distinct orders (Schooneveld *et al.* 1985). However, an antiserum said to be raised to the N–terminal part of AKH gives only a weak

Fig. 8. *Ostrinia nubilalis*. (Insulin antiserum). Immunoreactive brain perikarya in the pars intercerebralis. n, neuropile. × 530.

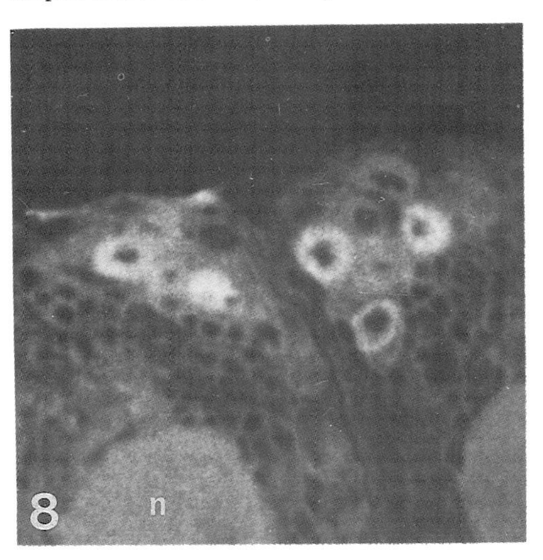

Table 7

Orders	Species	Localization	Authors
Proctolin antiserum			
Dic.	*Periplaneta americana*	Brain, SOG Ventral G.	Bishop *et al.*, 1981 Eckert *et al.*, 1981 Bishop & O'Shea, 1982 Agricola *et al.*, 1985
Or.	*Schistocerca nitens*	Ventral G.	Keshishian & O'Shea, 1985a
C.	*Leptinotarsa decemlineata*	Brain, SOG Ventral G.	Veenstra *et al.*, 1985
Adipokinetic hormone (AKH) antiserum			
Dic.	*Periplaneta americana*	Gland. CC	Raabe, 1985
Dic.	*Blattella germanica*	Gland. CC	Schooneveld *et al.*, 1985
P.	*Carausius morosus*	Gland. CC	Schooneveld *et al.*, 1985
P.	*Extatosoma tiaratum*	Gland. CC	Schooneveld *et al.*, 1985
Or.	*Acheta domesticus*	Gland. CC	Schooneveld *et al.*, 1985
Or.	*Locusta migratoria*	Brain, SOG Gland. CC	Schooneveld *et al.*, 1983 Schooneveld *et al.*, 1986
	Nineteen different species	Brain	Schooneveld *et al.*, 1985

Gland CC.: Glandular corpora cardiaca, SOG.: Suboesophageal ganglion, Ventral G.: Ventral ganglia.
C.: Coleoptera, Dic.: Dictyoptera, Or.: Orthoptera, P.: Phasmida

immunoreaction at the level of these neurons, thereby indicating that this related neuropeptide is probably different chemically from AKH (see also Chapters 5 and 7 this volume).

c. Molluscan cardio-excitatory peptide (FMRF-amide)

A neuropeptide related to the molluscan FMRF-amide (Price & Greenberg 1977) is present in insect nervous systems (Fig. 9). Positive immunohistological results are probably more numerous than is shown on Table 8 because, as already mentioned, the neuropeptide detected by PP antisera (Table 5) is likely to be more closely related to FMRF than it is to PP.

d. Crustacean hyperglycemic hormone (CHH)

Antibodies raised against the crustacean neuropeptide CHH (for a review see Keller, 1981 and Chapter 9, this volume) have permitted the demonstration of positive immunoreactions in various crustacean nervous systems. Using these antibodies, Jaros & Gäde (1982) could detect by immunocytochemistry a CHH-like

Table 8

Orders	Species	Localization	Authors
Molluscan cardio-excitatory tetrapeptide (FMRF-amide) antiserum			
Dic.	*Periplaneta americana*	Brain, CC, SOG	Boer *et al.*, 1980 Verhaert *et al.*, 1985
Or.	*Locusta migratoriaa*	Brain, SOG, CC	Boer *et al.*, 1980 Veenstra, 1984
Or	*Schistocerca gregaria*	SOG, Ventral G	Myers & Evans, 1985
L.	*Pieris brassicae*	Brain, CC, SOG	Boer *et al.*, 1980
C.	*Leptinotarsa decemlineata*	Brain, CC, SOG	Veenstra & Schooneveld, 1984
Crustacean hyperglycemic hormone (CHH) antiserum			
P.	*Carausius morosus*	Brain, CC	Jaros & Gäde, 1982

CC.: Corpora cardiaca, SOG.: Suboesophageal ganglion, Ventral G.: Ventral ganglia.
C.: Coleoptera, Dic.: Dictyoptera, L.: Lepidoptera, Or.: Orthoptera, P.: Phasmida

Fig. 9. *Locusta migratoria.* (FMRF-amide antiserum). Cluster of immunoreactive cells in the medioventral part of suboesophageal ganglion. n, neuropile. × 220.

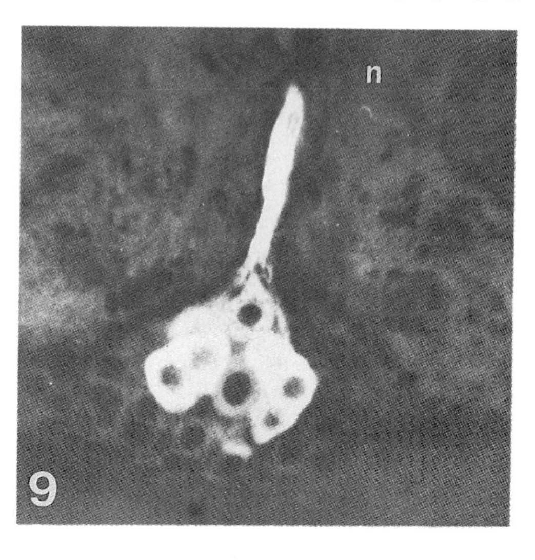

material in the pars intercerebralis and corpora cardiaca of the stick insect *Carausius morosus*. Radioimmunoassays also indicated that a pair of corpora cardiaca contains at least 7 pg of CHH-like neuropeptide.

II. The significance of immunocytochemical results

Prior to 1977, insect neurosecretory cells were studied using various specific staining methods. The numerous descriptions published allow these cells to be categorised into three major types: A, B and Cr with generally 2 subtypes within each type (Raabe 1982) – that is to say a limited number of cellular categories with similar tinctorial affinities. Immunocytochemical techniques allow a much greater discrimination, because the various antisera used have permitted the identification of different neurosecretory products with the same tinctorial affinity. For example SRIF-like substance is present in A1 type cells of the locust pars intercerebralis (Doerr–Schott *et al.* 1978) and vasopressin–neurophysin-like material in A1 type cells of locust suboesophageal ganglion (Rémy *et al.* 1979). They have also made it possible to visualize some neurosecretory cells which had never been observed before, because they did not show any specific tinctorial affinity. For example, locust protocerebral cells synthesize a met–enkephalin-like neuropeptide (Rémy & Dubois 1981). A study of Tables 1 to 8 shows that more than 10 neurosecretory cell categories could be identified among the most commonly studied insect species, such as *Periplaneta americana, Locusta migratoria* or *Manduca sexta*. In the *Manduca*

Fig. 10. *Locusta migratoria* third thoracic ganglion (Neurophysin antiserum). Neurophysin–vasopressin-like substance in terminal arborizations at the level of proximal parts of two segmental nerves. n, neuropile. × 300.

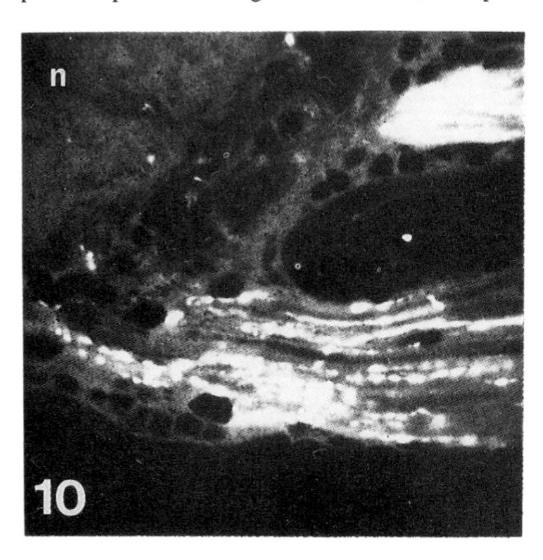

nervous system, El–Salhy *et al.* (1983) observe about 170 neurosecretory cells subdivided into 13 distinct types showing relationships with 13 different vertebrate neuropeptides.

The search for immunochemical relationships between insect neuropeptides and vertebrate neuropeptides has provided additional knowledge of insect neurosecretory systems.

Thus immunocytochemical investigations have shown that a relatively large number of distinct neuropeptides are involved in classical neurosecretory areas such as the protocerebral and corpus cardiacum systems (Figs. 3, 4 and 5). Eight neuropeptides were identified in *Manduca sexta* and 6 in *Leucophaea maderae* (Hansen *et al.* 1982).

Previously unknown neurosecretory systems were discovered using antisera to vertebrate neuropeptides, mainly in the ventral nerve cord. Thus, in *Locusta migratoria* (Rémy & Girardie 1980) a vasopressin–neurophysin-like material is synthesized by 2 suboesophageal perikarya (Table 1), the fibres of which run through the whole CNS and produce immunoreactive arborizations in the roots of all the segmental nerves where the neurosecretory product is released into the haemolymph (Fig. 10). A neurohaemal organ was discovered by Duve & Thorpe (1982) in the fused ventral ganglia of *Calliphora vomitoria*, where considerable amounts of PP-like substance in nerve fibres is collected within the dorsal sheath of the ganglia.

Figs. 11–12. *Locusta migratoria* (Met–enkephalin antiserum).

Fig. 11. Frontal section of protocerebrum. Two perikarya with met–enkephalin-like content (arrow) near mushroom bodies (mb) and their terminal arborizations (star). × 60.

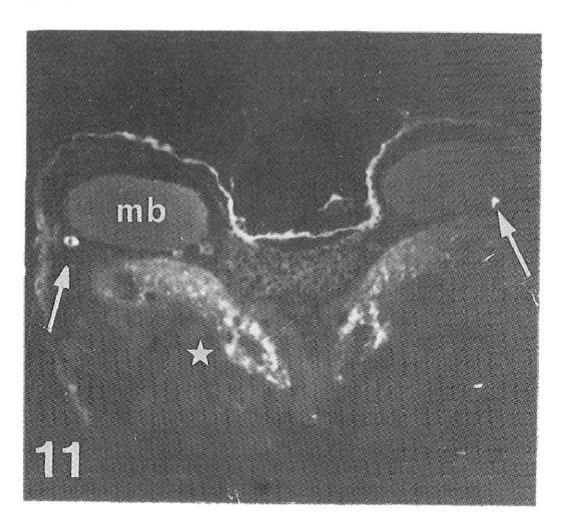

In most cases, however, related neuropeptides have only been described in perikarya, because axons and axon terminals are not visible or have not been investigated. It is therefore difficult to classify or categorize these immunoreactive cells. Yet, in more favourable cases, the whole neuron can be visualized; fibres and arborizations are observed in the neuropile, or in the vicinity of other neurosecretory cells, or of common neurons in *Locusta migratoria* (optic lobes, vasopressin–neurophysin-like substances – Rémy & Girardie 1980), *Aeshna cyanea* (optic lobes, GRF-like neuropeptide – Andriès *et al.* 1984), *Periplaneta americana* and *Locusta migratoria* (protocerebral neuropile, substance P-like material – Verhaert & De Loof 1985; Benedeczky *et al.* 1982) and *Eristalis aeneus* (protocerebral neuropile, gastrin/CCK-like neuropeptide – El Salhy *et al.* 1980). In some insects, the axons in certain neurosecretory cells of the pars intercerebralis do not leave the brain, but remain in the neighbourhood of other neurosecretory cells containing gastrin/CCK-like material in *Calliphora erythrocephala* (Duve *et al.* 1983).

In the locust brain (Rémy & Dubois 1981), the 4 perikarya synthesizing the met–enkephalin-like neuropeptide provide a striking example of an intracerebral neurosecretory system. The fibres coming from 2 of these perikarya cross the protocerebral neuropile and give rise to a network of ramifications in the two neuropilar zones surrounding the tractus to the corpora cardiaca which contain the fibres of the median neurosecretory cells (Figs. 11 and 12).

Fig. 12. High magnification of part of Fig. 11. Immunoreactive arborizations surrounding tractus to the corpora cardiaca (tcc) below pars intercerebralis (pi). Compare with Fig. 7. × 300.

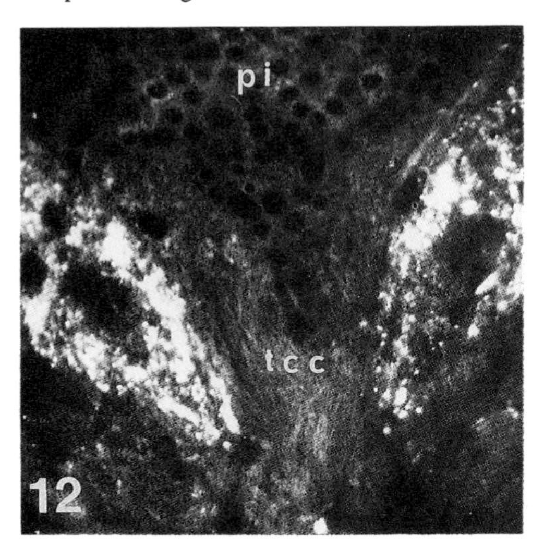

Although there is no tangible evidence to support the hypothesis, the logical conclusion drawn by the various authors who have studied these immunoreactive neuropeptides that do not leave the nervous system is that they might play a neurotransmitter or neuromodulator role.

Finally it should be pointed out that while antisera directed against vertebrate neuropeptides provide information about new neurosecretory systems in insects, it is necessary for the nature of the product to be fully characterized before any definitive functional conclusions may be reached.

The neurosecretory systems of insect embryos have not been widely studied by means of staining methods. However, immunocytochemical techniques allow the detection of very small amounts of neurosecretory product and particularly permit the identification of a wide range of neuropeptides. Only the embryos of *Schistocerca nitens* (Keshishian & O'Shea 1985b) and *Locusta migratoria* (Romeuf & Rémy 1984; Romeuf *et al.* 1986) have been studied.

In *Locusta migratoria*, embryonic development lasts for 12 days. Neurosecretory products such as SRIF-like and CHH-like neuropeptides are detected first in nervous lobes of the corpora cardiaca 4 days before hatching (Fig. 13). Simultaneously, the presence of AKH-like material in the glandular lobes of the corpora cardiaca becomes visible. Immunocytochemical detection of vasopressin–neurophysin-like or FMRF-amide-like substances in perikarya or fibres is possible 1 day later (Fig. 14). Met–

Fig. 13. *Locusta migratoria embryo.* 24 h before hatching (SRIF antiserum). Very brightly fluorescent immunoreactive material in neural part of corpora cardiaca surrounding aortic cavity (ac). hg, hypocerebral ganglion. x 250.

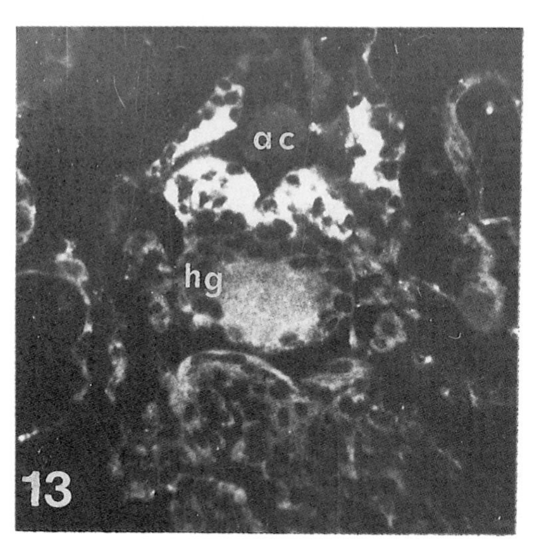

enkephalin-like and CCK-like neuropeptides are visible in the axons and perikarya respectively 2 days and 1 day before hatching.

The activity of embryonic neurosecretory cells probably begins before immunocytochemical detection is possible. For example radioimmunoassays show in *Locusta* that the vasopressin-like neuropeptide is present in the CNS one day before immunohistological detection (Romeuf *et al.* 1986).

Two suboesophageal cells of *Clitumnus* and *Locusta* react with antisera to vasopressin or neurophysin (Rémy *et al.* 1977, 1979) suggesting that, as in vertebrates, the neurosecretory product of some insect neurons could comprise a hormonal peptide (vasopressin-like) linked to a carrier protein (neurophysin-like). Biochemical studies of the vasopressin-like substance (Proux & Rougon–Rapuzzi 1980; Cupo & Proux 1983) and the neurophysin-like material (Camier *et al.* 1980) confirm the coexistence of 2 related polypeptides. A similar association has also been observed in some rare neurons of *Periplaneta americana* by Verhaert *et al.* (1984). It seems, however, that the concept cannot be applied to all insect species at present because such an association has not been widely investigated and contradictory results have been obtained in *Periplaneta*. In this dictyopteran, a large number of brain neurons synthesize a neurophysin-like substance but are not immunoreactive with an anti-vasopressin serum. Conversely, the neurosecretory product of other neurons is vasopressin-like, yet it does not show any reaction to antisera against neurophysin (Verhaert *et al.* 1984). Conceivably, the neurophysin-like material

Fig. 14. Neonate *Locusta migratoria* (Neurophysin antiserum). Two immunoreactive perikarya in the ventral part of the suboesophageal ganglion. ×580.

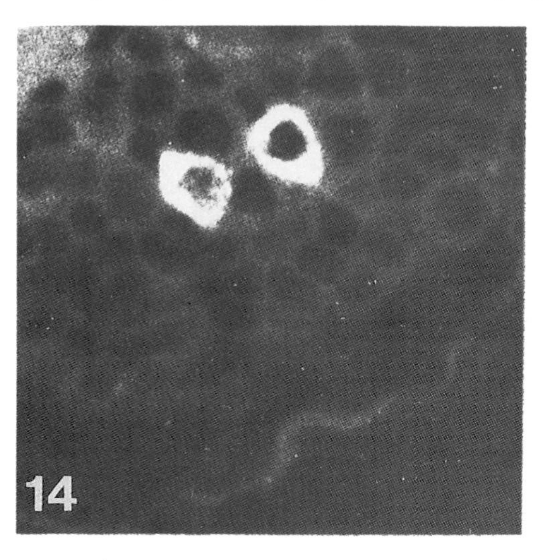

present in some cells might be linked to another neuropeptide distinct from vasopressin, or the specificity of antiserum is not appropriate.

In insects, most neurosecretory cells appear to produce only one peptide. In a few there is the suggestion of coexistence of at least 2 neuropeptides; as for example PP-like and gastrin-like substances in *Calliphora vomitoria* (Duve & Thorpe 1984), proctolin-like and PP-like neuropeptides in *Leptinotarsa decemlineata* (Veenstra *et al.* 1985). Similar examples have been found in other invertebrates, particularly in molluscs.

Immunocytochemical results are of prime importance as a starting point for biochemical investigations. Once the existence of insect neuropeptides related to vertebrate neuropeptides was known, it became possible to attempt isolation and purification from crude extracts by making use of such immunological relationships. This has resulted in the determination of the amino acid sequence of some insect neuropeptides (see Chapter 6 this volume).

Biochemical studies have been focused on insect neuropeptides related to 5 vertebrate brain-gut neuropeptides: vasopressin in *Locusta migratoria* (see Girardie *et al.* 1984); enkephalin in *Locusta migratoria* (Gros *et al.* 1978) and *Drosophila melanogaster* (Pages *et al.* 1983); gastrin/CCK and PP in *Calliphora vomitoria* (Dockray *et al.* 1981; Duve *et al.* 1981, 1982); and insulin in various insects (see Thorpe & Duve 1984). All the results obtained tend to show that the insect neuropeptides which could be purified bear immunochemical, physical and chemical resemblances to the related vertebrate neuropeptides, but appear never to be absolutely identical. Molecular weights can be very close (PP-like substance) or show differences of about 1000 (enkephalin-like material). When the amino acid determination was completed for PP-like and insulin-like neuropeptides in *Calliphora* the amino acids of the insect neuropeptide proved to be, with minor variation, present in similar proportions to those of the related vertebrate neuropeptide. Similarly the vasopressin-like neuropeptide of *Locusta* (J. Proux, personal communication) is closely related to vasopressin but with a difference in 2 amino acid residues.

B. Insect amine-like substances

Biochemical methods show that the insect CNS contains amines. Identification and quantitative measurements of amines in different parts of the brain have been undertaken recently using HPLC (Sloley & Owen 1982). The most convenient histological technique for locating amine-containing neurons has been the formaldehyde fluorescence method which induces endogenous indolylalkylamines and catecholamines to fluoresce after exposure to formaldehyde vapour. It is very sensitive for primary catecholamines but has not identified individual amines. It is less sensitive for indolylalkylamines because the fluorophores obtained from these fade rapidly and are masked by the brighter, more stable, fluorescence produced by catecholamine fluorophores. The introduction of antibodies to amines has provided a more sensitive method for the detection in tissues of amine-like antigens. The application of this method, however, is rather difficult and raises a few extra problems beyond those associated with any immunocytochemical technique. These problems probably explain why the results obtained in the same animal are sometimes variable, even though the same type of antibodies is used.

I. Immunocytochemical procedure and its associated problems

It is difficult to obtain suitable antibodies against amines because of their small size. Coupled to proteins as haptens, they elicit the production of antibodies which recognize primarily the monoamine coupled by the coupling reagent (essentially formaldehyde, but also recently glutaraldehyde) and to a lesser degree the original molecule. So, a correct immunochemical detection is achieved only if a similar chemical modification is present in both the immunogen used for producing antibodies and in the antigen fixed in the tissue. The coupling reagent may be present in the different fixatives used (paraformaldehyde in phosphate buffer, Bouin's fluid, paraformaldehyde + glutaraldehyde for ultrastructural studies).

Amine-like immunoreactivity cannot be taken as proof that a specific amine is present because as with peptides, a cross-reactivity of the antisera with other molecules is always possible (with other amines if the antibody concentration is too high, or with ring structure due in some cases to an atypical condensation product of amine and formaldehyde). Therefore, the specificity of the antibodies must be very carefully examined in solution as well as in the tissue, and their working dilution is also very important and must be as high as possible.

II. Immunohistological results

Using antibodies prepared recently (Steinbusch *et al.* 1978; Geffard *et al.* 1982), 2 amine-like substances have been visualized in the CNS of insects belonging to 6 different orders, and also in the stomatogastric nervous system, the neurohaemal organs, various somatic and visceral nerves, and in the gut. The results are summarized in Table 9.

1. Serotonin

a. CNS

Serotonin-immunoreactive cell bodies are observed in the whole brain, chiefly in the pars intercerebralis, near the mushroom bodies and in the optic lobes, but also in the lateral areas of the protocerebrum, in the deutocerebrum and the tritocerebrum. Fibres were located in unstructured neuropiles and structured neuropiles such as the central body, pedunculi, α and β lobes, calyces, olfactori glomeruli, lamina, medulla and lobula in *Periplaneta americana* (Bishop & O'Shea

Table 9

Orders	Species	Localization	Authors
Serotonin antiserum			
Dic.	*Periplaneta americana*	Brain, Ventral G.,	Bishop & O'Shea, 1983
		Nerves	Nässel & Klemm, 1983
			Klemm *et al.*, 1984
			Nishiitsutsuji–Uwo *et al.*, 1984
			Davis, 1985
Or.	*Schistocerca gregaria*	Brain, Ventral G	Klemm & Sundler, 1983
		Stom., Gut	Nässel & Klemm, 1983
			Taghert & Goodman, 1984
			Tyrer *et al.*, 1984
			Klemm *et al.*, 1986
Or.	*Melanoplus differentialis*	Brain, Ventral G.	Taghert & Goodman, 1984
Or.	*Locusta migratoria*	Brain, Ventral G.	Tyrer *et al.*, 1984
Or.	*Grillus bimaculatus*	Stom., Gut	Klemm *et al.*, 1986
Or.	*Acheta domesticus*	Stom., Gut	Klemm *et al.*, 1986
He.	*Rhodnius prolixus*	Ventral G.,	Flanagan, 1984, 1986
			Abd. N.O.
Dip.	*Calliphora erythrocephala*	Brain, Ventral G.	Nässel *et al.*, 1985
	Nerves		Nässel & Klemm, 1983
			Nässel & Elekes, 1984, 1985
			Nässel *et al.*, 1985
Hy.	*Apis mellifica*	Brain, SOG	Schürmann & Klemm, 1984
			Nässel *et al.*, 1985
Hy.	*Cataglyphis bicolor*	Brain	Nässel *et al.*, 1985
Dopamine antiserum			
Or.	*Locusta migratoria*	Brain	Vieillemaringe *et al.*, 1984
L.	*Bombyx mori*	Brain, SOG	Takeda *et al.*, 1986

Abd. N.O.: Abdominal neurohemal organ, SOG.: Suboesophageal ganglion, Stom.: Stomatogastric nervous system, Ventral G.: Ventral ganglia.

Dic.: Dictyoptera, Dip.: Diptera, He.: Hemiptera, Hy.: Hymenoptera, L.: Lepidoptera, Or.: Orthoptera.

1983; Klemm, Steinbusch & Sundler 1984; Nishiitsutsuji–Uwo *et al.* 1984), *Apis mellifera* (Schurmann & Klemm 1984) and *Locusta migratoria* and *Schistocerca gregaria* (Tyrer *et al.* 1984).

Some structured neuropiles have been investigated more extensively: the mushroom bodies, where the serotonin fibres visualized originate from extrinsic neurons in *Schistocerca gregaria* (Klemm & Sundler 1983), the lamina, the medulla and the lobula in *Calliphora erythrocephala* (Nassel *et al.* 1983; Nassel & Klemm 1983) and *Schistocerca gregaria* and *Periplaneta americana* (Nassel & Klemm 1983) where ultrastructural investigations show that immunoreactive terminals contain large granular vesicles (100 nm) and small clear vesicles (60 nm). These terminals are either presynaptic to second order neurons of the lamina in *Apis mellifera*, and *Cataglyphis bicolor* (Nassel *et al.* 1985), or distal to the synaptic layer of the lamina in a region of retinal photoreceptor axons where the neuroactive substance could be released non-synaptically from varicosities in *Calliphora erythrocephala* (Nassel & Elekes 1984; Nassel *et al.* 1985).

Serotonin-immunoreactive cell bodies occur mainly in segmentally repeated groups localized in the postero–lateral margin of the suboesophageal, thoracic and abdominal ganglia, even if other bilaterally symmetrical perikarya are found in the suboesophageal ganglion and sometimes in the thoracic ganglia and terminal abdominal ganglion in *Periplaneta americana* (Bishop & O'Shea 1983; Nishiitsutsuji–Uwo *et al.* 1984), *Apis mellifera* (Schurmann & Klemm 1984), *Schistocerca americana* (Taghert & Goodman 1984), *Schistocerca gregaria* and *Locusta migratoria* (Tyrer *et al.* 1984), *Melanoplus differentialis* (Taghert & Goodman 1984), *Calliphora erythrocephala* (Nassel & Elekes 1985), and *Rhodnius prolixus* (Flanagan 1984, 1986). Stained fibres were found in all ganglia and axon-like processes occurred in interganglionic connectives.

b. Neurohaemal organs

Relatively few serotonin–immunoreactive fibres are observed in the neurohaemal part of the corpora cardiaca in *Periplaneta americana* (Nishiitsutsuji–Uwo *et al.* 1984; Davis 1985), *Schistocerca gregaria, Gryllus bimaculatus* and *Acheta domesticus* (Klemm *et al.* 1986), but some emerge from the brain via the NCC I and the NCC II and/or from the stomatogastric nervous system; in other cases, the source of serotonin material is not clear. Endogenous serotonin within fibres and terminals is also detected with immunocytological techniques in the abdominal neurohaemal organs of *Rhodnius prolixus* (Flanagan 1984).

c. Stomatogastric nervous system

Most serotonin–immunoreactive neurons occur in the frontal ganglion (in which 30 to 40% of the total number of perikarya react with antisera). Few serotonergic cell bodies are located in hypocerebral (occipital), ventricular and ingluvial ganglia in *Periplaneta americana* (Davis 1985), *Schistocerca gregaria, Gryllus bimaculatus* and *Acheta domesticus* (Klemm *et al.* 1986).

The main processes of the serotonin immunoreactive cells in the frontal ganglion leaves the ganglion along the recurrent nerve. Several fibres descend within the nerves and ganglia of the stomatogastric nervous system in *Periplaneta americana* (Davis 1985), *Schistocerca gregaria, Gryllus bimaculatus* and *Acheta domesticus* (Klemm *et al.* 1986). In locusts and crickets, most immunoreactive fibres form a dense network on the periphery of the entire stomatogastric nervous system (Klemm *et al.* 1986).

d. Somatic and visceral nerves

Serotonin-like immunoreactivity has been demonstrated in and/or on the surface of nerves in the CNS: the labral nerve, suboesophageal nerves, thoracic and abdominal nerves in *Locusta migratoria* and *Schistocerca gregaria* (Tyrer *et al.* 1984); the maxillary–labial and labro–frontal nerves, cervical nerves, dorsal prothoracic nerves and median abdominal nerve in *Calliphora erythrocephala* (Nassel & Elekes 1984, 1985) and the nerves of the tritocerebrum and suboesophageal ganglion in *Periplaneta americana* (Davis 1985). Immunoreactivity in the serotonergic varicose fibres in the neural sheath is located in large granular vesicles (100 nm) and on the membranes of agranular vesicles (60 and 100 nm). In *Calliphora erythrocephala* these varicose fibres can constitute neurohaemal regions (Nassel & Elekes 1984, 1985).

Lateral branches of serotonin–immunoreactive fibres leave the nerves of the stomatogastric nervous system. They provide innervation to the surface of the visceral muscles of the foregut and the midgut in *Periplaneta americana* (Davis 1985): they form a plexus along the surface of the entire intestinal tract in *Schistocerca gregaria, Gryllus bimaculatus* and *Acheta domesticus* (Klemm *et al.* 1986), from where they supply all muscle layers of the muscularis and probably some somatic muscles. Electron microscopic analysis of nerves innervating the muscle layers of the alimentary tract reveals 1 type of serotonin–immunoreactive fibres in orthopterans (Klemm *et al.* 1986). As is found in the brain and the stomatogastric nervous system, the terminals contain large granular vesicles and large 'grainy' vesicles (80–120 nm). No synaptic membrane specializations are seen. The fibres are in close contact with the haemolymph in some regions which can thus be regarded as sites of neurohormonal release (Klemm *et al.* 1986).

Both in *Periplaneta americana* (Davis 1985) and orthopterans (Klemm *et al.* 1986), fibres from nerves of the stomatogastric nervous system supply the acini of the salivary glands.

2. Dopamine

Evidence for dopaminergic neurons in the CNS of *Locusta migratoria* (Vieillemaringe *et al.* 1984) and *Bombyx mori* (Takeda *et al.* 1986) was obtained immunocytochemically. In the brain of the locust, almost all the numerous immunoreactive perikarya are located in the pars intercerebralis (Fig. 15). Their processes can be traced to their endings in the neural region of the corpora cardiaca. Other dopamine–immunoreactive cell-bodies were observed near the mushroom bodies, in external areas of the protocerebrum and the tritocerebrum, and in the optic lobes. Among the structured neuropiles, the pedunculi, the central body, the α– and β–lobes, and the lamina contain immunoreactive fibres. In the silkworm, few dopaminergic perikarya are present in the median and lateral protocerebral areas of the brain and in the suboesophageal ganglion, but the dopamine immunoreactive cells in the larval suboesophageal ganglion (Fig. 16) also contain an α endorphin-like substance (Fig. 17)(Takeda *et al.* 1986).

III. Possible functions of serotonergic and dopaminergic terminals: contribution of the immunochemical method

The main role of selective mapping of serotonin– and dopamine–immunoreactive neurons is to complement studies on the distribution of monoamine-containing cells in the nervous system of insects. However, a few hypotheses concerning the potential roles of amines have also been put forward by a number of authors.

It appears from light and electron microscopic immunocytochemistry that amines may interact with other neurons or target cells in a variety of ways: by conventional

Fig. 15. *Locusta migratoria.* (Dopamine antiserum). Anterior area of the pars intercerebralis with perikarya and beginning of the tractus to the corpora cardiaca. × 220.

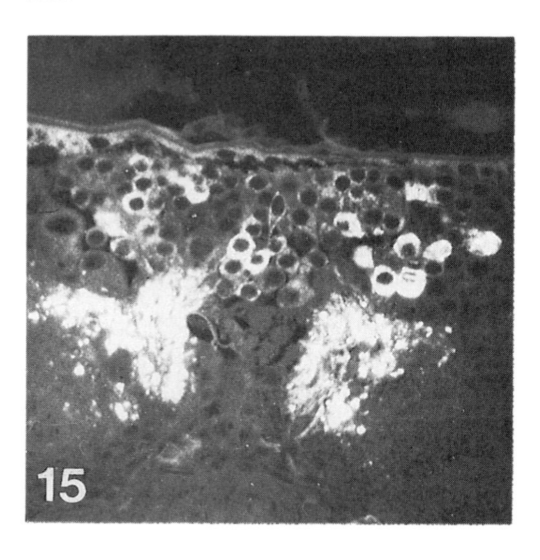

Figs. 16–17. *Bombyx mori*. Two transverse adjacent sections of same larval suboesophageal perikarya.

Fig. 16. Section treated with dopamine antiserum. × 450.

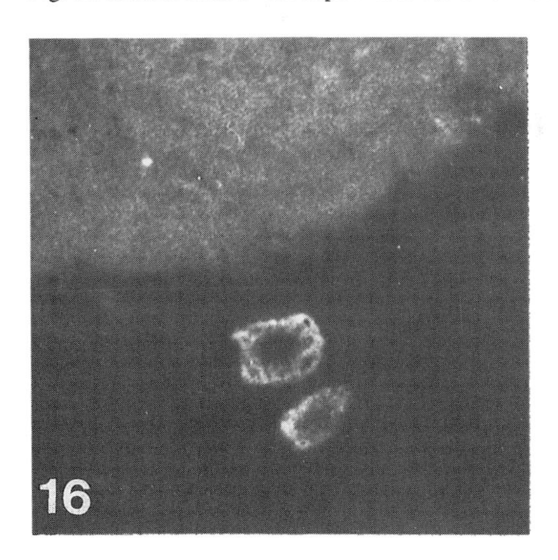

Fig. 17. Section treated with α–endorphin antiserum.

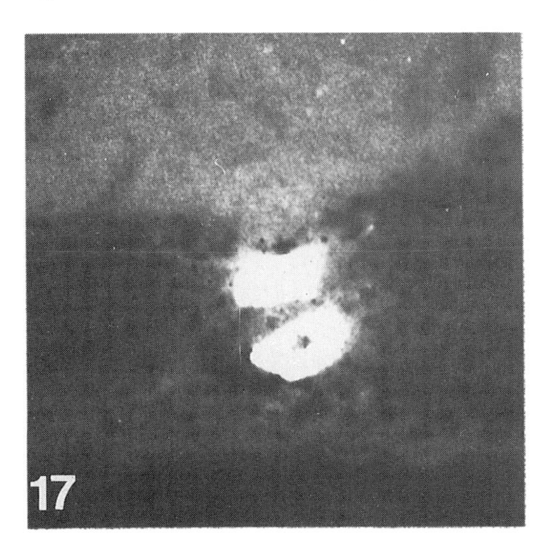

chemical synapses in many neuropiles of the CNS; by extracellular release from varicosities in some areas of the brain; and/or for those terminals in close contact with the haemolymph, by release into the body fluid.

The locations of serotonergic neurons in the optic lobes and in the antennal lobes of many insects suggest that they play a role in the control of information in sensory processing channels, or as level controllers, setting the responsiveness of the sensory channels. The findings that in all insects the optic lobes contain extensive serotonin–immunoreactive fibre layers derived from relatively few perikarya may suggest that the serotonin-containing neurons have an influence on many classes of visual neurons.

The distribution of serotonin–immunoreactive fibres in some intestinal and somatic muscles suggests that serotonin acts as a neurotransmitter or could have a modulatory effect on transmission.

The presence of serotonin and dopamine in the corpora cardiaca, as well as that of serotonin in the abdominal neurohaemal organs and in some regions which are thought to constitute neurohaemal sites (such as the neural sheath) suggests that these amines could be released directly into the haemolymph and act as neurohormones.

Conclusions

The use of antibodies raised against vertebrate neuropeptides or amines has allowed significant progress to be made in insect neuroendocrinology. The method also helps to further our understanding of the potential interactions between the aminergic and other functional systems, but, it is clear that immunochemistry is only a first step towards unravelling the possible functions of many peptides and amines.

Immunocytochemical techniques have shown that the various categories of neurosecretory cells are much more numerous than was formerly believed when conventional staining techniques alone were used. The existence of insect neurosecretory products related to almost all vertebrate neuropeptides known today has now been shown.

Previously unknown insect neurosecretory systems have been discovered and the co-localization of two neuropeptides or one peptide and one amine in individual insect neurosecretory cells has been demonstrated.

Biochemical studies tend to show that insect neurosecretory products which are immunochemically related to vertebrate neuropeptides possess very similar physico–chemical characteristics but are never, as yet, totally identical.The origin of vertebrate neuropeptides or amines probably dates back to an earlier period than that of vertebrates since related bioactive substances exist in higher invertebrates such as insects, and also at a very early stage of evolution as indicated by their presence in coelenterates.

References

Agricola, H., Eckert, M., Ude, J., Birkenbeil, H. & Penzlin, H. (1985). The distribution of a proctolin-like immunoreactive material in the terminal ganglion of the cockroach, *Periplaneta americana* L. *Cell Tissue Research*, **239**, 203–209.

Andries, J., Belemtougri, G. & Tramu, G. (1984). Immunohistochemical identification of growth hormone-releasing factor-like material in the nervous system of an insect, *Aeshna cyanea* (Odonata). *Neuropeptides*, **4**, 519–528.

Andries, J. & Tramu, G. (1984). Détection immunohistochimique de substances apparentées à des hormones peptidiques de Mammifères dans le mésentéron d'*Aeshna cyanea* (Insecte, Odonate). C.R. Académie Sciences Paris, **299**, Série III, 181–184.

Andries, J. & Tramu, G. (1985). Ultrastructural and immunohistochemical study of endocrine cells in the midgut of the cockroach *Blaberus craniifer* (Insecta, Dictyoptera). *Cell Tissue Research*, **240**, 323–332.

Benedeczky, I., Kiss, J. & Somogyi, P. (1982). Light and electron microscopic localization of substance P-like immunoreactivity in the cerebral ganglion of locust with a monoclonal antibody. *Histochemistry*, **75**, 123–131.

Bishop, C. & O'Shea, M. (1982). Neuropeptide proctolin (H–Arg–Tyr–Leu–Pro–Thr–OH): Immunocytochemical mapping of neurons in the central nervous sytem of the cockroach. *Journal of Comparative Neurology*, **207**, 223–238.

Bishop, C. & O'Shea, M. (1983). Serotonin immunoreactive neurons in the central nervous sytem of an insect (*Periplaneta americana*). *Journal of Neurobiology*, **14**, 251–269.

Bishop, C., O'Shea, M. & Miller, R. (1981). Neuropeptide proctolin (H–Arg–Tyr–Leu–Pro–Thr–OH): Immunological detection and neuronal localization in insect central nervous system. *Proc. National Academy Sciences* USA, **78**, 5899–5902.

Boer, H., Schot, L., Veenstra, J. & Reichelt, D. (1980). Immunocytochemical identification of neural elements in the central nervous sytems of a snail, some insects, a fish, and a mammal with an antiserum to the molluscan cardio-excitatory tetrapeptide FMRF-amide. *Cell Tissue Research*, **213**, 21–27.

Bounias, M. & Dubois, M. (1982). Inhibition par la somatostatine de l'action hypoglycémiante de l'insuline chez l'Abeille in vivo et localisation cytochimique d'un antigène apparenté à la somatostatine dans l'épithélium digestif ainsi que les tubes de Malpighi de l'Abeille. *C.R. Académie Sciences* Paris, **294**, Série III, 1029–1034.

Brown, B. & Starratt, A. (1975). Isolation of proctolin, a myotropic peptide from *Periplaneta americana*. *Journal of Insect Physiology*, **21**, 1879–81.

Camier, M., Girardie, J., Rémy, C., Girardie, A. & Cohen, P. (1980). Identification of immunoreactive neurophysin-like proteins in the central nervous system of an insect *Locusta migratoria*. *Biochemical and Biophysical Research Communications*, **93**, 792–796.

Cupo, A. & Proux, J. (1983). Biochemical characterization of a vasopressin-like neuropeptide in *Locusta migratoria*. Evidence of high molecular weight protein encoding vasopressin sequence. *Neuropeptides*, **3**, 309–318.

Davis, N. (1985). Serotonin–immunoreactive visceral nerves and neurohemal system in the cockroach *Periplaneta americana* L. *Cell Tissue Research*, **240**, 593–600.

Dockray, G., Duve, H. & Thorpe, A. (1981). Immunochemical characterization of gastrin/cholecystokinin-like peptides in the brain of the blowfly, *Calliphora vomitoria*. *General and Comparative Endocrinology*, **45**, 491–496.

Doerr–Schott, J., Joly, L. & Dubois, M. (1978). Sur l'existence dans la Pars intercerebralis d'un insecte (*Locusta migratoria* R. et F.) de cellules neurosécrétrices fixant un antisérum antisomatostatine. *C.R. Académie Sciences* Paris, **286**, Série D, 93–95.

Duve, H. & Thorpe, A. (1979). Immunofluorescent localization of insulin-like material in the median neurosecretory cells of the blowfly, *Calliphora vomitoria* (Diptera). *Cell Tissue Research*, **200**, 187–191.

Duve, H. & Thorpe, A. (1980). Localisation of pancreatic polypeptide (PP) like immunoreactive material in neurones of the brain of the blowfly, *Calliphora erythrocephala* (Diptera). *Cell Tissue Research*, **210**, 101–109.

Duve, H. & Thorpe, A. (1981). Gastrin/Cholecystokinin (CCK)-like immunoreactive neurones in the brain of the blowfly, *Calliphora erythrocephala* (Diptera). *General and Comparative Endocrinology*, **43**, 381–391.

Duve, H. & Thorpe, A. (1982). The distribution of pancreatic polypeptide in the nervous system and gut of the blowfly, *Calliphora vomitoria* (Diptera). *Cell Tissue Research*, **227**, 67–77.

Duve, H. & Thorpe, A. (1983). Immunocytochemical identification of α endorphin-like material in neurones of the brain and corpus cardiacum of the blowfly, *Calliphora vomitoria* (Diptera). *Cell Tissue Research*, **233**, 415–426.

Duve, H. & Thorpe, A. (1984). Immunocytochemical mapping of gastrin/CCK-like peptides in neuroendocrine system of the blowfly *Calliphora vomitoria* (Diptera). *Cell Tissue Research*, **237**, 309–320.

Duve, H., Thorpe, A. & Lazarus, N. (1979). Isolation of material displaying insulin-like immunological and biological activity from the brain of the blowfly *Calliphora vomitoria*. *Biochemical Journal*, **184**, 221–227.

Duve, H., Thorpe, A., Lazarus, N. & Lowry, P. (1982). A neuropeptide of the blowfly *Calliphora vomitoria* with an amino acid composition homologous with vertebrate pancreatic polypeptide. *Biochemical Journal*, **201**, 429–432.

Duve, H., Thorpe, A., Neville, R. & Lazarus N. (1981). Isolation and partial characterization of pancreatic polypeptide-like material in the brain of the blowfly *Calliphora vomitoria*. *Biochemical Journal*, **197**, 767–770.

Duve, H., Thorpe, A. & Strausfeld, N. (1983). Cobalt-immunocytochemical identification of peptidergic neurons in *Calliphora* innervating central and peripheral targets. *Journal of Neurocytology*, **12**, 847–861.

Eckert, M., Agricola, H. & Penzlin, H. (1981). Immunocytochemical identification of proctolinlike immunoreactivity in the terminal ganglion and hindgut of cockroach *Periplaneta americana* (L.). *Cell Tissue Research*, **217**, 633–645.

El–Salhy, M. (1981). Immunohistochemical localization of pancreatic polypeptide (PP) in the brain of the larval instar of the hoverfly, *Eristalis aeneus* (Diptera). *Experientia*, **37**, 1009.

El–Salhy, M., Abou–El–Ela, R., Falkmer, S., Grimelius, L. & Wilander, E. (1980). Immunohistochemical evidence of gastro–entero–pancreatic neurohormonal peptides of vertebrate type in the nervous system of the larva of a dipteran insect, the hoverfly, *Eristalis aeneus*. *Regulatory Peptides*, **1**, 187–204.

El–Salhy, M., Falkmer, S., Kramer, K. & Speirs, R. (1983). Immunohistochemical investigations of neuropeptides in the brain, corpora cardiaca, and corpora allata of an adult lepidopteran insect, *Manduca sexta* (L.). *Cell Tissue Research*, **232**, 295–317.

El–Salhy, M., Falkmer, S., Kramer, K. & Speirs, R. (1984). Immunocytochemical evidence for the occurrence of insulin in the frontal ganglion of a lepidopteran insect, the tobacco hornworm moth, *Manduca sexta* L. *General and Comparative Endocrinology*, **54**, 85–88.

Endo, Y., Iwanaga, T., Fujita, T. & Nishiitsutsuji–Uwo, J. (1982). Localization of pancreatic polypeptide (PP)-like immunoreactivity in the central and visceral nervous systems of the cockroach *Periplaneta. Cell Tissue Research,* **227**, 1–9.

Evans, P. (1980). Biogenic amines in the insect nervous system. *Advances in Insect Physiology,* **15**, 317–473.

Flanagan, T. (1984). Cytological evidence for serotonin-containing fibers in an abdominal neurohemal organ in a Hemipteran. *Brain Research,* **306**, 235–242.

Flanagan, T. (1986). Serotonin-containing, catecholamine-containing and dopamine-sequestering neurones in the ventral nerve cord of the Hemipteran *Rhodnius prolixus. Journal of Insect Physiology,* **32**, 17–26.

Frontali, N. & Norberg, K. (1966). Catecholamine-containing neurons in the cockroach brain. *Acta physiologica scandinavica,* **66**, 243–244.

Geffard, M., Kah, O., Chambolle, P., Le Moal, M. & Delaage, M. (1982). Première application immunocytochimique d'un anticorps anti-dopamine à l'étude du système nerveux central. *C.R. Académie Sciences Paris,* **295**, Série III, 797–802.

Girardie, A. & Girardie, J. (1967). Etude histologique, histochimique et ultrastructurale de la pars intercerebralis chez *Locusta migratoria* L. (Orthoptère). *Zeitschrift für Zellforschung,* **78**, 54–75.

Girardie, A., Girardie, J., Proux, J., Remy, C. & Vieillemaringe, J. (1984). Neurosecretion in insects: multisynthesis, mode of action, and multiregulation. *Biosynthesis, Metabolism and Mode of Action of Invertebrate hormones, Springer–Verlag Berlin,* 97–105.

Greenberg, M. & Price, D. (1983). Invertebrate neuropeptides: native and naturalized. *Annual Review of Physiology,* **45**, 271–288.

Grimm–Jorgensen, Y. & McKelvy, J. (1975). Immunoreactive thyrotrophin releasing factor in gastropod circumoesophageal ganglia. *Nature,* **254**, 620.

Gros, C., Lafon–Cazal, M. & Dray, F. (1978). Présence de substances immunoréactivement apparentées aux enképhalines chez un insecte, *Locusta migratoria. C.R. Académie Sciences, Paris,* **287** Série D: 647–650.

Guillemin, R., Ling, N. & Burgus, R. (1976). Endorphines peptides d'origine hypothalamique et neurohypophysaire à activité morphinomimétique. Isolement et structure moléculaire de l'α endorphine. *C.R. Académie Sciences*, Paris, **282**, Série D: 783–785.

Hansen, B., Hansen, G. & Scharrer, B. (1982). Immunoreactive material resembling vertebrate neuropeptides in the corpus cardiacum and corpus allatum of the insect *Leucophaea maderae. Cell Tissue Research,* **225**, 319–329.

Iwanaga, T., Fujita, T., Nishiitsutsuji–Uwo, J. & Endo, Y. (1981). Immunohistochemical demonstration of PP-, somatostatin-, enteroglucagon- and VIP-like immunoreactivities in the cockroach midgut. *Biomedical Research,* **2**, 202–207.

Jaros, P. & Gade, G. (1982). Evidence for a crustacean hyperglycemic hormone-like molecule in the nervous system of the stick insect, *Carausius morosus. Cell Tissue Research,* **227**, 555–562.

Keller, R. (1981). Purification and animoacid composition of the hyperglycemic neurohormone from the sinus gland of *Orconectus limosus* and comparison with the hormone of *Carcinus maenas. Journal of Comparative Physiology,* **141**, 445–450.

Keshishian, H. & O'Shea, M. (1985a). The distribution of a peptide neurotransmitter in the postembryonic grasshopper central nervous system. *The Journal of Neuroscience,* **5**, 992–1004.

Keshishian, H. & O'Shea, M. (1985b). The acquisition and expression of a peptidergic phenotype in the grasshopper embryo. *Journal of Neuroscience,* **5**, 1005–1015.

Klemm, N. (1976). Histochemistry of putative transmitter substances in the insect brain. *Progress in Neurobiology*, **7**, 99–169.

Klemm, N., Hustert, R., Cantera, R. & Nassel, D. (1986). Neurons reactive to antibodies against serotonin in the stomatogastric nervous system and in the alimentary canal of locust and crickets (Orthoptera, Insecta). *Neuroscience*, **17**, 247–261.

Klemm, N., Steinbusch, H. & Sundler, F. (1984). Distribution of serotonin-containing neurons and their pathways in the supraoesophageal ganglion of the cockroach *Periplaneta americana* (L) as revealed by immunocytochemistry. *Journal of Comparative Neurology*, **225**, 387–395.

Klemm, N. & Sundler, F. (1983). Organization of catecholamine and serotonin-immunoreactive neurons in the corpora pedunculata of the desert locust, *Schistocerca gregaria* Forsk. *Neuroscience Letters*, **36**, 13–17.

Lavenseau, L., Gadenne, C. & Trabelsi, M. (1984). Immunofluorescent localization of a substance immunologically related to insulin in the protocerebral neurosecretory cells of the European corn borer. *Cell Tissue Research*, **238**, 207–208.

Marshall, J. (1951). Localization of adrenocorticotropic hormone by histochemical and immunochemical methods. *Journal of experimental medicine*, **94**, 21–30.

Myers, C. & Evans, P. (1985a). An FMRFamide antiserum differentiates between populations of antigens in the ventral nervous system of the locust, *Schistocerca gregaria*. *Cell Tissue Research*, **242**, 109–114.

Myers, C. & Evans, P. (1985b). The distribution of Bovine pancreatic polypeptide/FMRFamide-like immunoreactivity in the ventral nervous system of the locust. *Journal of Comparative Neurology*, **234**, 1–16.

Nassel, D. & Elekes, K. (1984). Ultrastructural demonstration of serotonin–immunoreactivity in the nervous system of an insect (*Calliphora erythrocephala*). *Neuroscience Letters* **48**, 203–210.

Nassel, D. & Elekes, K. (1985). Serotonergic terminals in the neural sheath of the blowfly nervous system: electron microscopical immunocytochemistry and 5,7–dihydroxytryptamine labelling. *Neuroscience*, **15**, 293–307.

Nassel, D., Hagberg, M. & Seyan, H. (1983). A new, possibly serotonergic, neuron in the lamina of the blowfly optic lobe: an immunocytochemical and Golgi–EM study. *Brain Research*, **280**, 361–367.

Nassel, D. & Klemm, N. (1983). Serotonin-like immunoreactivity in the optic lobes of three insect species. *Cell Tissue Research*, **232**, 129–140.

Nassel, D., Meyer, E. & Klemm, N. (1985). Mapping and ultrastructure of serotonin–immunoreactive neurons in the optic lobes of three insect species. *Journal of Comparative Neurology*,**232**, 190–204.

Nishiitsutsuji–Uwo, J., Takeda, M. & Saito, H. (1984). The production of an antiserum to serotonin and serotonin-like immunoreactivity in the cockroach brain–midgut system. *Biomedical Research*, **5**, 211–224.

Pages, M., Jimenez, F., Ferrus, A., Peralta, E., Ramirez, G. & Gelpi, E. (1983). Enkephalin-like immunoreactivity in *Drosophila melanogaster*. *Neuropeptides*, **4**, 87–98.

Price, D. & Greenberg, M. (1977). Structure of a molluscan cardioexcitatory neuropeptide. *Science*, **197**, 670–671.

Proux, J. & Rougon–Rapuzzi, G. (1980). Evidence for vasopressin-like molecule in migratory locust. Radioimmunological measurements in different tissues: correlation with various states of hydration. *General and Comparative Endocrinology*, **42**, 378–383.

Raabe, M. (1982). Insect neurohormones. *Plenum Press* - New York.

Raabe, M. (1985). Réactions immunocytochimiques au niveau de cellules neurosécrétrices périphériques, des aires neurohémales des organes périsympathiques et des corpora cardiaca chez quelques insectes. *C.R. Académie Sciences*, Paris, **301**, série III, 407–412.

Remy, C. & Dubois, M. (1981). Immunohistological evidence of methionine Enkephalin-like material in the brain of the migratory locust. *Cell Tissue Research*, **218**, 271–278.

Remy, C. & Girardie, J. (1980). Anatomical organization of two vasopressin–neurophysin-like neurosecretory cells throughout the central nervous system of migratory locust. *General and Comparative Endocrinology*, **40**, 27–35.

Remy, C., Girardie, J. & Dubois, M. (1977). Exploration immunocytologique des ganglions cérébroïdes et sous-oesophagien du phasme *Clitumnus extradentatus:* existence d'une neurosécrétion apparentée à la vasopressine–neurophysine. *C.R. Académie Sciences Paris*, **285**, Série D, 1495–1497.

Remy, C., Girardie, J. & Dubois, M. (1978). Présence dans le ganglion sous-oesophagien de la chenille processionnaire du pin (*Thaumetopoea pityocampa* Schiff) de cellules révélées en immunofluorescence par un anticorps anti–α–endorphine. *C.R. Académie Sciences Paris*, **286**, Série D, 651–653.

Remy, C., Girardie, J. & Dubois, M. (1979). Vertebrate neuropeptide-like substances in the suboesophageal ganglion of two insects: *Locusta migratoria* R. and F. (Orthoptera) and *Bombyx mori* L. (Lepidoptera). Immunocytological investigation. *General and Comparative Endocrinology*, **37**, 93–100.

Romeuf, M., Proux, J. & Remy, C. (1986). The neurophysin–vasopressin-like system in migratory locust embryos. Immunohistological and radioimmunological studies. *Neurochemistry International*, **8**, 299–302.

Romeuf, M. & Remy, C. (1984). Early immunohistochemical detection of somatostatin-like and methionine–enkephalin-like neuropeptides in the brain of the migratory locust embryo. *Cell Tissue Research*, **236**, 289–292.

Schooneveld, H., Romberg–Privee, H. & Veenstra, J. (1985). Adipokinetic hormone-immunoreactive peptide in the endocrine and central nervous system of several insect species: a comparative immunocytochemical approach. *General and Comparative Endocrinology*, **57**, 184–194.

Schooneveld, H., Romberg–Privee, H. & Veenstra, J. (1986). Immunocytochemical differentiation between adipokinetic hormone (AKH)-like peptides in neurons and glandular cells in the corpus cardiacum of *Locusta migratoria* and *Periplaneta americana* with C—erminal and N–terminal specific antisera to AKH. *Cell Tissue Research*, **243**, 9–14.

Schooneveld, H., Tesser, G., Veenstra, J. & Romberg–Privee, H. (1983). Adipokinetic hormone and AKH-like peptide demonstrated in the corpora cardiaca and nervous system of *Locusta migratoria* by immunocytochemistry. *Cell Tissue Research*, **230**, 67–76.

Schurmann, F. & Klemm, N. (1984). Serotonin–immunoreactive neurons in the brain of the Honeybee. *Journal of Comparative Neurology*, **225**, 570–580.

Simantov, R., Goodman, R., Aposhian, D. & Snyder, S. (1976). Phylogenetic distribution of a morphine-like peptide "enkephalin". *Brain Research*, **111**, 204–211.

Sloley, B. & Owen, M. (1982). The effects of reserpine on amine concentrations in the nervous system of the cockroach (*Periplaneta americana*). *Insect Biochemistry*, **12**, 469–476.

Snyder, S. (1980). Brain peptides as neurotransmitters. *Science*, **209**, 976–983.

Stefano, G. & Scharrer, B. (1981). High affinity binding of an enkephalin analog in the cerebral ganglion of the insect *Leucophea maderae* (Blattaria). *Brain Research*, **225**, 107–114.

Stefano, G., Scharrer, B. & Assanah, P. (1982). Demonstration, characterization and localization of opioid binding sites in the midgut of the insect *Leucophaea maderae* (Blattaria). *Brain Research*, **253**, 205–212.

Steinbusch, H., Verhofstad, A. & Joosten, H. (1978). Localization of serotonin in the central nervous system by immunohistochemistry: description of a specific and sensitive technique and some applications. *Neuroscience*, **3**, 811–819.

Stone, J., Mordue, W., Batley, K. & Morris, H. (1976). Structure of locust adipokinetic hormone, a neurohormone that regulates lipid utilisation during flight. *Nature*, **263**, 207–211.

Strambi, C., Rougon–Rapuzzi, G., Cupo, A., Martin, N. & Strambi, A. (1979). Mise en évidence immunocytologique d'un composé apparenté à la vasopressine dans le système nerveux du Grillon *Acheta domesticus*. *C.R. Académie Sciences Paris*, **288**, Série D, 131–133.

Straus, E. & Yalow, R. (1979). Gastrointestinal peptides in the brain. *Federation Proceedings*, **38**, 2320–2324.

Sundler, F., Hakanson, J., Alumets, J. & Walles, B. (1977). Neuronal localization of pancreatic polypeptide (PP) and vasoactive intestinal peptide (VIP) immunoreactivity in the earthworm (*Lumbricus terrestris*). *Brain Research Bulletin*, **2**, 61–65.

Taghert, P. & Goodman, C. (1984). Cell determination and differentiation of identified serotonin–immunoreactive neurons in the grasshopper embryo. *Journal of Neuroscience*, **4**, 989–1000.

Takeda, S., Vieillemaringe, J., Geffard, M. & Remy, C. (1986). Immunohistological evidence of dopamine cells in the cephalic nervous system of the silkworm *Bombyx mori*. Coexistence of dopamine and α endorphin-like substance in neurosecretory cells of the suboesophageal ganglion. *Cell Tissue Research*, **243**, 125–128.

Thorpe, A. & Duve, H. (1984). Insulin- and glucagon-like peptides in insects and molluscs. *Molecular Physiology*, **5**, 235–260.

Tyrer, N., Turner, J. & Altman, J. (1984). Identifiable neurons in the locust central nervous system that react with antibodies to serotonin. *Journal of Comparative Neurology*, **227**, 313–330.

Veenstra, J. (1984). Immunocytochemical demonstration of a homology in peptidergic neurosecretory cells in the suboesophageal ganglion of a beetle and a locust with antisera to bovine pancreatic polypeptide, FMRFamide, vasopressin and αMSH. *Neuroscience Letters*, **48**, 185–190.

Veenstra, J., Romberg–Privee, H. & Schooneveld, H. (1984). Immunocytochemical localization of peptidergic cells in the neuro-endocrine system of the Colorado potato beetle *Leptinotarsa decemlineata*, with antisera against vasopressin, vasotocin and oxytocin. *Histochemistry*, **81**, 29–34.

Veenstra, J., Romberg–Privee, H. & Schooneveld, H. (1985). A proctolin-like peptide and its immunocytochemical localization in the Colorado potato beetle, *Leptinotarsa decemlineata*. *Cell Tissue Research*, **240**, 535–540.

Veenstra, J. & Schooneveld, H. (1984). Immunocytochemical localization of neurons in the nervous system of the Colorado potato beetle with antisera against FMRFamide and bovine pancreatic polypeptide. *Cell Tissue Research*, **235**, 303–308.

Veenstra, J. & Yanaihara, N. (1984). Immunocytochemical localization of gastrin-releasing peptide/bombesin-like immunoreactive neurons in insects. *Histochemistry*, **81**, 133–138.

Verhaert, P. & De Loof, A. (1985a). Substance–P-like immunoreactivity in the central nervous system of the blattarian insect *Periplaneta americana* L revealed by a monoclonal antibody. *Histochemistry*, **83**, 501–507.

Verhaert, P. & De Loof, A. (1985b). Immunocytochemical localization of a Methionine–Enkephalin-Resembling neuropeptide in the central nervous system of the american cockroach, *Periplaneta americana* L. *Journal of Comparative Neurology*, **239**, 54–61.

Verhaert, P., Geysen, J., De Loof, A. & Vandesande, F. (1984). Immunoreactive material resembling vertebrate neuropeptides and neurophysins in the brain, suboesophageal ganglion, corpus cardiacum and corpus allatum of the dictyopteran *Periplaneta americana* L. *Cell Tissue Research*, **238**, 55–59.

Verhaert, P., Grimmelikhuijzen, C. & De Loof, A. (1985). Distinct localization of FMRFamide and bovine pancreatic polypeptide-like material in the brain, retrocerebral complex and suboesophageal ganglion of the cockroach *Periplaneta americana* L. *Brain Research*, **348**, 331–338.

Verhaert, P., Marivoet, S., Vandesande, F. & De Loof, A. (1984). Localization of CRF immunoreactivity in the central nervous system of three vertebrate and one insect species. *Cell Tissue Research*, **238**, 49–53.

Vieillemaringe, J., Duris, P., Geffard, M., Le Moal, M., Delaage, M., Bensch, C. & Girardie, J. (1984). Immunohistochemical localization of dopamine in the brain of the insect *Locusta migratoria migratorioides* in comparison with the catecholamine distribution determined by the histofluorescence technique. *Cell Tissue Research*, **237**, 391–394.

Yui, R., Fujita, T. & Ito, S. (1980). Insulin-, Gastrin-, Pancreatic polypeptide-like immunoreactive neurons in the brain of the silkworm, *Bombyx mori*. *Biomedical Research*, **1**, 42–46.

G. MARTIN

Immunocytochemistry and ultrastructure of crustacean endocrine cells

The crustacean neuro-endocrine system has an extensive and diverse organisation (GABE 1966). It includes the Y–organs, mandibular organs and gonads as well as all parts of the nervous sytem (NS) (Echalier 1954; Le Roux 1968; Charniaux–Cotton 1954; Juchault 1966).

This chapter will concentrate on the peptidergic and aminergic areas of the NS, utilizing the isopod model as a framework upon which to base comparative data obtained from decapods.

The neurosecretory cells

The Isopod brain comprises 3 major areas: protocerebrum, innervating the compound eyes; deutocerebrum, innervating the antennule; and tritocerebrum, innervating the antennary.

Neurosecretory products are often revealed by standard histological techniques such as paraldehyde fuchsin (PF) and chrome hematoxylin phloxine (CHP). We have used these techniques to locate 4 types and subtypes of neurosecretory cells (NSC) in the NS (Martin 1981), and the ones described here will be restricted to those terminating in the sinus gland (SG), following the pattern established by the cobalt back-filling experiments of Chiang and Steel (1985a). NSC are classified according to the terminology introduced by Matsumoto (1959).

β–cells: These comprise a group of heterogeneous cells, located in the anterior part of the protocerebrum, on each side of the midline (Fig. 1a; 1c).

Subgroup β_1 contains 6 cells, which are polygonal, 14–28 µm in diameter which stain intensely with both PF and CHP. Ultrastructure shows an ovoid nucleus, stacks of RER, Golgi-derived electron dense granules (160–180 nm) and glycogen particles associated with electron lucent vacuoles (Fig. 1d).

Subgroup β_2 consists of 4 tear-drop shaped cells, 10–20 µm in diameter which stain intensely with both PF and CHP. Ultrastructure reveals a spherical nucleus, sparse RER and large accumulations of ovoid granules (87–169 nm diam) (Fig. 1d).

B–cells: These polygonal cells of 10–20 µm diameter are located more posteriosly than β_1 cells, and stain weakly with both PF and CHP. Electron microscopy shows polygonal cells with electron lucent nucleus with dominant nucleolus. RER is concentrically arranged around the nucleus and the spherical granules (12.8 nm ± 3.8 diameter) accumulate near the Golgi complex (Fig. 2a). Both β and B cells have lysosomes associated with the granules. These may be involved in autophagy of excess granules or processing of pro–hormones (Morris & Nordmann 1979).

γ-cells: These ovoid cells are 7 by 13 μm and are located solely in the optic lobe between each medulla; they have a high affinity for PF and CHP. As visualized by EM, each cell shows a spherical nucleus surrounded by a narrow fringe of cytoplasm littered with irregular or elongate granules (diameter ranging from 100 to 130 nm)(Fig. 2b). In terrestrial isopods the NSC–types are easily located (Fig. 4). Light microscopic data on Decapods NSC (See Bressac 1978) however, do not always agree; recent studies using cobalt back-filling or micro iontophoresis with Lucifer yellow have shown that most of the axons ending in the SG of *Macrura* originate from NSC located in the medulla terminalis (Medulla terminalis ganglionic X–organ: MTGX)(Andrew *et al.* 1978; Jaros 1978; Kallen 1985). With the description of collaterals from axons of X–organ cells, one may expect that synapses are to be found on other dendritic elements.

The sinus gland

The sinus gland (SG) located adjacent to the optic lobes, is an area of storage and release for several hormones; it is the major neurohaemal organ of crustaceans and neurosecretory granules (NSG) are found in a mosaic of secretory axon dilatations or 'terminals' (Nordmann & Morris 1980). Decapod SGs have received most attention, probably due to the large size of their axon terminals (see Cooke & Sullivan 1982). Isopods have also been quite widely investigated (Martin 1972; 1981; Chataigner *et al.* 1978; Chiang & Steel 1985) while to date there is only a single report on Amphipod sinus glands (Brodie & Halcrow 1977). Generalizations that can be drawn from these studies are that terminals of several types can be distinguished by morphological appearance of granules and cytoplasm, and that the number of types varies from 2 to 7, depending on investigator, conditions of fixation, and species. This section will focus mainly on the ultrastructure of SG from terrestrial Isopods. In woodlice the SG appears as a bulb-like outgrowth (Fig. 3a), 100 by 70 μm, attached by a short stalk to the ventro-lateral side of the optic lobe. The gland comprises 3 main components: the neurilemma, the glial cells, and the axons and axon terminals (Fig. 3b). The neurilemma forms a 60 nm layer separating the axon endings from haemolymph.

Fig. 1. (a) Semithin section with 2 perikarya β_1. × 680. (b) Ultrastructural aspect of a β_1-perikaryon. × 9680. (c) Semithin section with a β_2–perikaryon; notice the stainability.× 820. (d) Ultrastructural aspect of a β_2–perikaryon. Note the granule load. × 7320.

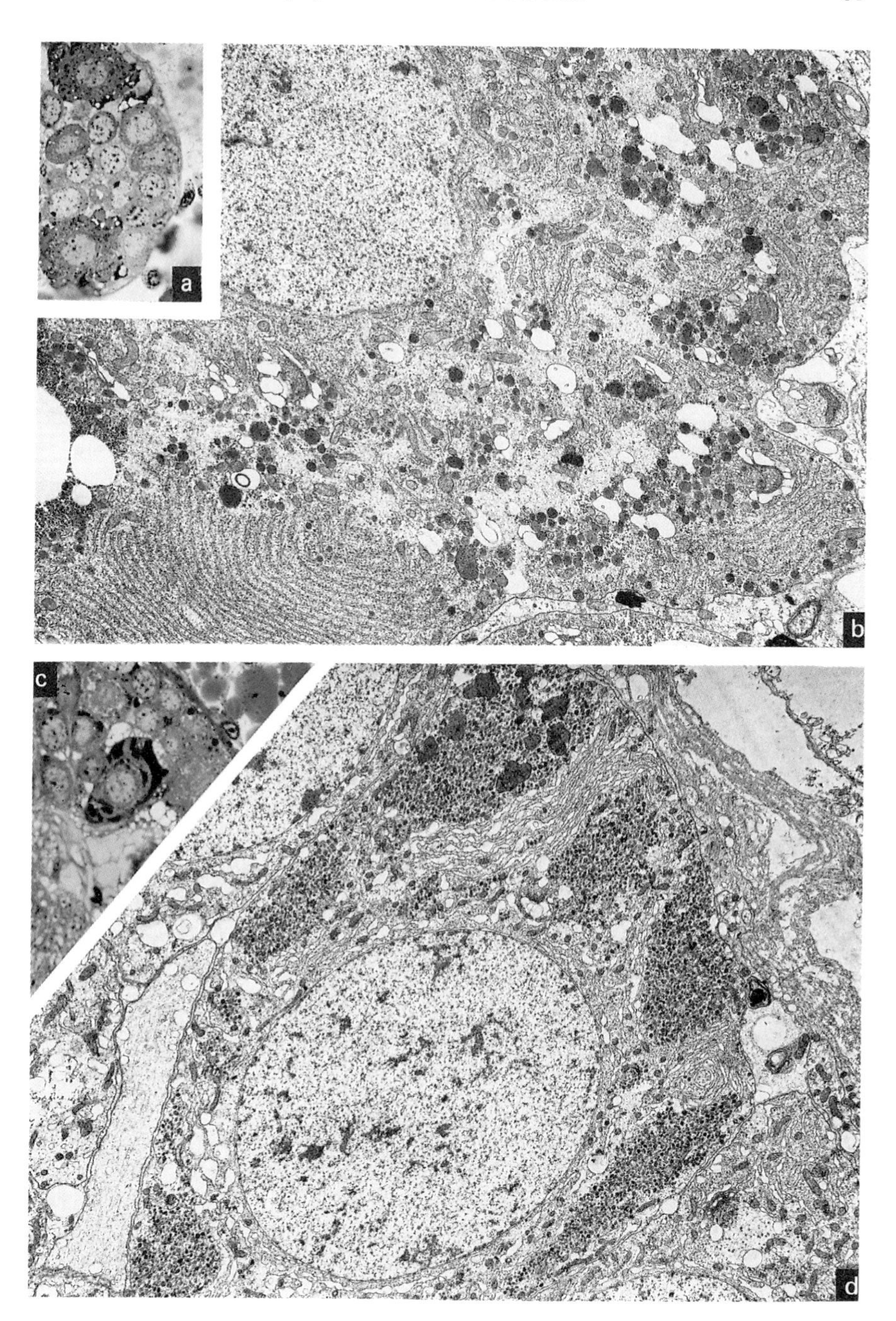
a
b
c
d

Nuclei in the SG (Fig. 3c) belong to glial cell processes which enwrap the axons but not the endings (Gabe 1954). The morphology (size, shape and electron density) of the NSG enable us to distinguish the following three types of terminal:
type 1, with large, spherical, electron dense granules with a diameter between 200 and 270 nm;
type 2, with medium sized, spherical, electron dense granules with a diameter between 100 and 165 nm; and type 3, with irregularly shaped granules of varying size, and varying degrees of electron density (Maximum granule size 100–135 nm).

Some terminals (Fig. 3e) do not enter the stalk of the SG but run ventrally along the optic nerve producing a thin secondary neurohaemal organ extended towards the ommatidia. This region, with 2 terminal types, is called "lateral extension of the SG" by Chiang and Steel (1985). The single type of terminal here is likely to be represented by the phloxine positive, PF negative A cells, containing large granules described by Martin (1972a). The release of hormones from the SG occurs by exocytosis (Fig. 3d) and although the precise mechanism of membrane retrieval is still unclear, Nordmann and Morris (1980) propose that cytoplasmic vacuoles larger than the microvesicles are responsible for the recapture of membrane following exocytosis. In an alternative proposal, Nordmann and Chevallier (1980) and Shaw and Morris (1980) consider that microvesicles are involved in the sequestration of calcium ions and do not participate in the endocytotic process.

Lateral cephalic nerve plexus (LCNP)

The LCNP is characteristic of Isopoda, first described in *Porcellio* and *Oniscus* by Messner (1963), and soon afterwards identified as a neurohaemal organ based upon its tinctorial affinities and connections with the nerve cord (Besse & Legrand 1964). Subsequently, more information has become available from studies of a number of *Anilocra frontalis, Sphaeroma serratum, Helleria brevicornis, Ligia oceanica, Idotea balthica* (Chaigneau 1983; Martin *et al.* 1983).

It is thought that in ancestral forms, SG and LCNP were linked, and in extant forms this link persists in the genera *Anilocra, Helleria* and *Ligia*. Demassieux (1976) reports a paired lateral sinus nerve running from the brain to the abdominal ganglion with metameric neurohaemal areas. This peculiarity is unique among the neurohaemal organs of Peracarida (Chaigneau 1983).

Optical microscope studies showed that the LCNP in *P. dilatatus* is formed in the tritocerebrum on the ventral edge of the antennary root nerve; nMd, nMX_1, nMx_2 and nMxp which come from the sub-oesophageal ganglion; and nR_1, which comes from the ipsilateral cardiac nerve and is a link with pericardial organs described by

Fig. 2. (a) Ultrastructural aspect of a B perikaryon. × 9590. (b) Ultrastructural aspect of a γ-perikaryon. × 9960.

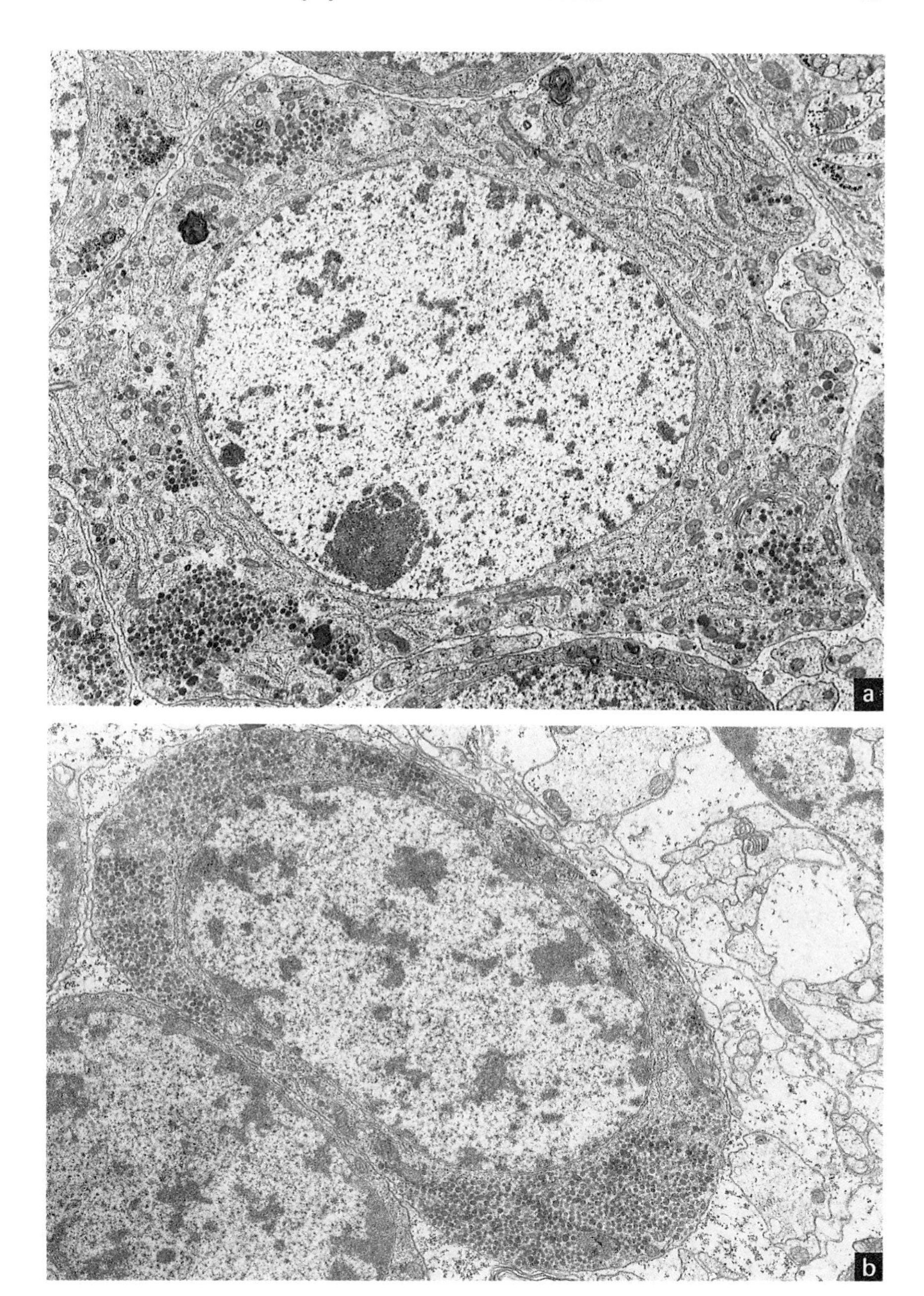

a
b

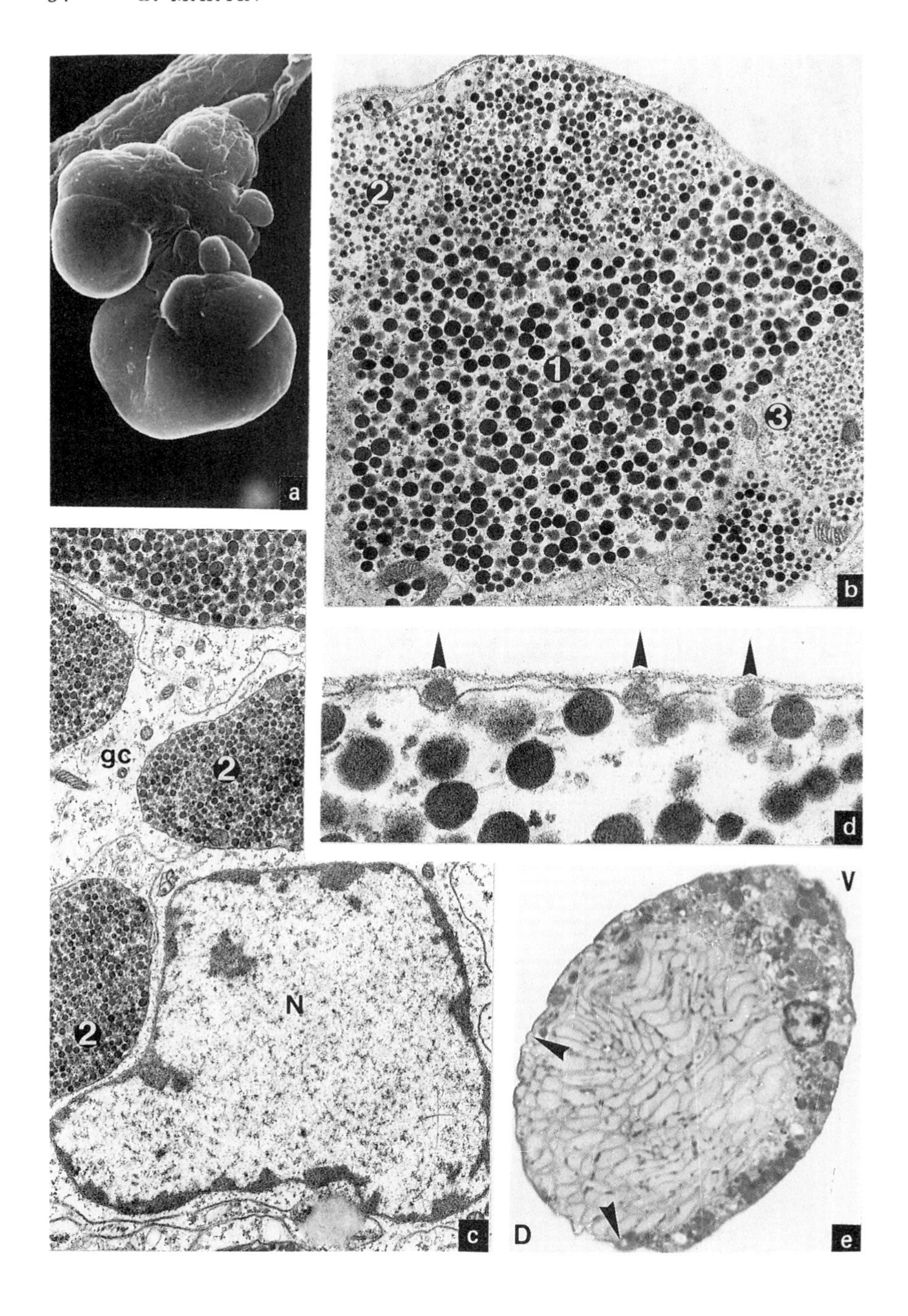
a
2
1
3
b
gc
2
d
N
2
c
V
D
e

Fig. 4. Left side, localization of the different types of NSCS in the NS of *Porcellio dilatatus*. Right side, drawing of LCNP from Besse & Legrand (1964). nR_1 and nR_2, nerves linked with the stomatogastric system. nA_2, nMd, nMx, nMx_2, nPMx, segmental nerves.

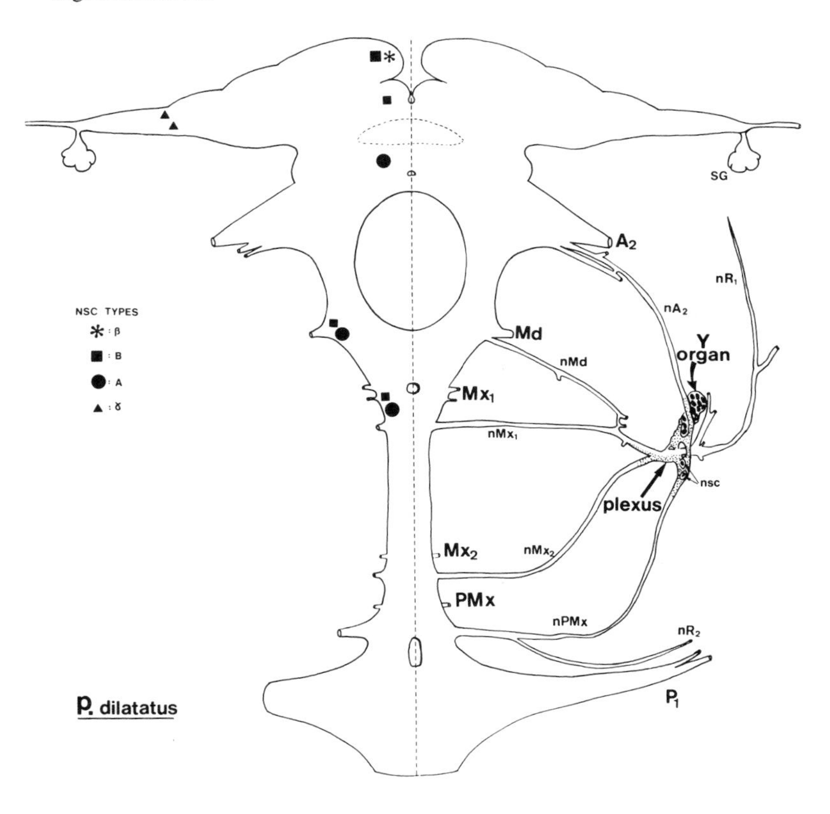

Fig. 3. (a) SG a bulb-like structure (SEM). × 570. (b) 3 types of terminals in the SG. × 16000. (c) Nucleus of a glial cell (gc). × 9500. (d) Release by exocytosis. × 48000. (e) Semithin section showing the "lateral extensions" towards the optic peduncle. Arrow heads dorsal limits of the extensions. D, dorsal side; V, ventral side. × 2600. (Facing page)

Delaleu (1974). The enlarged part (blister) of the LCNP, adjacent to the Y–organ was seen to contain 2 intrinsic neurosecretory cells and a cluster of axon terminals (Fig. 5a).

Ultrastructural investigation of the LCNP blister shows a large number of axon terminals of various sizes, sometimes with glial processes running along them, enclosed in a thin neurilemma-like sheath. At release sites the glial sheath is absent. If axon terminals are classified according to the size and density of their elementary granules, the LCNP may be said to include 5 types of terminals (Fig. 5b): type I, spherical electron dense granules, diameter 172–240 nm; type II, spherical granules less electron dense than type I, diameter 135–176 nm; type III, similar to type II but diameter 100–150 nm; type IV, granules subspherical with pale appearance and found only in a few small terminals diameter 82–100 nm; and type V, oblong granules, density as in type II, long axis ranging from 165 to 170 nm, narrow axis from 85–90 nm.

The intrinsic neurosecretory cells comprise pear-shaped perikarya measuring 18 × 13 μm, and are stained more easily with PF than CHP. The staining affinity and morphological appearance of these cells suggests they are type B according to the Matsumoto classification (1959).

At the ultrastructural level, the two 9 to 10 μm diameter cells with clear ovoid nuclei display very diffuse chromatin. They also contain many small RER cisternae (mostly dilated and dispersed throughout the cytoplasm), numbers of lysosomal structures (Fig. 5d), and only few dense neurosecretory granules (107–132 nm diameter)(Fig. 5c).

During the moulting cycle, granule concentration is always very low. At present, ultrastructural data about LCNP are also available for the species *S. serratum* (Maissiat *et al.* 1970) and *L. oceanica* (Martin *et al.* 1983).

Post-commissural organs

Little is known of the role of post-commissural organs. They are unknown in Isopoda, although they have been described by Carlisle and Knowles (1959) in

Fig. 5. (a) Semithin section at the level of the LCNP blister. insc, Intrinsic neurosecretory cell. × 550. (b) 3 types of terminals within the LCNP blister. × 10000. gc, glial cell. (c) Localization of CHH containing axons on the edge of the LCNP blister. × 280. Y, y organ. (d) Dense bodies of the plexus cell. × 16000. (c) Localization of CHH containing axons on the edge of the LCNP blister. × 280. Y, y organ.(e) Plexus cell. × 15000. Arrow heads, neurosecretory granules; N, nucleus.

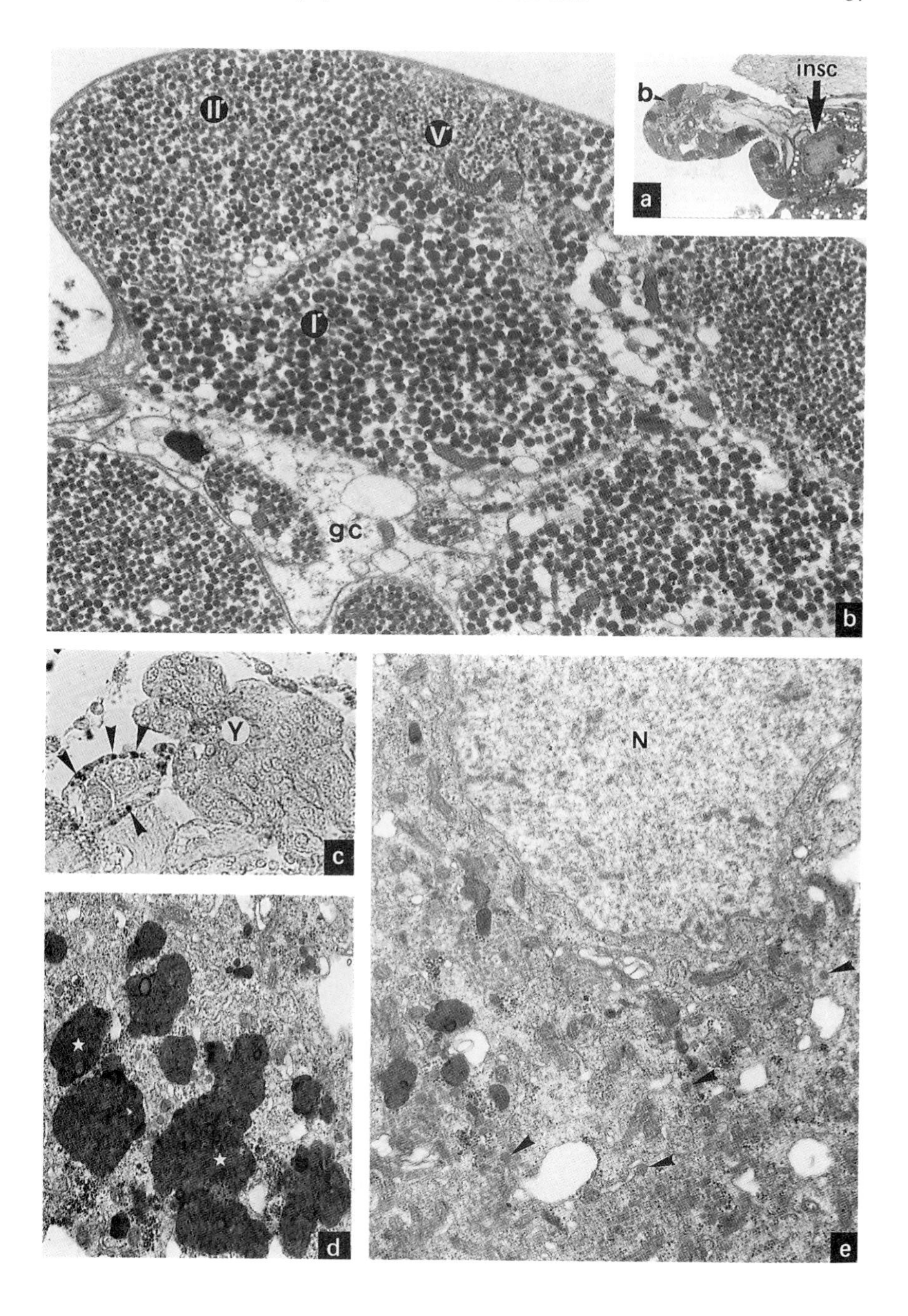
insc
b
a
II
V
I
gc
b
Y
c
N
d
e

Peneida and in a stomatopod. Maynard (1961a,b) was the first to find them in a crab where they are linked with the tritocerebral commissure.

Pericardial organs

Pericardial organs are neurohaemal organs lying in the venous cavity surrounding the crustacean heart. In *P. dilatatus* dorsolateral nerves run along the dorsal musculature and at fourth thoracic segment level, they enter into the pericardial septum which surrounds the heart as far as its posterior limit. In the head, each of these nerves joins the suboesophageal ganglion by way of 2 small nerves, the most anterior of which leads close to the blister of the LCNP. In the 3 anterior segments of the abdomen, on either side of the heart, the dorsolateral nerves send out branching processes which form the intrapericardic nerve plexus. Some authorities consider this plexus to be homologous to the decapod pericardial organ (Delaleu 1974).

Preliminary work on the dorsolateral nerves in *Armadillidium vulgare* reveals that 9 of the axons enclosed in each nerve carry 2 types of NSG, with a diameter of 85 nm and 140 nm respectively. Materials in these NSG_S are thought to modulate heart activity. In decapods the pericardial organs are more variable (Cooke & Sullivan 1982). Ultrastructurally, up to 7 ending types have been described and each may contain a specific hormone, for example octopamine (type 3), serotonin (type 4), and dopamine (type 7). Proctolin-containing endings are still unknown (Schwarz *et al.* 1984). Pericardial organs may be thought of as neurohaemal organs for the synthesis and storage of catecholamines, indolamines and phenolamines, together with a number of peptide hormones.

Biogenic amines in crustacean NS

Early work, utilizing the formaldehyde-induced fluorescence technique (Falck 1962) established the presence of both catecholamines and indolamines in the crustacean nervous system (Kushner & Maynard 1977; Elofsson & Klemm 1972; Elofsson *et al.* 1977; Aramant & Elofsson 1976). Octopamine is not detectable by this method and later Molinoff *et al.* (1969) introduced a sensitive radioenzymatic assay which has revealed octopamine in the lobster (Baker *et al.* 1972) and in the brain (1.2 μg/g, w/w) and nerve cords (0.24 μg/g w/w) of *P. dilatatus*. (M. Lafon Cazal & Martin, unpublished observation).

High performance liquid chromatography (HPLC) with electrochemical detection (ED) has been particularly useful for the identification and quantitative measurement of biogenic amines and dehydroxyphenylalamine in the CNS of Crustacea. Elofsson *et al.* (1982) and Laxmyr (1984) described dopamine, noradrenaline, octopamine, serotonin (5–HT) and the amino acid dehydroxyphenylalamine; these substances occur in concentrations between 0.1 and 0.7 μg/g, w/w. In *P. dilatatus* levels of serotonin varied with the range 0.4–0.5 ng/g, w/w (Picard & Martin, unpublished).

Immunochemical methods have also proved useful in the detection of amines (Seguela *et al.* 1984). Elofsson (1983), using 3 different antisera against serotonin, produced a map of bundle fibres and perikarya in the CNS of the crayfish *Pacifastacus leniusculus*. Immunoreactive areas coincided in general with areas of the brain where synapses are thought to be concentrated; for example, the central body, protocerebral bridge, and the olfactory lobes. Immunoreactivity in the medulla externa agreed with the pattern (3 layers) of the catecholamine fluorescence (Falck–Hillarp), but in the medulla interna a difference was noted; immunoreactive neurons branched in 3 layers, whereas the catecholaminergic neurons branch in 5 layers. The lamina ganglionaris did not possess any immunoreactive neurons. In addition a diffuse network of immunoreactive nerve fibres is present in the central NS.

In *P. dilatatus* we have found two bundles of fibres that traverse the whole length of the nerve cord (Fig. 6a); on the ventral side of the segmental ganglia two symmetrical immunoreactive perikarya are located whose axons join these bundles. At present classical monoaminergic synapses are unknown in crustaceans.

Amines as releasing factor

As early as 1968, Keller & Beyer demonstrated that serotonin will trigger the release of the hyperglycemic hormone in *Orconectes limosus*. Later Strolenberg and Van Herp (1977) noted an increase in exocytosis in crayfish SG and a hyperglycemic response 2 h after injection of serotonin. We confirmed this result in *P. dilatatus* SG (Fig. 6g)(Martin 1978) and reported an unexpected effect on glial cells.

The release of chromatophorotropins from neurohaemal organs (SG and PCO) is thought to be controlled by neural signals that may occur in areas adjacent to neurosecretory fibre tracts, especially the X–organ–SG tract. Thus serotonin stimulates the release of red pigment dispersing hormone, dopamine triggers the release of red pigment concentrating hormone and black pigment concentrating hormone and norepinephrine stimulates the release of black pigment/melanin dispersing hormone (Staub & Fingerman 1984).

Other evidence suggests that release of specific neurotransmitters by neurons acting on NSC results in release of specific neurohormones (Rao & Fingerman 1983; Quackenbush & Fingerman 1984). In addition studies by Mattson & Spaziani (1985a,b) show that serotonin-containing neurons provide excitatory input to moult inhibiting hormone-containing NSC.

Native peptides

The regulation of physiological and biochemical processes in crustaceans by endogenous peptides has been known for some time (Kleinholz & Keller 1979). These effects, formalized as bioassays, have allowed the purification and characterization of the peptide molecules themselves (see Chapter 9). An antiserum raised against a *Carcinus* peptide by Jaros and Keller (1979a) was used to develop a RIA to measure blood titres of CHH (Crustacean Hyperglycemic Hormone) and to locate the cells that synthesize CHH. Jaros and Keller (1979b) reported the direct demonstration of the CHH in the MTGX–organ–SG complex by means of a double antibody fluorescence technique. Jaros (1979) and Gorgels–Kallen and Van Herp (1981) have extended the immunocytochemical studies to the electron microscopical level and here there is good evidence that CHH is associated with the large predominant type 5 granules.

The isolation and characterization of CHH from *Porcellio* SGs (Martin *et al.* 1984a) and its crossreactivity with the antiserum raised against *Carcinus* CHH has also allowed the localization of its sites of production (Martin *et al.* 1984b).

The β_1–cells show strong specific CHH-like immunoreactivity and their axons can be traced towards the SG. These investigations have recently been extended and immunoreactivity has been found in axon terminals of the LCNP (Fig. 5e). Plexus cells are not stained and we speculate that terminals of type I contain CHH; based on granule morphology. Type III was assigned to plexus cells and types II, IV and V to neurosecretory perikarya of the suboesophageal ganglion.

Use of the immunogold method has shown specific labelling of the NSG at the ultrastructural level, other areas such as RER, Golgi complex, which might be expected to contain CHH–prohormone, were essentially unlabelled (Fig. 6f). Axon profiles of the SG with the biggest granules (type 1) are intensely labelled (Fig. 6d). This confirms that the β_1–cells of *Porcellio* are homologous to the cells of the X–organ in Decapods.

Proctolin

In the lobster, proctolin-like immunoreactivity (see Chapter 8) is detectable in nearly every portion of the NS. The amount of proctolin-like material tends to be greater in the anterior ganglia and the largest amount of immunoreactivity (varicosities) is found in the pericardial organs, although no cell bodies in this area are positive (Schwarz *et al.* 1984). The nerve endings in the pericardial organs are

Fig. 6. (a) Whole mount immunocytochemical preparation of *Porcellio* ventral cord. (antiserum 5–HT, d = 1/350). × 45. (b) Falck Hillarp method; yellow fluorescence of the central body (5–HT). × 720. (c) PAP method immunoreactivity in the central body (antiserum Leu–enk, d = 1/500). × 720. (d) Immunogold staining (IGS) of type 1 granules in the SG (antiserum CHH, d = 1/600). × 32400. IGS is performed on non-osmicated tissues embedded in glycolmetacrylate (GMA). (e) Control antiserum preadsorbed with an extract of SGs. × 32000. (f) Immunogold staining of granules in β_1 cell. x 32000. (g) Aspect of terminals type in the SG. Note the microvesicles (open circles) and exocytosis aftermaths (arrow heads). × 36000. (h) Immunoreactive cell in the protocerebrum (antiserum hpGRF, d = 1/500. × 480.

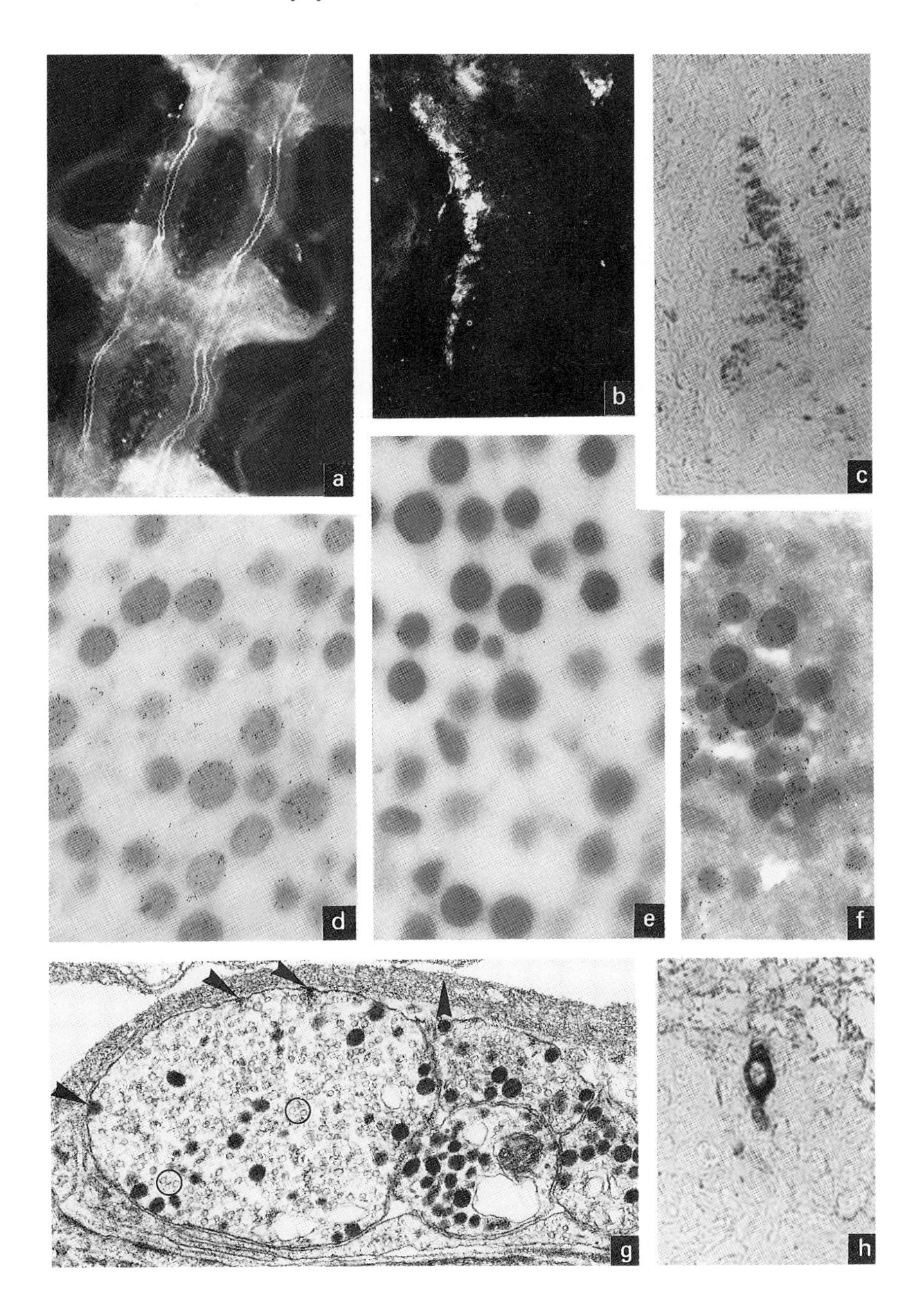
a
b
c
d
e
f
g
h

capable of releasing proctolin-like material when depolarized in the presence of Ca^{2+}, and while these findings suggest that proctolin is a neurohormone in the lobster, no proctolin-like immunoreactivity has been detected in the haemolymph.

In *Porcellio* a few perikarya with proctolin-like immunoreactivity (IM) are found in the optic lobe near the cells and in the cortical area where β_1–cells are located.

'Vertebrate-related' peptides

Van Deijnen and Van Herp (1985) have used a variety of antisera raised against both vertebrate and invertebrate peptides to reveal the complexity of the neurosecretory system in the eyestalk of the crayfish. These authors emphasize the close relationship between the aminergic and peptidergic systems, but are unable to decide whether amines and peptides co-exist in the same fibre.

In Isopods, somatostatin-like (Martin & Dubois 1981) and Leu–enkephalin-like immunoreactivity is found in the central body, where serotonin is also found (Fig. 6b,c). In the upper and lower parts of the protocerebron, 2 perikarya (Fig. 6h) were intensely labelled with hp/GRF (human pancreatic growth hormone releasing factor, which is very similar to hypothalamic GRF; hpGRF, Proctolin, Leu–enk, antisera were kindly provided by Dr. G. Tramu Lille).

Leucine–enkephalin-like immunoreactivity has been detected in retinular cells of *Panulirus interruptus* by Mancillas *et al.* (1981) while Enkephalin-like and substance–P-like immunoreactivity were demonstrated in the same cells in *Uca pugilator* by Fingerman *et al.* (1985). The same authors show that SP-like, enkephalin-like, and FMRF–amide-like reactivity is widely distributed in the eystalks of different Decapods and it is conceivable that they are involved in visual processing. By immunostaining ultrathin section, Jaros *et al.* (1985), depicted small leu–enkephalin positive profiles containing granules (82 ± 23 nm). These terminals are more numerous in the areas close to haemolymph lacunae.

Met–enkephalin has been shown to release both red and black pigment concentrating hormones (Quackenbush & Fingerman 1984). Subsequently, following the demonstration of vasopressin-like immunoreactivity in *Palaemon serratus* (Van Herp & Bellon–Humbert 1982), Mattson and Spaziani (1985) studied the effect on steroidogenesis: in *Cancer antennarius* Y–organ activity *in vitro*.is inhibited The relative protencies of the tested peptides were: lysine vasopressin, arginine vasopressin, vasotocin, oxytocin which suggest that MIH may be most closely related to lysine vasopressin.

In Decapods, as in Isopods, the availability of antisera raised to vasopressins will permit the identification of axon terminals in the SG or other neurohaemal organs that could regulate the rate of ecdysteroid secretion by the Y–organ. However, confirmation of the identity of MIH must ultimately depend upon sequence analysis of the native molecules whose amino acid composition is very similar to CHH (Webster & Keller 1986; see also Chapter 9, this volume).

References

Andrew, R.D., Orchard, I. & Saleuddin, A.S. 1978 Structural re-evaluation of the neurosecretory system in the crayfish eyestalk. *Cell Tiss. Res.*, vol. 290, 235–246.

Aramant, R. & Elofsson, R. 1976 Distribution of monoaminergic neurons in the nervous system of non-malacostracan crustaceans. *Cell Tiss. Res.*, vol. 166, 1–24.

Barker, D.L., Molinoff, P.B. & Kravitz, E.A. 1972 Octopamine in the lobster nervous system. *Nature*, **236**, 61–63.

Besse, G. & Legrand, J.J. 1964 Contribution à l'étude d'un organe neurohemal en relation avec l'organe Y. chez l'Oniscoïde *Porcellio dilatatus* Brandt. *C.R. Acad. Sc. Paris*, **239**, 3858–3861.

Bressac, C. 1978 Etudes cytologiques et ultrastructurales des organes neuro-endocrines et endocrines de *Pachygrapsus marmoratus* (Fabricius 1787)(Crustacé, Décapode). *Thèse Etat Université des Sciences et Techniques du Languedoc, Montpellier*, 179p.

Brodie, D. & Halcrow, K. 1977 The ultrastructure of the sinus gland of *Gammarus oceanicus* (Crustacea:Amphipoda). *Cell Tiss. Res.*, **182**, 557–564.

Carlisle, D.B. & Knowles, F.G. (eds.) 1959 *Endocrine control in Crustaceans.* Cambridge University Press, Cambridge, New York.

Chaigneau, J. 1983 Neurohemal organs in Crustacea; In: *Structures and Functions*, ed. A.P. Gupta. Charles C. Thomas.

Charniaux–Cotton, H. 1954 Découverte chez un Crustacé Amphipode (*Orchestia gammarella*) d'une glande endocrine responsable de la différenciation des caractères sexuls primaires et secondaires mâles. *C.R. Acad. Sci. Paris*, **239**, 780–782.

Chataigner, J.P., Martin, G. & Juchault, P. 1978 Etude histologique cytologique et expérimentale des centres neurosécréteurs céphaliques du flabellifère *Sphaeroma serratum* (Crustacé, Isopode). *Gen. Comp. Endocrinol.*, **35**, 52–69.

Chiang, R. & Steel, C. 1985a Structural organization of neurosecretory cells terminating in the sinus gland of the terrestrial isopod, *Oniscus asellus*, revealed by paraldehyde fuchsin and cobalt backfilling. *Can. J. Zool.*, **63**, 543–549.

Chiang, R. & Steel, C. 1985b Ultrastructure and distribution of identified neurosecretory terminals in the sinus gland of the terrestrial isopod *Oniscus asellus. Tiss. and Cell*, **17**, 405–415.

Cooke, I.M. & Sullivan, R.E. 1982 Hormones and Neurosecretion. In: *The Biology of Crustacea.* Ed. H.L. Atwood & D.C. Sandeman. Academic Press, New York, London. Vol. 3, 205–290.

Delaleu, J.C. 1974 Structure et physiologie d'un coeur à automatisme neurogène (Crustacé, Isopode); myocarde, relations neuromusculaires, facteurs de la régulation. *Thèse Etat Université Cl. Bernard, Lyon I*, p.226

Demassieux, C. 1976 Le système neurosécréteur du crustacé Isopode, *Asellus aquaticus* Linné. Etude morphologique et mise en évidence de variations cycliques des cellules de type β en fonction de la physiologie sexuelle femelle. *Thèse 3ème cycle, Universite de Nancy I*, 96 p.

Echalier, G. 1959 L'organe Y et le déterminisme de la croissance et de la mue chez *Carcinus maenas. Ann. Sc. Nat. Zool.*, **12**, no.1, 1–59.

Elofsson, R. 1983 5–HT-like immunoreactivity in the central nervous sytem of the crayfish, *Pacifastacus leniusculus. Cell Tiss. Res.*, **232**, 221–236.

Elofsson, R. & Klemm, N. 1972 Monoamine-containing neurons in the optic ganglia of crustaceans and insects. *Z. Zellforsch.*, **133**, 475–499.

Elofsson, R., Näaael, D. & Myhrberg, H. 1977 A catecholaminergic neuron connecting the first two optic neuropiles (lamina ganglionaris and medulla externa) of the crayfish *Pacifastacus leniusculus*. *Cell Tiss. Res.*, **182**, 287–297.

Elofsson, R., Laxmyr, L., Rosengren, E. & Hansson, C. 1982 Identification and quantitative measurements of biogenic amines and dopa in the central nervous system and haemolymph of the crayfish *Pacifastacus leniusculs* (Crustacea). *Comp. Biochem. Physiol.*, **71C**, 195–201.

Falck, B. 1962 Observations on the possibilities of the cellular localization of monoamines by a fluorescence method. *Acta physiol. scand.*, **56**, 1–25.

Fingerman, M., Hanumante, M., Kulkarni, G., Ikeda, R. & Vacca, L. 1985 Localization of substance P-like, Leucine–enkephaline-like, methionine–enkephaline-like, and FMRF–amide-like immunoreactivity in the eyestalk of the fiddler crab, *Uca pugilator*. *Cell Tiss. Res.*, **241**, 111–117.

Gabe, M. 1954 La neurosécrétion chez les Invertébrés. *Année Biol.*, t.30, 5–62.

Gabe, M. 1956 *Neurosécrétion*. Pergamon Press, Oxford.

Gorgels–Kallen, J. & Van Herp, F. 1981 Localization of crustacean hyperglycemic hormone (CHH) in the X–organ sinus gland–complex in the eyestalk of the crayfish, *Astacus leptodactylus* (Nordmann 1842). *J. Morph.*, **170**, 347–355.

Jaros, P.P. 1978 Tracing of neurosecretory neurons in crayfish pptic ganglia by cobalt iontophoresis. *Cell Tiss. Res.*, **194**, 297–302.

Jaros, P.P. 1979 Immunocytochemical demonstration of the neurosecretory X–organ complex in the eyestalk of the crab, *Carcinus maenas*. *Histochemistry*, **63**, 303–310.

Jaros, P.P. & Keller, R. 1979a Radioimmunoassay of an invertebrate peptide hormone – the custacean hyperglycemic hormone. *Experientia*, **35**, 1252–1253.

Jaros, P.P. & Keller, R. 1979b Immunocytochemical identification of hyperglycemic hormone-producing cells in the eyestalk of *Carcinus maenas*. *Cell Tiss. Res.*, **204**, 379–385.

Jaros, P.P., Dircksen, H. & Keller, R. 1985 Occurrence of immunoreactive enkephalins in a neurohemal organ and other nervous structures in the eystalk of the shore crab, *Carcinus maenas* L. (Crustacea, Decapoda). *Cell. Tiss. Res.*, **241**, 111–117.

Juchault, P. 1966 Contribution à la différenciation sexuelle mâle chez les Crustacés Isopode. *Thèse Doctorat d'Etat*, 1–111. Université Poitiers, France.

Kallen, J. 1985 The hyperglycemic hormone producing system in the eyestalk of the crayfish *Astacus Leptodactylus*. *Thèse Nijmegen*, 127 p.

Kleinholz, L.H. & Keller, R. 1979 Endocrine regulation in crustacea. In: E.J.W. Barrington (ed.) *Hormones and Evolution*, **1**, 159–214. Academic Press, New York.

Keller, R. & Beyer, J. 1968 Zur hyperglykamischen wirkung von serotonin und Augenstielextrakt beim Fluβkrets *Orconectes limosus*. *Z. Vergl. Physiol.*, **59**, 78–85.

Keller, R., Jaros, P.P. & Kegel, G. 1985 Crustacean hyperglycemic neuropeptides. *Amer. Zool.*, **25**, 207–221.

Kushner, P.D. & Maynard, E.A. 1977 Localization of monoamine fluorescence in the stomatogastric nervous system of lobsters. *Brain Res.*, **129**, 13–28.

Laxmyr, L. 1984 Biogenic amines and dopa in the central nervous sytem of Decapod crustaceans. *Comp. Biochem. Physiol.*, **77C**, 139–143.

Le Roux, A. 1968 Description d'organes mandibulaires nouveaux chez les Crustacés Décapodes. C.R. Acad. Sci. Paris, **266**, 1414–1417.

Maissiat, R., Martin, G., Maissiat, J. & Juchault, P. 1979 Ultrastructural development of the neurohemal organ joined to the ecdysial gland after imaginal moulting in the male Isopod *Sphaeroma serratum* Fab. (Crustacea Flabellifera). *Cell Tiss. Res.*, **203**, 403–414.

Mancillas, J.R., McGinty, J.F., Selverston, A.I., Karten, H. & Bloom, F.E. 1981 Immunocytochemical localization of enkephalin and substance P in retina and eyestalk neurones of lobster. *Nature*, **293**, 576–578.

Martin, G. 1972a Analyse ultrastructurale des cellules neurosécrétrices du protocérébron de *Porcellio dilatatus* (Brandt)(Crustacé Isopode Oniscoïde). *C.R. Acad. Sc. Paris.* **274**, 243–246.

Martin, G. 1972b Contribution à l'étude ultrastructurale de la glande du sinus de l'Oniscoïde *Porcellio dilatatus* Brandt. *C.R. Acad. Sc. Paris*, **275**, 839–842.

Martin, G. 1978 Action de la sérotonine sur la glycémie et la libération des neurosécretions contenues dans la glande du sinus de *Porcellio dilatatus* Brandt (Crustacé, Isopode, Oniscoïde). *C.R. Soc. Biol.*, **172**, 304–307.

Martin, G. (1981) Contribution à l'étude cytologique et fonctionnelle des systèmes de neurosécrétion des crustacés isopodes. *Thèse Etat, Université de Poitiers*, 374p.

Martin, G. 1982 Etude ultrastructurale de la régéneration de la glande du sinus chez l'Oniscoïde *Porcellio dilatatus* Brandt, données complémentaires sur l'origine des terminaisons de cet organe neurohemal. *J. Physiol., Paris*, **78**, 558–565.

Martin, G. & Dubois, M.P. 1981 A somatostatin-like antigen in the nervous system of an isopod *Porcellio dilatatus* Brandt. *Gen. Comp. Endocrinol.*, **45**, 125–130.

Martin, G., Maissiat, R. & Girard, P. 1983 Ultrastructure of the sinus gland and lateral cephalic nerve plexus in the Isopod *Ligia oceanica* (Crustacea Oniscoïdea). *Gen. Comp. Endocrinol.*, **52**, 38–50.

Martin, G., Keller, R., Kegel, G., Besse, G. & Jaros, P.P. 1984a The hyperglycemic neuropeptide of the terrestrial isopod, *Porcellio dilatatus*. I. Isolation and characterization. *Gen. Comp. Endocrinol.*, **55**, 208–216.

Martin, G., Jaros, P.P., Besse, G. & Keller, R. 1984b The hyperglycemic neuropeptide of the terrestrial isopod, *Porcellio dilatatus*. II. Immunocytochemical demonstration in neurosecretory structures of the nervous system. *Gen. Comp. Endocrinol.*, **55**, 217–226.

Matsumoto, K. 1959 Neurosecretory cells of an Isopod, *Armadillidium vulgare* (Latreille). *Biol. J. Okayama Univ.*, **5**, 43–50.

Mattson, M.P. & Spaziani, E. 1985a 5–hydroxytryptamine mediates release of molt inhibiting hormone activity from isolated crab eyestalk ganglia. *Biol. Bull.*, **169**, 246–255.

Mattson, M.P. & Spaziani, E. 1985b Functional relations of crab molt-inhibiting hormone and neurohypophysial peptides. *Peptides*, **6**, 635–640.

Maynard, D.P. 1961a Thoracic neurosecretory structures in Brachyura. I. Gross anatomy. *Biol. Bull.*, **121**, 316–329.

Maynard, D.M. 1961b Thoracic neurosecretory structures in Brachyura. II. Secretory neurons. *Gen. Comp. endocrinol.*, **1**, 237–263.

Messner, B. 1963 Neue histologische Befunde zur Neurosekretion bei terrestrichen Isopoden (*Porcellio scaber* latr. und *Oniscus asellus* L.). *Sond. Naturwiss.*, **8**, 338–339.

Molinoff, P.B., Landsberg, L. & Axelrod, J. 1969 An enzymatic assay for octopamine and other β-hydroxylated phenyl–ethylamines. *J. Pharmac. exp. Ther.*, **170**, 253–261.

Morris, J.F. & Nordmann, J.J. 1979 Lysosomal Control on Neurosecretion: the Roles of Cellular Compartmentation and Granule Age. *Biol. Cell.*, **36**, 193–200.

Nordmann, J.J. & Chevallier, J. 1980 The role of microvesicles in buffering Ca^{2+} in neurohypophysis. *Nature*, **287**, 46–54.

Nordmann, J.J. & Morris, J.F. 1980 Depletion of neurosecretory granules and membrane retrieval in the sinus gland of the crab. *Cell Tiss. Res.*, **205**, 31–42.

Quackenbush, L.S. & Fingerman, M. 1984 Regulation of the release of chromatophorotropic neurohormones from the isolated eyestalk of the fiddler crab, *Uca pugilator. Biol. Bull.*, **166**, 237–250.

Rao, K.R. & Fingerman, M. 1983 Regulation of release and mode of action of crustacean chromophorotropins. *Amer. Zool.*, **23**, 517–527.

Schwarz, T., Lee, M.–H., Siwiki, K., Standaert, D. & Kravitz, E. 1984 Proctolin in the lobster: the distribution, release and chemical characterization of a likely neurohormone. *J. Neurosci.*, **4**, 1300–1311.

Seguela, P., Geffard, M., Buijs, R. & Le Moal, M. 1984 Antibodies against γ–aminobutyric acid = specificity studies and immunocytochemical rests. *Proc. Natl. Aca. Sci. USA.*, **81**, 3888–3892.

Shaw, F.D. & Morris, J.F. 1980 Calcium localization in the rat neurohypophysis. *Nature*, **287**, 56–58.

Staub, G.C. & Fingerman, M. 1984 Effect of Naphthalene on Color changes of the Sand Fiddler Crab, *Uca pugilator. Comp. Biochem. Physiol.*, **77C**, 7–12.

Van Deijnen, J.E., Vek, F. & Van Herp, F. 1985 An immunocytochemical study of the optic ganglia of the crayfish *Astacus leptodactylus* (Nordmann 1842) with antisera against biologically active peptides of vertebrates and invertebrates. *Cell Tiss. Res.*, **240**, 175–183.

Van Herp, F. & Bellon–Humbert, C. 1982 Localisation immunocytochimique de substances apparentées à la neurophysine et à ka vasopressine dans le pédoncule oculaire de *Palaemon serratus* Pennant (Crustacé Décapode Natantia). *C.R. Acad. Sci. Paris*, **295**, 97–102.

Webster, S.G. & Keller, R. 1986 Purification, characterisation and amino acid composition of the putative moult-inhibiting hormone (MIH) of *Carcinus maenas* (Crustacea, Decapoda). *J. Comp. Physiol., B.*, **156**, 617–62.

PART II

Arthropod Neurohormones

WILLIAM MORDUE & KARL J. SIEGERT

Characterization of insect neuropeptides

Introduction

In insects a variety of functions (e.g. intermediary metabolism, ion and osmoregulation, developmental and neuronal processes) are regulated by peptides from different parts of the nervous system. This diversity necessitates the restriction of the present chapter to a particular group of peptides. We have been interested in the structure and biological functions of peptide hormones regulating intermediary metabolism and fluid secretion in insects and these studies are used to highlight the problems encountered in characterizing these peptides.

Lipid and carbohydrate metabolism in insects can be regulated by adipokinetic (AKH) and so-called hyperglycaemic (HGH) hormones present in the corpora cardiaca (Mayer & Candy 1969; Steele 1961). These peptides increase concentrations of haemolymph lipids or carbohydrates. The physiological functions and modes of action of these hormones are reviewed in Chapter 7. Diuretic hormones (DH) present in the corpora cardiaca but also in other parts of the nervous system, e.g. brain, suboesophageal and thoracic ganglia (Proux *et al.* 1982; Morgan & Mordue 1984a; Aston & Hughes 1980) regulate fluid secretion.

In recent years investigations have been conducted to isolate and characterize prothoracicotropic hormones (PTTH) which stimulate the release of ecdysone from the prothoracic glands (Williams 1947). In *Bombyx mori*, the PTTHs have been found to exhibit significant homologies in their N–terminal sequences with insulin and insulin-like growth factors (Nagasawa *et al.* 1984). These peptides are discussed in Chapter 6 which is devoted to the presence of 'vertebrate' peptides in insects. The involvement of peptides in the functioning of neurones as neurotransmitters and/or neuromodulators is covered in Chapter 8.

The corpora cardiaca of insects are neurohaemal organs (Scharrer 1952) analogous to the crustacean sinus glands which are known to contain metabolically active materials (Carlisle & Knowles 1959). Therefore, early studies on hormonal regulation of intermediary metabolism in insects focused on the corpora cardiaca as source of similar materials.

Isolation of insect neurohormones

The bioassay. A first step in the elucidation of the hormonal regulation of a particular metabolic pathway is to bring about changes in this pathway as a response to extracts from the tissue(s) containing the putative hormone, and to establish a reliable and sensitive bioassay. For adipokinetic, hyperglycaemic and diuretic peptides, such bioassays measure the increase in concentration of haemolymph diacylglycerols (Mayer & Candy 1969) or trehalose (Steele 1961), or of fluid secretion by the Malpighian tubules (Maddrell 1962) respectively. Very sensitive assays can be developed by monitoring changes in enzyme activities in the target tissues. For example, one DH from *Locusta migratoria* mediates its action through the adenylate cyclase system (Rafaeli *et al.* 1984; Morgan & Mordue 1985), and the use of enzyme activities to measure hormonal effects provides a very sensitive assay and facilitates detection of active material during isolation procedures. The introduction of an assay based on the measurement of cAMP production has proved especially helpful in the isolation of DH from *L. migratoria* (Morgan *et al.* 1987): a membrane preparation from the Malpighian tubules is incubated with corpus cardiacum extract or

Fig. 1. A methanol extract of 5 *L. migratoria* corpora cardiaca was chromatographed on a HPSEC column using a flow rate of 1 ml 0.1% TFA min^{-1} (wavelength 206 nm). One-ml-fractions were collected and consecutively numbered from injection of extract (arrow). DH activity measured using the fluid secretion assay eluted in fractions 7/8 (stipple) and 11/12 (hatching); AKHI in 12/13 (filled box) and AKHII in 14/15 (open box).

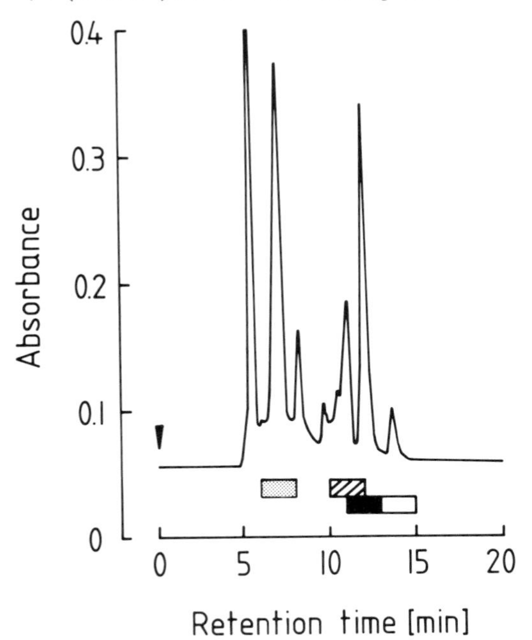

isolated DH fractions, and the formation of cAMP measured. For assay of other peptides, the fat body glycogen phosphorylase is more sensitive to corpus cardiacum extracts than release of trehalose into the haemolymph: 0.001 of a pair of corpora cardiaca produces no hypertrehalosaemic response at all, but induces half-maximum activation of phosphorylase (Siegert *et al*, 1986). The adipokinetic response in locusts is very sensitive: 0.005 of a pair gives half-maximum hyperlipaemia (Goldsworthy *et al.* 1972), but until now no detailed studies of the lipase(s) have been conducted successfully. AKHI, however, activates glycogen phosphorylase in the fat body and this assay is as sensitive as that based on the adipokinetic response (Van Marrewijk *et al.* 1983).

Nerve tissues other than the corpora cardiaca also contain biologically active peptides. The brain, suboesophageal and thoracic ganglia contain DH activity. It is thus necessary to establish the distribution of hormonal activity throughout the nervous system and to ascertain whether these activities from different sources are linked to the same molecules. AKH and HGH have only been found in the corpora cardiaca of locusts or cockroaches, respectively, but Ziegler (1980) reported that in adult *Manduca sexta* AKH can also be detected in the corpora allata. In *Periplaneta americana* (Siegert & Mordue, unpublished observations) and in *Calliphora erythrocephala* (Vejbjerg & Normann 1974) the corpora allata are free of hyperglycaemic activity.

Corpus cardiacum extracts contain a mixture of compounds and the hormonal action of one compound might be modulated by the presence of another. The presence of such modulators may be of physiological importance but makes detailed and reliable studies difficult. The presence of two different AKHs (Carlsen *et al.* 1979; Gäde *et al.* 1984; Siegert *et al.* 1985), of two DHs (Morgan & Mordue 1984b) in locust and of two HGHs in cockroach corpora cardiaca (Scarborough *et al.* 1984) clearly shows the necessity to isolate and identify individual compounds, and to study their biological activities separately before more complex experiments using two peptides tested at known doses can follow.

Extraction of biologically active material. After the identification of the tissues containing biological activities, an optimum extraction procedure is required. Extractions with 0.1 M acetic or 0.1 N hydrochloric acid have been used for the extraction of invertebrate peptides (Traina *et al.* 1976). AKHs and HGHs are very hydrophobic and therefore organic solvents such as 80% aqueous methanol are used. Sufficient biological activity can be obtained from corpora cardiaca using salt solutions (Steele 1961) or distilled water (Natalizi & Frontali 1966; Ziegler 1979). It may prove necessary to introduce a peptidase inhibitor into aqueous extraction media to prevent enzymatic breakdown of certain peptides such as DH which are rather unstable.

The homogenization of tissues can be performed with tissue grinders of the Potter–Elvejhem type (glass–teflon)(Steele 1963) or glass–glass homogenizer (Ziegler 1979); in other studies sonication has often been used to disrupt tissues. Three to four repeated sonications are sufficient to extract AKHI and AHKII from a single pair of locust corpora cardiaca (Siegert & Mordue 1986a); using batches of 100 corpora cardiaca or more, however, may require a larger number of extractions. The number of repetitions necessary to remove all the biologically active material from the tissue should be determined using the bioassay in pilot experiments.

In locusts, the corpora cardiaca comprise two structurally defined lobes, the glandular lobes containing AKH (Goldsworthy *et al.* 1969) and the storage lobes containing DH activities (Mordue & Goldsworthy 1972). Thus, it is possible to reduce the heterogeneity of peptides present in a tissue extract by dissecting the corpora cardiaca into separate lobes. In other species, such as *P. americana* and *M. sexta*, however, the corpora cardiaca are not structurally differentiated into distinct lobes.

Early attempts to isolate hyperglycaemic hormones. The hypertrehalosaemic response in cockroaches (Steele 1961) was demonstrated some 8 years before the adipokinetic response (Mayer & Candy 1969), but the HGH peptides eluded isolation and characterization for almost a quarter of a century.

Early attempts in the late 1960s and early 1970s to isolate the HGHs from cockroach corpora cardiaca used a variety of chromatographic techniques such as gel

Fig. 2. A methanol extract of 10 cockroach corpora cardiaca was chromatographed on a HPSEC column. Fractions 13/14 contained HGH activity and were injected onto a RP-column (start 21% B (arrow), gradient 0.3% B min $^{-1}$; flow rate 1 ml min^{-1}. HGHI eluted after 11.7 min, HGHII after 20.1 min.

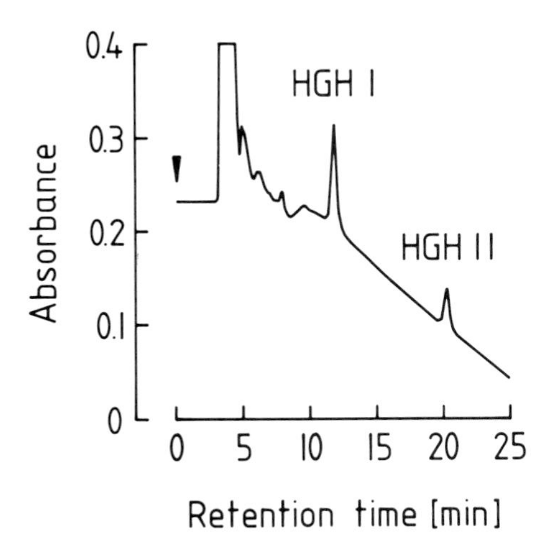

filtration on e.g. Sephadex G–10 and G–15 and ion exchange on Sephadex DEAE A–25 (the principles of separation are the same as for the appropriate HPLC type; see below), sometimes in combination with other methods such as electrofocusing or ultrafiltration. In most of these studies the fractions with biological activity were either not subjected to amino acid analysis, or showed non-integral ratios for the amino acids present. Thus, the isolation procedures never produced reliable data on the characteristics of the peptide studied. These studies will not be discussed further here. Because these separation techniques did not yield material of sufficient purity to determine the molecular structure, a more sensitive separation technique(s) was needed to isolate and analyze the HGHs from cockroach corpora cardiaca.

The isolation of adipokinetic hormones. AKHI was the first insect neurohormone to be isolated and fully characterized in its structure. Although the isolation methods used have been superseded (see below), the strategy followed by Stone *et al.* (1976) was successful and will be described in more detail than those used for the isolation of HGH.

Methanolic extracts of the corpora cardiaca were chromatographed on controlled-pore glass beads (G–75–50 beads, mesh size 120–200, Sigma Chemicals), eluted in distilled water, the active fractions collected, and chromatographed on thin-layers (Woelm 20×20 cm silica gel F254 Rapid-plates, ICN Pharmaceuticals) in isopropanol–water–acetic acid (25:10:1, V/V). The AKH activity travelled with an R_f

Fig. 3. HGHII was treated with pyroglutamate aminopeptidase to cleave the pyroglytamyl residue from the N–terminus. The shorter peptide (HGHII*) was isolated from the original peptide using RP–HPLC (start 0% B (arrow), gradient 1% B min^{-1}; 1 ml min^{-1}).

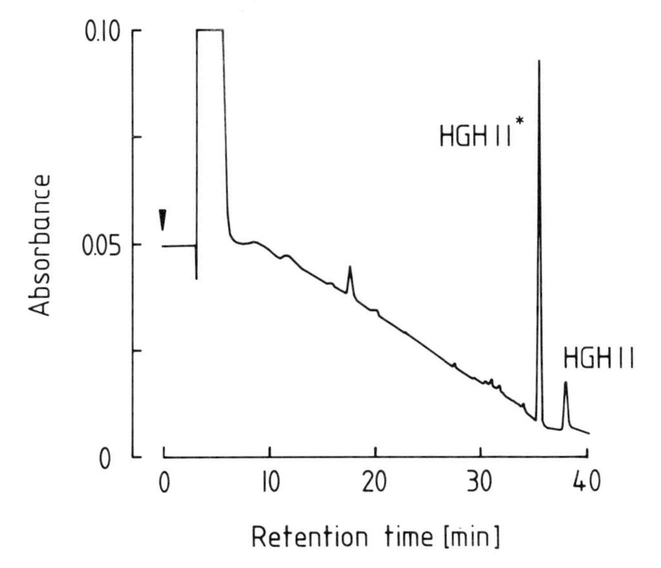

value of 0.67. This fraction was determined as pure by amino acid analysis (showing only integral residue numbers), and the structure of this decapeptide was eventually elucidated fully using mass spectrometry (see below) in combination with other techniques (Stone & Mordue 1980).

Using gel filtration on Sephadex G–25 and LH–20 Carlsen *et al.* (1979) separated a second AKH from the corpora cardiaca of *S. gregaria* (AKHII–S); this peptide was determined as an octapeptide with a similar amino acid composition to AKHI and another invertebrate peptide hormone, the crustacean red–pigment concentrating hormone (RPCH; Fernlund 1974). The AKHII–S was probably the last insect peptide hormone to be isolated using only conventional gel-permeation techniques.

The appearance of new methodologies. In the late 1970s a technique was developed which eventually led to increased speed of separations of complex mixtures of peptides, sugars and nucleic acids. This method is called high-performance liquid chromatography (HPLC) as it possesses increased resolving power when compared with previous liquid chromatography. Sometimes this technique is also referred to as high-pressure liquid chromatography; it can operate at high pressures (as high as 3000–8000 psi) because of the more rigid material, small column diameters and pore sizes used for the separation. It is beyond the scope of this chapter to discuss the details of HPLC methodology, and the reader is advised to consult specialized reviews or handbooks (e.g. Walker & Gaastra 1983). There are basically 3 different types of HPLC: high-performance size exclusion chromatography (HPSEC), reversed-phase (RP–HPLC) and ion exchange HPLC. In the present context only the former two are of major importance, because none of the AKHs and HGHs are charged.

The principle of HPSEC is that a complex mixture can be separated according to the molecular diameters of the compounds present (this is identical to the principle of gel filtration). Columns are packed with a silica base; although the pore size of the packing material is of critical importance in influencing the separation, the silica base also acts as a weak cation–exchanger because silanol residues are present. This function can be reduced by the use of capping agents, or eluants with high ionic strengths or pH values in the acid range. If these disturbing factors can be minimized, separation is achieved on the basis of molecular size, and large molecules elute first. Using HPSEC, peptide separations can be obtained in less than 1 h when operating at flow rates of 1–2 ml min^{-1}.

In RP–HPLC the silica base has been reacted with hydrocarbons, e.g. C–8 (octyl) or C–18 (octadecyl) derivatives. Peptides are injected onto the column in a predominantly aqueous solution and hydrophobic parts of the molecule bind to the C–8 or C–18 residues present in the column. Then an elution gradient in which the aqueous solvent is gradually replaced by a water-miscible organic modifier (e.g. acetonitrile or methanol) is applied to the column. At a particular percentage of organic modifier the peptide detaches from the stationary phase, elutes from the column and is usually detected with a UV spectrophotometer at wavelengths around 210 nm: at such wavelengths peptide bonds absorb maximally, but so also do aromatic amino acids such as phenylalanine and tryptophan. The peptides elute from the columns, and are detected as peaks in *c.* 1–2 ml depending on the type of gradient applied and the flow rate. The peak volumes in conventional gel filtration separations can be up to 10–20× this volume. But as has been mentioned above some secondary effects can occur such as interaction with non-reacted silica residues (again, these can be reduced by a capping agent) which may produce a secondary RP effect. Another effect may result from the pore-size of the batch material. RP columns such as the widely used μBondapak C–18 (Waters) have pore sizes of *c.* 60–100, which only allows the separation of small molecules, but a wider range of peptides can be studied when wide pore columns such as the Aquapore RP–300 (300Å; Brownlee) are used. Molecules or particles exceeding the pore size of the column will accumulate on top of the column and eventually block it; it is therefore necessary to have a guard column (of identical material) ahead of the main column to prevent contamination of the latter.

Chromatographic solvents used in chromatography must be chosen very carefully for two reasons: first, the solvent system is partly responsible for the interaction

Fig. 4. AKHII–L and angiotensin II (ANGII) were treated with carboxypeptidase A and the reaction mixture chromatographed on a RP column (start 10% B (arrow), gradient 2% B min^{-1}; 1 ml min^{-1}). ANGII was converted to $ANGII^{-}$. Neither the retention time nor the peak area of AKHII–L were changed during treatment. AKHII–L eluted after 17.4 min, ANGII after 15.1 min (open arrow), $ANGII^{-}$ after 9.9 min.

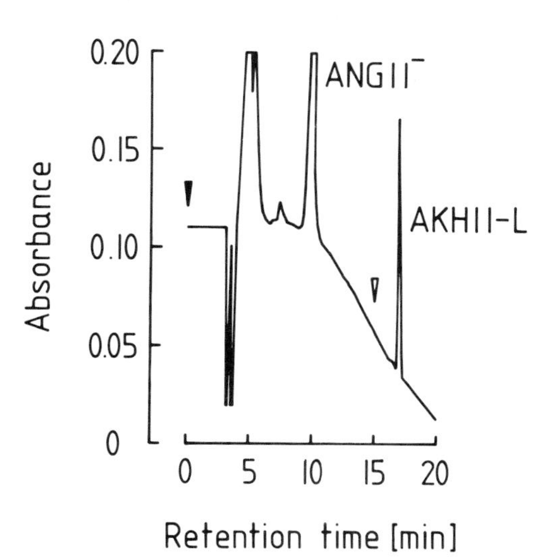

between the molecules to be separated and the stationary phase; second, the eluant should be removed completely prior to the bioassay. Therefore, volatile solvents are recommended. If a low pH (*c.* 2) is required, 0.1% trifluoroacetic acid (TFA) can be used; if only a slightly acidic pH is needed, ammonium acetate buffers of appropriate molarity can be employed. Both substances can be removed with the water during freeze-drying. Organic solvents frequently used for the isolation of insect peptides are acetonitrile and methanol which can be easily removed by gentle heating under reduced pressure.

The isolation of HGH, AKH and DH using HPSEC and RP–HPLC

The introduction of HPLC has greatly facilitated the isolation of insect peptide hormones, and a number of groups isolated identical peptides independently. The methodologies used by different authors can be very similar, but the resulting chromatograms may look different. As is clear from the literature, independent studies on the same peptides always revealed identical primary structures (see below). In the following we concentrate on our own work concerning the isolation of HGH, AKH and DH and highlight the work of other authors where necessary.

We have used a combination of HPSEC and RP–HPLC in our attempts to isolate HGHs, AKHs and DHs from the corpora cardiaca of cockroaches and locusts. In a first step we used HPSEC to separate peptides on a TSK 2000 SW (predicted fractionation range for peptides 500–70000 d; Toya Soda) according to their molecular masses. We employed 0.1% TFA as eluant at a flow rate of 1 ml min^{-1}. Under these conditions UV-absorbing molecules from locust and cockroach corpora cardiaca eluted between 5 and 18 min after injection. Fractions of 1 ml were taken and numbered consecutively. DH activity from *L. migratoria* eluted in fractions 7/8 (apparent molecular weight *c.* 3500) and 11/12 (*c.* 2000) (Morgan *et al.* 1987), while AKH activity both in *L. migratoria* and *S. gregaria* was present in fractions 12–15 (*c.* 2000–1000) (Siegert & Mordue 1986a). Hyperglycaemic activity from cockroach corpora cardiaca was found in fraction 13/14 (*c.* 1000) (Siegert *et al.* 1986). By selecting a wide pore column (Aquapore RP–300) we were able to identify a peak containing DH activity; for convenience we used the same column for the isolation of the HGHs and AKHIIs. Individual gradients were designed and optimized for the different peptides. In pilot experiments wide-ranging gradients were used because the elution characteristics of the peptides were unknown. Thus, gradients starting at 100% 0.1% TFA (A) and 0% acetonitrile (B) with an increase of 2% B min^{-1} were used. Four-ml-fractions were collected, dried and bioassays run for the individual fractions. The volume of the fractions collected was reduced until individual peaks with biological activities were identified. Then the initial percentage for B was increased, and the slope of the gradient decreased, to make the separation system efficient with the peptides eluting as quickly as possible.

The AKHII–S and AKHII–L were eventually isolated using an initial percentage B of 23% with an increase of 0.3% B min^{-1} (Siegert *et al.* 1985). It must be said that these values are only valid for our system and they might be different in other laboratories, even if identical columns from the same manufacturer are used. Using the same equipment the isolation of the HGHs was achieved with an initial percentage B of 21% with a gradient increasing at a rate of 0.3% B min^{-1} (Siegert & Mordue 1986b). These cockroach peptides were later found to be identical with MI and MII (O'Shea *et al.* 1984). DHI was isolated eventually using a gradient starting at 30% B with an increase of 0.5% B min^{-1} (Morgan *et al.* 1987).

Analysis of primary structures

After the identification and purification of individual peptides the structure analysis can be attempted; this includes the establishment of the amino acid sequence and whether the N– and/or the C–terminals are modified or blocked.

Amino acid composition. The first step in the analysis of a primary structure of a peptide must be the determination of its amino acid composition. From this, the purity of the preparation can be judged, and the likely molecular mass of the peptide calculated; the composition data are also necessary for the interpretation of the results from the structure analysis (see below). Prior to the identification and quantification of amino acids the peptides must be hydrolyzed. This is a crucial step because amino acids such as tryptophan or proline can be destroyed or derivatized when the peptide is hydrolyzed in 6 M HCl. Mild hydrolysis in 4 M methanesulfonic acid containing 0.2% tryptamine protects tryptophan and allows its determination (Ziegler *et al.* 1984). But even under these conditions relatively large amounts of the tryptophan can sometimes be lost. This is an important point because the composition data are used to calculate the likely number of amino acids present in the peptide. If in a given peptide sample the presence of 1 nmol X, 1 nmol Y, 1 nmol Z and 0.5 nmol A was determined, then the peptide could be either a tetrapeptide, if it is assumed that 50% of A was lost during hydrolysis, or a heptapeptide containing 2 X, 2 Y, 2 Z and 1 A.

After hydrolysis, the amino acids are reacted with e.g. o–phthalaldehyde (OPA; Turnell & Cooper 1982) or phenylthiohydantoin (PTH; Zimmermann *et al.* 1977); both are more or less equally sensitive, but PTH derivatives of amino acids are more stable. Amino acids in the hydrolysate are then identified and quantified using RP–HPLC and reference standards.

Fast-atom-bombardment mass spectrometry. The structure of peptides can be assigned by two different approaches. Fast-atom-bombardment mass spectrometry (FABMS) elucidates the complete primary structure directly. The other approach determines the amino acid sequence using the Edman degradation (see below) and modifications of the termini are shown in separate experiments.

In recent years FABMS has become the predominant technique to elucidate the structure of peptides containing 10–20 amino acids. The molecule is ionized and the ions subsequently separated and analyzed according to their mass-charge-ratio. The theory and practice of this technique are not within the scope of this article, and the reader should refer to specialist books or reviews (Rinehart 1982) for detailed information. FABMS elucidates the entire structure of a peptide including any modification to the N– and C–terminals.

The hyperglycaemic/myotropic peptides from cockroach corpora cardiaca have been characterized using this technique (Scarborough *et al.* 1984; Witten *et al.* 1984; Table 1). The data obtained from the FABMS in conjunction with the amino acid composition enabled the entire structure to be elucidated. The structure of the *M. sexta* AKH was elucidated using FAB tandem mass spectrometry from only 2 nmols of material (Table 1; Ziegler *et al.* 1985).

Edman degradation. The second way to establish the primary structure of peptides is not as convenient as FABMS and has a slight draw-back (see below). Edman (1950) described a convenient way to establish the amino acid sequence of peptides and proteins. There are basically three kinds of automated sequencers which differ only in the way the solvents are delivered to the reaction chamber, but they all use the Edman degradation: solid-, liquid- and gas-phase sequencers, the most efficient method being the latter, because all reagents are delivered to the reaction chamber as gases and thus the proportion of the peptide in solution and lost is minimized. The Edman degradation requires a free unmodified N–terminus. This is coupled to the matrix of the sequencer using a special coupling agent (first step of the degradation; see Kia–Ki Han *et al.* 1985). The conditions are then changed to cleave the first amino acid from the N–terminus of the peptide chain (second step; cleavage agent). The released amino acid is converted to a PTH derivative because these are very stable (third step; conversion reagent), and the derivatives are subsequently determined using conventional amino acid analysis. The cleavage of amino acids is repeated and, from the sequence of appearance of amino acids, the sequence in the peptide is established. Ideally, the cleavage only removes one amino acid at a time, but especially after a few cycles more than one residue may be cleaved, and carry-over may occur which can make interpretation of the analysis difficult. In the event it is necessary to have reliable data for the amino acid composition of the peptide.

Determination of the N–terminus. The Edman–degradation cannot be performed on peptides which contain a blocked N–terminus, e.g. a pyroglutamyl residue. As the locust AKHI, RPCH, HGHI and HGHII are all N–terminally blocked, AKHII from *S. gregaria* (AKHII–S) and from *L. migratoria* (AKHII–L) were expected also to be blocked, because amino acid analysis demonstrated a single glutamine/glutamic acid

Table 1. *Known structures of insect neurohormones and related peptide hormones*

		1 2 3 4 5 6 7 8 9 10
AKHI	(*S. gregaria*) (*S. nitans*) (*L. migratoria*)	pGlu–Leu–Asn–Phe–Thr–Pro–Asn–Trp–Gly–ThrNH$_2$
AKHII	(*S. gregaria*) (*S. nitans*)	pGlu–Leu–Asn–Phe–Ser–Thr–Gly–TrpNH$_2$
AKII	(*L. migratoria*)	pGlu–Leu–Asn–Phe–Ser–Ala–Gly–TrpNH$_2$
AKH	(*M. sexta*) (*H. zea*)	pGlu–Leu–Thr–Phe–Thr–Ser–Ser–Trp–GlyNH$_2$
HGHI/MI	(*P. americana*)	pGlu–Val–Asn–Phe–Ser–Pro–Asn–TrpNH$_2$
HGHII/MII	(*P. americana*)	pGlu–Leu–Thr–Phe–Thr–Pro–Asn–TrpNH$_2$
HTH	(*B. discoidalis*)	pGlu–Val–Asn–Phe–Ser–Pro–Gly–Trp–Gly–ThrNH$_2$
HTH	(*N. cinerea*)	pGlu–Val–Asn–Phe–Ser–Pro–Gly–Trp–Gly–ThrNH$_2$
HTHII	(*C. morosus*)	pGlu–Leu–Thr–Phe–Thr–Pro–Asn–Trp–Gly–ThrNH$_2$
RPCH	(*P. borealis*)	pGlu–Leu–Asn–Phe–Ser–Pro–Gly–TrpNH
Glucagon	(3–12)	–Gln–Gly–Thr–Phe–Thr–Ser–Asp–Tyr–Ser–Lys–
DH Lys–	(*L. migratoria*)	? –Gly– ? –Gly–Ile–Gln–Ala– ? –V/M–Tyr–

residue and a composition very similar to the above mentioned peptides. We cleaved the pyropglutamyl residues of AKHII–S and AKHII–L with pyroglutamate amino-peptidase, and isolated the resulting shorter peptides. These, which were no longer blocked, were sequenced conventionally (Siegert *et al.* 1985). The sequence data confirmed the amino acid compositions obtained by Carlsen *et al.* (1979) and Gäde *et al.* (1984): the peptides were both octapeptides (Table 1).

Determination of the C–terminus. One of the most widely occurring post-translational modifications of peptide hormones is the amidation of the C–terminal, involves dehydrogenation and hydrolysis of the prohormone (Bradbury *et al.* 1982). The presence of an amide group cannot be concluded from the Edman degradation.

absence of such an amide group. Its biological function seems to be to protect peptides from cleavage by carboxypeptidases. This property can be used to show the presence of the blocking group, but not its nature. The amide group is, however, the only known modification of the C–terminus (Schulz & Schirmer 1979). There are different types of carboxypeptidases with different specifities for C–terminal amino acids: carboxypeptidase A does not cleave arginine, lysine and proline; carboxypeptidase B removes arginine and lysine with high specificity; whereas carboxypeptidase C and Y release all C–terminal amino acids. Thus, if a peptide is treated with a carboxypeptidase and is not cleaved it must carry an amide group. In the case of the AKHIIs, an unmodified C–terminal tryptophan would be cleaved by all of these carboxypeptidases. Because it was expected that the enzymatic treatment would be unable to cleave the AKHIIs, it was necessary to show that the enzyme was active under the conditions used. This was achieved using angiotension II (human) as a substrate which contains a C–terminal phenylalanine residue; angiotensin II was added therefore to the same reaction mixture as the AKHIIs as an internal control. No conversion of the AKHIIs was observed when treated with carboxypeptidase A, but angiotensin II was cleaved. It was concluded that both AKHIIs must indeed by C–terminally blocked, presumably by an amide group (Siegert *et al.* 1985). Recently FABMS has been used to confirm the structures of AKHIIs from *S. gregaria* (AKHII–S) and from *L. migratoria* (AKHII–L) (Gäde *et al.* 1986).

The structure of insect neurohormones

As shown in Table 1, all adipokinetic and hyperglycaemic hormones characterized so far are very similar and obviously form a family of peptides. They all contain a N–terminal pyroglutamyl residue, phenylalanine at position 4, tryptophan at 8 and are C–terminally amidated. These features may be of general importance for receptor binding. The amino acids between these positions may determine the species-specific interactions between hormone and receptors. It is very interesting that AKHII–S and AKHII–L are different from the crustacean RPCH in only one amino acid at position 6 which demonstrates that these molecules are, in evolutionary terms, closely related.

The cockroach hyperglycaemic hormones and AKHII–S and AKHII–L are all octapeptides, whereas AKHI is a decapeptide. The *M. sexta* AKH, however, contains 9 amino acids (Ziegler *et al.* 1985) as was reported for the hypertrehalosaemic hormone II from the stick insect, *Carausius morosus* (Gäde 1985). The structural analysis, however, revealed that this compound is a decapeptide (Table 1, Gäde & Rinehart 1987).

Scarborough *et al.* (1984) pointed out the slight similarity between HGHII and the N–terminal part of glucagon (residues 3–12; Table 1). The *M. sexta* AKH shows a slightly higher degree of similarity.

The DH isolated from *L. migratoria* (Morgan *et al.* 1987) does not seem to belong to this group of peptides. Its molecular weight is higher (*c.* 3500) than that of the above mentioned peptides. Additionally, the N–terminus is not blocked by a pyroglutamyl residue, because a part of the amino acid sequence can be revealed using Edman degradation; this fragment does not show any similarity with the adipokinetic or hyperglycaemic hormones (Table 1). The N–terminal amino acid could not be distinguished amongst Phe, Ser, Gly, Ala, Tyr, Val/Met, and Lys. Gaps are also present at positions 3 and 8, which might be occupied by cysteine; this must be derivatized before sequencing to enable determination. Tryptophan, threonine or serine might also be present, but as a consequence of the small amount of initial material (c. 130 pmol), and the low yield of these amino acids when derivatized with PTH, the amounts present in these sequencing cycles could have been too small to be detected by the amino acid analysis.

Acknowledgements

Original research reported here was supported by grants from the Deutsche Forschungsgemeinschaft (Si 139/1–1) and by AFRC.

References

Aston, R.J. & Hughes, L. (1980) Diuretic hormone – Extraction and chemical properties. In: Neurohormonal techniques in insects. (Ed: T.A. Miller). pp. 91–115. Springer–Verlag, New York.

Bradbury, A.F., Finnie, M.D.A. & Smyth, D.G. (1982) Mechanism of C–terminal amide formation by pituitary enzymes. *Nature*, **298**, 686–688.

Carlisle, D.B. & Knowles, F. (1959) *Endocrine control in crustaceans.* 120 pp. Cambridge University Press, Cambridge.

Carlsen, J., Herman, W.S., Christensen, M. & Josefsson, L. (1979) Characterization of a second peptide with adipokinetic and red pigment-concentrating activity from locust corpora cardiaca. *Insect Biochem.* **9**, 497–501.

Edman, P. (1950) A method for the determination of the amino acid sequence in peptides. *Acta Chem. Scand.* **4**, 283–293.

Fernlund, P. (1974) Structure of the red-pigment concentrating hormone of the shrimp, *Pandalus borealis. Biochim. Biophys. Acta*, **371**, 304–311.

Gäde, G. (1985) Isolation of the hypertrehalosaemic factors I and II from the corpus cardiacum of the Indian stick insect, *Carausius morosus*, by reversed-phase high-performance liquid chromatography, and amino-acid composition of factor II. *Biol. Chem. Hoppe–Seyler*, **366**, 195–199.

Gäde, G., Goldsworthy, G.J., Kegel, G. & Keller, R. (1984) Single step purification of locust adipokinetic hormones I and II by reversed-phase high-performance liquid chromatography, and amino-acid composition of the hormone II. Hoppe–Seyler's Z. *Physiol. Chem.* **365**, 393–398.

Gäde, G., Goldsworthy, G.J., Schaffer, M.H., Cook, J.C. & Rinehart, K.L. (1986) Sequence analyses of adipokinetic hormones II from corpora cardiaca of *Schistocerca nitans, Schistocerca gregaria*, and *Locusta migratoria* by fast atom bombardment mass spectrometry. *Biochem. Biophys. Res. Commun.* **134**, 723–730.

Gäde, G. & Rinehart, K.L. .(1986)..Amino acid sequence of a hypertrehalosaemic neuropeptide from the corpus cardiacum of the cockroach, *Nauphoeta cinerea. Biochem. Biophys. Res. Commun.*, **141**, 774–781.

Gäde, G. & Rinehart, K.L. ..(1987)..Primary structure of a hypertrehalosaemic factor II from the corpus cardiacum of the Indian stick insect, *Carausius morosus*, determined by fast atom bombardment mass spectrometry. *Biol. Chem. Hoppe–Seyler*, **368**, 67–75.

Goldsworthy, G.J., Mordue, W. & Guthkelch, J. (1972) Studies on insect adipokinetic hormones. *Gen. Comp. Endocrinol.*, B18, 545–551.

Hayes, T.K., Keeley, L.L. & Knight, D.W. (1986) Insect hypertrehalosaemic hormone: Isolation and primary structure from *Blaberus discoidalis. Biochem. Biophys. Res. Commun*, **140**, 674–679.

Jaffe, H., Raina, A.K., Riley, C.T., Fraser, A.B., Holman, G.M., Wagner, R.M., Ridgway, R.L. & Hayes, D.K. (1986) Isolation and primary structure of a peptide from the corpora cardiaca of *Heliothis zea* with adipokinetic activity. *Biochem. Biophys. Res. Commun.*, **135**, 622–828.

Kia–Ki Han, Belaiche, D, Moreau, O & Briand, G. (1985) Current developments in stepwise Edman degradation of peptides and proteins. *Int. J. Biochem.* **17**, 429–445.

Maddrell, S.H.P. (1962) A diuretic hormone in *Rhodnius prolixus* stål. *Nature* **194**, 605–606.

Mayer, R.J. & Candy, D.J. (1969) Control of haemolymph lipid concentration during locust flight: An adipokinetic from the corpora cardiaca. *J. Insect Physiol.* **15**, 611–620.

Migliori Natalizi, G. & Frontali, N. (1966) Purification of insect hyperglycaemic and heart accelerating hormones. *J. Insect Physiol.* **12**, 1279–1287.

Mordue, W. & Goldsworthy, G.J. (1969) The physiological effects of corpus cardiacum extracts in locusts. *Gen. Comp. Endocrinol.* **12**, 360–369.

Morgan, P.J. & Mordue, W. (1984a) Diuretic hormone: Another peptide with wide-spread distribution within the insect CNS. *Physiol. Entomol.* **9**, 197–206.

Morgan, P.J. & Mordue, W. (1984b) Separation and characteristics of diuretic hormone from the corpus cardiacum of *Locusta. Comp. Biochem. Physiol.* **75B**, 75–80.

Morgan, P.J. & Mordue, W. (1985) Cyclic AMP and locust diuretic hormone action. Hormone induced changes in cAMP levels offers a novel method for detecting biological activity of uncharacterized peptide. *Insect Biochem.* **15**, 247–257.

Morgan, P.J., Siegert, K.J. & Mordue, W. (1987) Preliminary characterisation of locust diuretic peptide (DP–1) and another corpus cardiacum peptide (LCCP). *Insect Biochem.* **17**, 383–387.

Nagasawa, H., Kataoka, H., Isogai, A., Tamaura, S., Suzuki, A., Ishizaki, H., Mizoguchi, A., Fujiwara, Y. & Susuki, A. (1984) Amino-terminal amino acid sequence of the silkworm prothoracicotrophic hormone: Homology with insulin. *Science* **226**, 1344–1345.

O'Shea, M., Witten, J. & Schaffer, M. (1984) Isolation and characterization of two myoactive neuropeptides: Further evidence of an invertebrate peptide family. *J. Neurosci.* **4**, 521–529.

Proux, J., Rougon, G. & Cupo, A. (1982) Enhancement of excretion across locust Malpighian tubules by a diuretic vasopressin-like hormone. *Gen. Comp. Endocrinol.* **47**, 449–457.

Rafaeli, A., Pines, M., Stern, P.S. & Applebaum, S.W. (1984) Locust diuretic hormone-stimulated synthesis and excretion of cyclic-AMP: A novel Malpighian tubule bioassay. *Gen. Comp. Endocrinol.* **54**, 35–42.

Rinehart, K.L. (1982) Fast atom bombardment mass spectrometry. *Science*, **218**, 254–260.
Scarborough, R.M. Jamieson, G.C., Kalish, F., Kramer, S.J., McEnroe, G.A., Miller, C.A. & Schooley, D.A. (1984) Isolation and primary structure of two peptides with cardioacceleratory and hyperglycemic activity from the corpora cardiaca of *Periplaneta americana. Proc. Natl. Acad. Sci. USA.* **81**, 5575–5579.
Scharrer, B. (1952) Neurosecretion XI. The effects of nerve section on the intercerebralis–cardiacum–allatum system of the insect *Leucophaea maderae. Biol. Bull. Woods Hole* **102**, 261–272.
Schulz, G.E. & Schirmer, R.H. (1979) *Principles of protein structure*. Springer–Verlag, Berlin.
Siegert, K.J. & Mordue, W. (1986a) Quantification of adipokinetic hormones I and II in the corpora cardiaca of *Schistocerca gregaria* and *Locusta migratoria. Comp. Biochem. Physiol.*, **84A**, 279–284.
Siegert, K.J. & Mordue, W. (1986b) Elucidation of the primary structures of the cockroach hyperglycaemic hormones I and II using enzymatic techniques and gas-phase sequencing. *Physiol. Entomol.*, **11**, 205–211.
Siegert, K.J., Morgan, P. & Mordue, W. (1985) Primary structures of locust adipokinetic hormones II. *Biol. Chem. Hoppe–Seyler*, **366**, 723–727.
Siegert, K.J., Morgan, P.J. & Mordue, W. (1986) Isolation of hyperglycaemic peptides from the corpus cardiacum of the American cockroach, *Periplaneta americana. Insect Biochem.* **16**, 365–371.
Steele, J.E. (1961) Occurrence of hyperglycemic factor in the corpus cardiacum of an insect. *Nature* **192**, 680–681.
Steele, J.E. (1963) The site of action of insect hyperglycaemic hormone. *Gen. Comp. Endocrinol.* **3**, 46–52.
Stone, J.V. & Mordue, W. (1980) Adipokinetic hormone. In: Neurohormonal techniques in insects (ed: T.A. Miller), pp. 31–80. Springer–Verlag, New
Stone, J.V., Mordue, W., Batley, K.E. & Morris, H.R. (1976) Structure of locust adipokinetic hormone, a neurohormone that regulates lipid utilisation during flight. *Nature*, **263**, 207–211.
Traina, M.E., Bellino, M., Serpietri, L., Massa, A. & Frontali, N. (1976) Heart–accelerating peptides from cockroach corpora cardiaca. *J. Insect Physiol.* **22**, 323–329.
Turnell, D.C. & Cooper, J.D.H. (1982) Rapid assay for amino acids in serum or urine by pre-column derivatization and reversed-phase liquid chromatography. *Clin. Chem.* **28**, 527–531.
Van Marrewijk, W.J.A., Van den Broek, A.Th.M. & Beenakkers, A.M.Th. (1983) Regulation of glycogen phosphorylase activity in fat body of *Locusta migratoria* and *Periplaneta americana. Gen. Comp. Endocrinol.* **50**, 226–234.
Vejbjerg, K. & Normann, T.C. (1974) Secretion of hyperglycaemic hormone from the corpus cardiacum of flying blowflies, *Calliphora erythrocephala. J. Insect Physiol.* **20**, 1189–1192.
Walker, J.M. & Gaastra, W. (1983) *Techniques in molecular biology*. pp. 333. Groom Helm, London and Canberra.
Williams, C.A. (1947) Physiology of insect diapause. II. Interaction between the pupal brain and prothoracic glands in the metamorphosis of the giant silkworm, *Platysamia cecropia. Biol. Bull.* **93**, 89–98.
Witten, J.L., Schaffer, M.H., O'Shea, M., Cook, J.C., Hemling, M.E. & Rinehart, K.L. (1984) Structures of two cockroach neuropeptides assigned by fast atom bombardment mass spectrometry. *Biochem. Biophys. Res. Comm.* **124**, 350–358.
Ziegler, R. (1979) Hyperglycaemic factor from the corpora cardiaca of *Manduca sexta* (L)(Lepidoptera: Sphingidae). *Gen. Comp. Endocrinol.* **39**, 350–357.

Ziegler, R. (1980) Hypertrehalokämische and adipokinetische Aktivität bei *Manduca sexta. Verh. Dtsch. Zool. Ges.* 1980, 309.
Ziegler, R., Eckart, K., Schwarz, H. & Keller, R. (1985) Amino acid sequence of Manduca sexta adipokinetic hormone elucidated by combined fast atom bombardment (FAB)/tandem mass spectrometry. *Biochem. Biophys. Res. Comm.* **133**, 337–342.
Ziegler, R., Kegel, G. & Keller, R. (1984) Isolation and amino acid composition of the adipokinetic hormone of *Manduca sexta.* Hoppe–Seyler's *Z. Physiol. Chem.* **365**, 1451–1456.
Zimmerman, C.L., Appella, E. & Pisano, J.J. (1977) Rapid analysis of amino acid phenylthiohydantoins by high-performance liquid chromatography. *Anal. Biochem.* **77**, 569–573.

Note added in proof

Since this paper was originally finalized more insect neuropeptides of the AKH/RPCH family have been characterized. These peptides are the AKH from *Heliothis zea* (Jaffe *et al.* 1986), the HTH from *Blaberus discoidalis* (Hayes *et al.* 1986) and the HTH from *Nauphoeta cinerea* (Gäde & Rinehart 1986) (Table 1).

H. DUVE & A. THORPE

The isolation and characterisation of vertebrate-type peptides in insects

Introduction

Vertebrates produce a variety of molecules that are important in controlling and regulating the life processes. Initially, a clear distinction was made between the peptide hormones produced by the endocrine system, and the other major class of regulatory substances, the neurotransmitters, produced by the nervous system. Emphasis was placed on the contrasts between these two types of molecules, and comparisons were made of the differences in the sites of production, methods of release, transportation within the body and the duration of action. On the basis of these comparisons the endocrine and nervous systems were considered to be completely different.

Gradually, however, these two systems have come to be seen as the extremes in a wide spectrum of control mechanisms over biological processes. In between, we have a subtle arrangement and interplay of molecules which cannot, usefully, be categorised as either hormonal or nervous.

A major factor in the shift away from this traditional and separatist view of biological control systems is the finding that the brain produces a great variety of peptides. Indeed, the term 'endocrine brain' has been coined (Motta 1980). The discovery of the peptide hormones of the hypothalamus, firstly oxytocin and vasopressin and shortly afterwards the releasing hormones, provided the first clear evidence that certain neurones were peptidergic. Subsequently the phenomenon of neurosecretion has been studied in a wide range of vertebrates and invertebrates.

The speed at which 'new' neuropeptides are being discovered is testimony to the considerable interest in this field of neurobiology. In a review article in 1980, Hökfelt *et al.* referred to more than 20 peptides in neurones of the mammalian brain, spinal cord and peripheral nervous system. This number has increased yearly largely due to the widespread application of the techniques of immunocytochemistry to neural tissues.

If the great variety of neuropeptides is one interesting facet of the subject, another of equal importance is the ubiquitous nature of their occurrence. Thus, several peptides are now known to be produced in a variety of organs and tissues, within the vertebrate body. As examples, cholecystokinin (CCK) is found in the brain as well as in the gut, the organ where it was discovered initially and somatostatin, originally localised within the hypothalamus, is also produced by a specific cell type of the endocrine pancreas. The so-called brain/gut peptides have come to be recognised as a well-defined group of substances with, presumably, quite different functions in their specific localities.

Another radical revision of a traditionally held belief in biology has been necessitated by the discovery that one nerve cell may produce more than one type of regulatory substance. A frequent arrangement appears to be the co-existence of peptides with amine neurotransmitters, but several other combinations have also been reported.

This brief introduction to the regulatory peptides of vertebrates describes some of the interest and the problems concerning this important group of molecules. One particular aspect, however, and the main subject of this chapter, has, until recently received little attention: exactly where and when did these molecules appear in evolution?

Biological research has primarily been concerned with understanding mammalian systems. Studies on other types of vertebrates are often of a comparative nature where the 'definitive' animal is a mammal (more often than not, the rat), and comparisons are made down to the level of the fishes, but usually no further.

Invertebrates have usually been considered to be so different from vertebrates, that they have not been considered likely to contain the same types of peptide. The very title of this chapter, "vertebrate peptides in invertebrates" is, in itself, an expression of the conceived uniqueness of a particular group of molecules.

There is no rationale for this thinking. Unless one believes that invertebrates and vertebrates are two completely different groups of animals that arose independently in evolution, then there is no reason for believing that they must have different and independent types of regulatory peptides. Indeed, it is extremely difficult to promote such an argument.

The problem lies in the fact that the molecules in question, the gastro-enteropancreatic peptides such as insulin, CCK and secretin, and the neuropeptides such as the endorphins and enkephalins were discovered in vertebrates and they are commonly known and named by their role in these organisms. Cholecystokinin is a prime example, named by virtue of its action on the gallbladder. Such nomenclature makes it somewhat difficult conceptually to accept that it is also a neuropeptide of the brain, although it is present there in larger quantities than in the gut (Rehfeld 1978). Why should a gallbladder-contracting peptide be produced by the brain? If we extend this argument to lower levels of evolutionary organisation, even more bizarre questions arise. Why, for example, should the coelenterates possess CCK when they have neither gallbladder nor brain?

The answers to these conundrums, of course, lie in the framing of the questions. We should not attempt to transpose the functions (or the names) of vertebrate peptides to lower levels of organisation. We should simply ask questions such as: for how long has the amino acid sequence of the molecule recognised as CCK been in existence? and: is the presence of this molecule in vertebrates the end-product of long term evolutionary conservation from pre-vertebrate times? It is important to obtain answers to these questions, not only from the inherently interesting standpoint of biochemical evolution, but also because such answers may provide vital clues to unresolved problems of fundamental regulatory processes.

The mammalian brain is complex to a degree that makes the elucidation of the roles of neuropeptides a formidable task. Some progress has been made for certain of the molecules, but their role in the normal and abnormal functioning of the brain remains a key issue in neurobiology.

Invertebrate nervous systems, on the other hand, contain many fewer neurones and are correspondingly several orders of magnitude less complex. If, therefore, the same or slightly modified, homologous amino acid sequences are found at these lower levels of organisation it may well prove easier to use invertebrates to study some of the unresolved fundamental problems of neurobiology.

This Chapter reviews some of the progress that has been made in identifying and characterising the vertebrate-type peptides in insects.

Methods

A. Immunocytochemistry

This technique, in which either polyclonal or monoclonal antibodies directed against peptide antigens are applied to fixed or frozen tissue sections, is widely used to identify the cellular localisation of the peptides against which the antibodies were produced. Over the last decade, numerous studies using this technique have appeared in the literature, and there is no doubt that immunocytochemistry has greatly contributed to an awareness of the ubiquitous nature of many of the regulatory peptides of vertebrates. Workers on invertebrates have also used a wide range of antibodies directed against vertebrate peptides on a large selection of invertebrate tissues. Studies of this type on insects are described separately in this volume (see Chapter 3). In general the results of such studies have been extremely interesting if, at times, a little difficult to interpret. Nevertheless, the conclusion is that there are cells present within many invertebrate tissues that react positively to vertebrate peptide antibodies. It is essential to emphasise that cross reactivity with antibodies raised against a peptide does not constitute evidence *per se* that the identical peptide exists in the reacting tissue. The antigenic determinant (epitope) comprises 3–8 residues only (Larsson 1980; 1983) and, therefore, even when all necessary controls have been carried out, the most one can claim is that the sequence of amino acids of a particular antigenic determinant residing in the parent peptide molecule under study, is likely to be present within the cell or cells reacting positively. If the antibodies are raised against, for example, insulin, then the term insulin-like is often used to denote the degree of similarity demonstrated by these techniques.

However, in studies of certain peptides, it is possible to use a series of antibodies that are region- or sequence-specific (i.e. specific for different regions of the immunizing peptide used to prepare the antiserum). If several, many, or all of these antisera react positively, there is a high probability that a greater part of the amino acid sequence is present within the tissue (Duve *et al.* 1985).

Immunocytochemistry provides clues that peptides of vertebrate type may be present in invertebrate tissues and, in our laboratory we have used the technique to localise peptide-containing cells in the blowfly, *Calliphora vomitoria* (Fig. 1). The results provide us with a hard starting point for biochemical studies on *Calliphora*, in which the specific objective is the purification and characterisation of vertebrate-type peptides. The goal of these studies is the determination of the amino acid sequence of the insect peptides in order that a direct, unequivocal comparison can be made with the corresponding vertebrate peptides. Once this has been achieved it will be possible to obtain sufficient quantities of synthetic peptides with which to study their biological function in *Calliphora*.

B. Purification and characterisation studies

The results of our studies on the purification of vertebrate-type peptides of the blowfly, *Calliphora* will provide the basis for the remainder of this Chapter, but other insect studies will be discussed in relation to our findings.

Fig. 1. (A, B) Median neurosecretory cells (MNC) of pars intercerebralis of the brain of *Calliphora* showing (A) insulin-immunoreactive cells (anti-bovine insulin antiserum, 1:1000, PAP technique) and (B) gastrin/CCK immunoreactive cells (C–terminal gastrin/CCK antiserum, 1:1000, PAP technique). n, neuropile; v, vacuole. (C) Transverse section of the brain of *Calliphora* showing bovine pancreatic polypeptide (PP)-immunoreactive cells and axons (arrows) within the dorsal part of the protocerebrum (bovine PP antiserum, 1:2000, PAP technique). (D) Sagittal section through part of MNC and median nerve bundle showing α–endorphin-immunoreactive perikarya and axons. (Indirect immunofluorescence technique with a rhodamine-conjugated antibody and an antibody to synthetic α–endorphin). (E) Sagittal section through the corpus cardiacum (CC) showing cells and neuropile containing gastrin/CCK-immunoreactive material (C–terminal gastrin/CCK antiserum, 1:500, indirect immunofluorescence technique with a rhodamine-conjugated antibody). (F, G) Transverse section of the brain showing (F) Leu–enkephalin and (G) Met–enkephalin-immunoreactive cells within the dorsal part of the protocerebrum. Arrows indicate transverse section through immunoreactive axons within the neuropile (n). (Leu– and Met–enkephalin antisera, 1:500, PAP technique). opl. optic lobe. ×460.

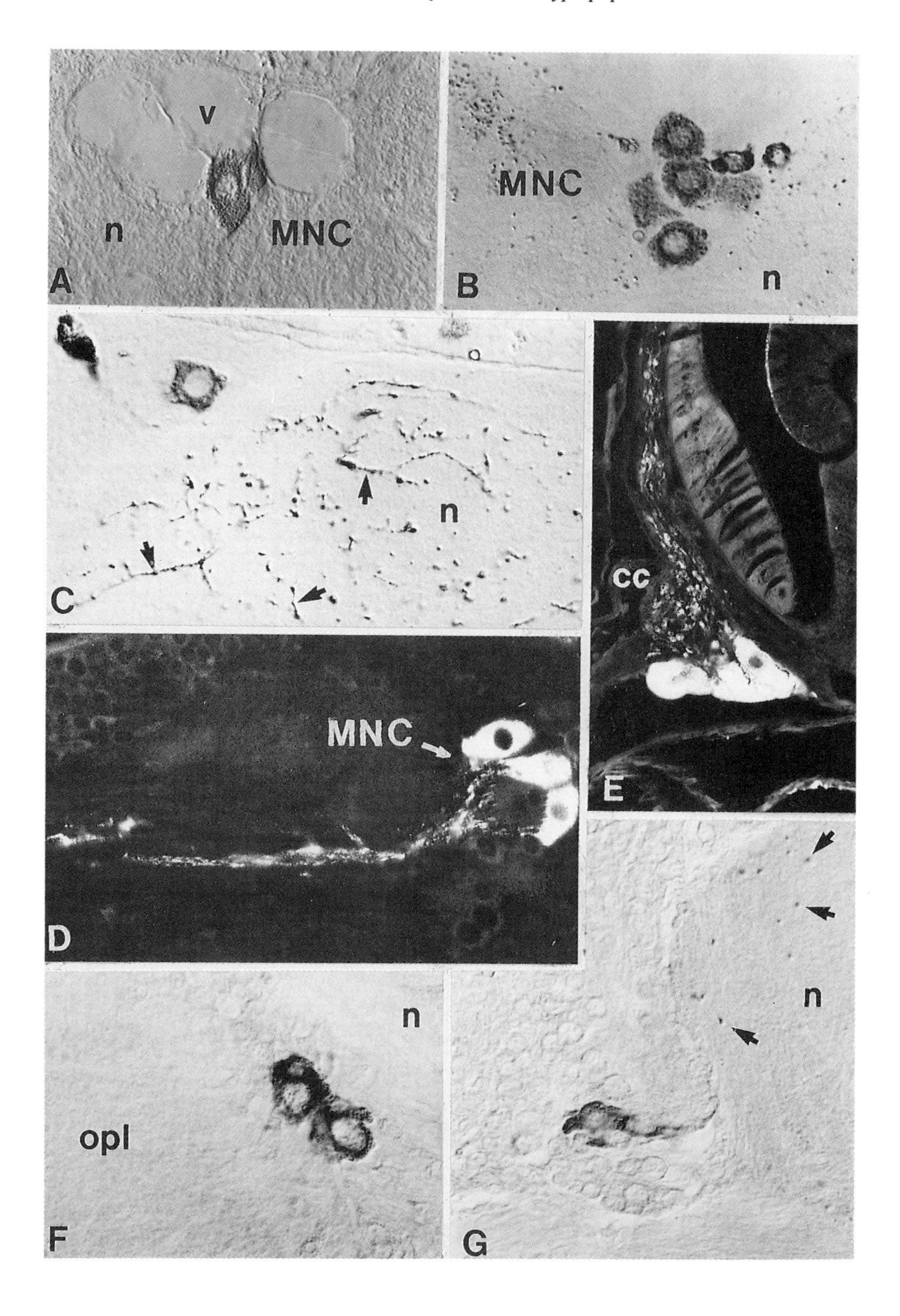
A
v
n
MNC
B
MNC
n
C
n
D
MNC
E
cc
F
n
opl
G
n

(i) General comment: There are several problems inherent in isolating and characterising vertebrate-type peptides in insects, but one prime consideration is exactly how to identify and monitor the material through the stages of purification. It is common practice to rely upon radioimmunoassay (RIA) for peptide identification, but bioassay is also a possibility if a specific function can be ascribed to the substance. The RIAs are, of necessity, based on the known vertebrate peptides, and generally have the advantages of speed and specificity. They are specific in that reliance is placed upon competition between the unknown peptides and the standard, radiolabelled peptide for binding sites on the antibodies. This is unlike immunocytochemistry which lacks this important element of competition in the binding. It is preferable to combine RIA and bioassay during the purification process and an example of this approach will be seen in the discussion of the isolation of the insulin-like peptide of *Calliphora.*

A second major problem in trying to assess the degree to which vertebrate-type peptides exist in invertebrates, and it applies particularly to insects, is the small size of the animals concerned and the often small amounts of the peptide present within the tissues. In the isolation and purification of a mammalian brain or pancreatic peptide, the investigator is presented with a clear-cut choice of starting material which can normally be obtained in large amounts. The need for large amounts of material, even in mammalian studies, is seen in the Nobel prize-winning work of R. Guillemin and A.V. Schally who obtained the first pure hypophysiotrophic hormone, thyrotrophic releasing factor, TRF, from more than 500 tons of sheep brain collected over 3–4 years from some 5 million sheep. In their definitive study in 1968, they used 300,000 hypothalami, a quantity that yielded just 1 mg of TRF.

In insects, the vertebrate-type peptides appear to be present largely within cells of the nervous system, a point of interest to which we will return later. In practical terms, however, this means that the investigator is faced either with trying to remove specific parts of the nervous system (as, for example, is done in the isolation of the peptides of the corpus cardiacum – see Chapter 5) or, as an alternative, to separate the heads from the rest of the body in order to obtain a higher brain content/volume. As a last resort it is possible to work with whole animals. Each approach carries with it particular difficulties.

The final problem is a natural corollary of the first two points. Is it possible to identify and purify *sufficient* of the vertebrate-type peptide(s) in order to be able to carry out the sequencing of the amino acid residues and so determine the degree of relatedness to the vertebrate peptide?

(ii) Gastroenteropancreatic peptides

(a) *Insulin:* Of the peptides studies from the viewpoint of biochemical evolution, perhaps most work has been carried out on insulin. The great interest in this molecule undoubtedly stems from the key roles that it plays in the metabolism of protein, fat and carbohydrate, the vital metabolites found in all living organisms.

Certain insects, such as the one used exclusively as an experimental animal in our laboratory, *Calliphora vomitoria*, have considerable quantities of glucose, glycogen and another disaccharide, trehalose, either within their tissues or in the haemolymph. A hyperglycaemic factor (hormone?) exists in *Calliphora* (Normann & Duve 1969) and questions as to whether or not there is a corresponding *hypoglycaemic* factor, led naturally enough to the search for a possible insulin-related peptide.

The history of studies on the insulin-like peptides of insects has recently been reviewed (Thorpe & Duve 1984) and will not be repeated in great detail here. The studies of Tager *et al.* (1975, 1976) on a lepidopteran *Manduca sexta*, provided the first real evidence that insulin (and glucagon, the main insulin antagonist) may be present at least in modified form, in insects. A detailed account of these and other studies has been presented by Kramer (1979). The experiments relied on corpus cardiacum/corpus allatum complexes extracted in aqueous guanidine hydrochloride and fractionated on Bio–Gel–P10. Eluted fractions contained both insulin-like, and glucagon-like peptides as measured by RIA. Bioassays showed that glucagon-like fractions could promote glycogenolysis in the fat body whereas the insulin-like fractions could promote hypotrehalocaemia. Subsequently, Kramer *et al.* (1980) showed the presence of both insulin- and glucagon-like peptides in the haemolymph of *Manduca* larvae and pupae and, more recently, Kramer (1985) has published a preliminary amino acid composition of the *Manduca* insulin-like peptide.

In *Calliphora*, we showed earlier (by means of immunocytochemistry) that a small number of the median neurosecretory cells (MNC) of the pars intercerebralis of the brain contain peptidergic material that cross-reacts with mammalian insulin antibodies (Duve & Thorpe 1979). Subsequently, in a number of separate experiments we isolated and partially purified an insulin-like peptide from separated heads of flies. In the last of these investigations more than 1 million heads were used.

In brief, the method adopted was a modified version of that normally used in the isolation of mammalian insulins from pancreatic tissue. Extraction of 5.5 kg of heads was carried out in acidified 74% ethanol overnight at 4°C followed by gel filtration on Sephadex G25, mainly for the purpose of de-salting. Insulin immunoreactive fractions, monitored at this and all subsequent stages in the purification by a standard bovine or porcine insulin RIA, were pooled, freeze-dried and subjected to further gel filtration (Sephadex G50–SF), cellulose ion-exchange (DE–52 DEAE) and a series of reversed-phase high performance liquid chromatographic (rp–hplc) steps (Fig. 2). (For full experimental details see Duve *et al.* 1979, 1981, 1982; Duve & Thorpe 1984a; and Thorpe & Duve 1984).

Sufficient pure material was obtained for amino acid composition analyses, and the results show a partial degree of similarity to the mammalian insulins. For the main peak of insulin immunoreactivity from the final analytical step, the following amino acid composition was determined (figures in parentheses are for bovine insulin):

Asx 6(3); Thr 3 (1); Ser 3 (3); Glx 7 (7); Pro 1 (1); Gly 4 (4); Ala 3 (3); Cys not determined (6); Val 3 (5); Met 0 (0); Ile 1 (1); Leu 5 (6); Tyr 2 (4); Phe 4 (3); His 2 (2); Lys 2 (1); Arg 4 (1).

We do not yet have the amino acid sequence of the *Calliphora* insulin-like material, so the degree of homology with mammalian insulin cannot be stated with certainty. The material in its pure state, adsorbs to glass and polypropylene surfaces and considerable losses are experienced because of this phenomenon.

Despite the lack of specific structural data, we have tested the partially purified material in three types of bioassay, each of which strongly suggests that it is truly insulin-like. Two of the bioassays are mammalian; the incorporation of glucose into lipid by rat adipocytes and the displacement of ^{125}I-labelled porcine insulin from insulin receptors of rat hepatocytes.

The slopes of the two curves, in both the displacement and incorporation assays are not identical with those for known insulins and we assume from this that the *Calliphora* insulin-like molecule is conformationally different. In addition to the two mammalian assays, we have tested the peptide in a *Calliphora* bioassay in which flies made hyperglycaemic through the surgical removal of their MNC were normalised following administration of 1 ng bovine insulin equivalents of the purified material. This experiment is particularly important since it links the removal of specific cells, shown to contain insulin immunoreactive material in immunocytochemistry, with hyperglycaemia, and then a return to normoglycaemia with purified extracts (presumably from the same cells?). Full details of these bioassays are given in an earlier review (Thorpe & Duve 1984).

In addition to the experiments described for *Manduca* and *Calliphora*, work has been carried out on at least three other insect species. Thus, Le Roith *et al.* (1981) have demonstrated by means of RIA an insulin-like peptide in acid alcohol extracts of the fruit fly, *Drosophila* and Maier *et al.* (1978, 1985) and O'Connor and Baxter (1985) have likewise shown the presence of insulin-like material in extracts of the

Fig. 2. (Top Panel) DE–52 DEAE-cellulose chromatography of *Calliphora* insulin-like and pancreatic polypeptide-like immunoreactive peptides. The peptides were extracted from heads in acid alcohol and gel-filtered using Sephadex G–15 followed by G–50. Freeze dried immunoreactive fractions from Sephadex G–50 were reconstituted and applied to the DEAE column. Column size, 1.5 cm × 30 cm; gradient 20–400 mM ammonium acetate in the presence of 25% acetonitrile; pH 8.5; flow rate 20 ml/h; 4 ml fractions. Samples (10 μl) were immunoassayed for insulin and PP.

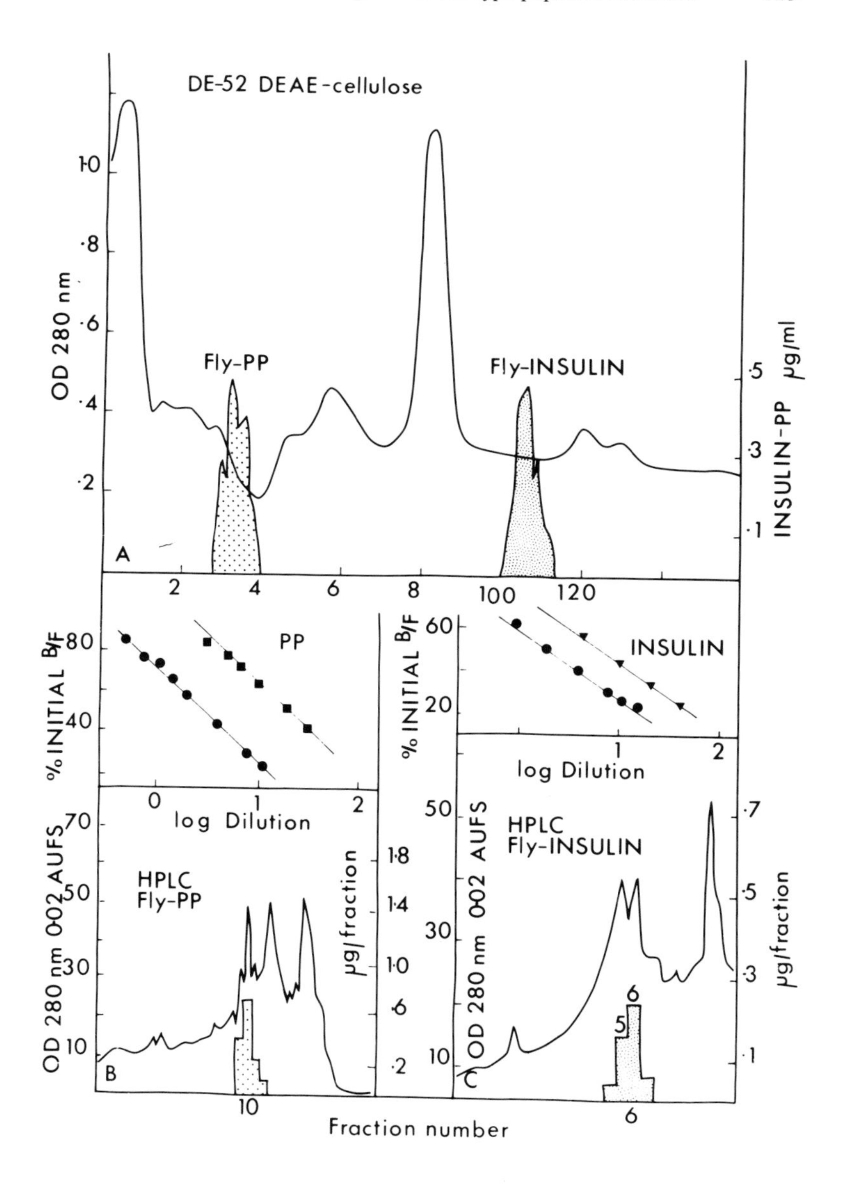

(Lower Panel) Final analytical HPLC of PP (B) and insulin (C) from the immunoreactive peaks from DEAE. Ultrasphere ODS (octadecylsilane); column size 0.4 cm × 25 cm; gradient 40–80% methanol/1% TFA. Fraction 10 was used for amino acid analysis of PP and both fractions 5 and 6 for insulin analysis. The insets show the linearity and parallelism of the dilution curves, indicating common antigenic determinants. (Reproduced from Duve & Thorpe (1984*a*) with the kind permission of Plenum Press.)

```
                              1   2   3   4   5    6      7    8   9  10  11   12  13  14  15  16  17  18  19

4K PTTH-I                  H - Gly-Val-Val-Asp-Glu-(Cys)-(Cys)-Phe-Arg-Pro-(Cys)-Thr-Leu-Asp-Val-Leu-Leu-Ser-Tyr

      -III                 H - Gly-Val-Val-Asp-Glu-(Cys)-(Cys)-Leu-Gln-Pro-(Cys)-Thr- ? -Asp-Val-Val-Ala-Thr-Tyr

      -II                  H - Gly-Ile-Val-Asp-Glu- Cys - Cys -Leu-Arg-Pro- Cys -Ser-Val-Asp-Val-Leu-Leu-Ser-Tyr

Codfish Insulin A chain        Gly-Ile-Val Asp Gln- Cys - Cys -His-Arg-Pro- Cys -Asp-Ile-Phe-Asp-Leu-Gln-Asn-Tyr

Human Insulin A chain

(residues 1 - 19)              Gly-Ile-Val-Glu-Gln- Cys - Cys -Thr-Ser-Ile- Cys -Ser-Leu-Tyr-Gln-Leu-Glu-Asn-Tyr

IGF - I

(residues 42 - 60)             Gly-Ile-Val-Asp-Glu- Cys - Cys -Phe-Arg-Ser- Cys -Asp-Leu-Arg-Arg-Leu-Glu-Met-Tyr
```

(Cys) - indicates that cysteine residues at positions 6, 7 and 11 in 4K PTTH-I and -III not identified fully.

Fig. 3. NH_2-terminal amino acid sequence of 4K–PTTH I, II and III from *Bombyx mori* compared with codfish and human insulin A chains (residues 1–19) and insulin-like growth factor residues 42–60. (From Nagasawa *et al.* 1984.)

heads of the honeybee, *Apis mellifera*. The two latter studies included bioassays of the purified material using the rat adipocyte assay.

The investigations discussed so far were all initiated with the specific objective of ascertaining whether or not a mammalian insulin-like peptide is present within the tissues of insects. Perhaps the most convincing evidence for this, however, has come indirectly, from a study designed to purify and characterise the prothoracicotropic hormone (PTTH) of the silkwork *Bombyx mori*. This hormone, secreted by certain of the MNC, acts on the prothoracic glands to stimulate the synthesis and release of ecdysone, which, in turn, initiates moulting and metamorphosis. Much is known about the physiological function of PTTH, and Japanese and Chinese groups have been working for a long time on a large-scale purification of the peptide with the specific objective of elucidating its primary structure.

From a starting weight of 4.86 kg of adult male *Bombyx* heads (648,000 individuals) and a 15-step purification procedure, Nagasawa *et al.* (1984) have finally succeeded in determining the sequence of the first 19 amino acid residues of the amino terminus of three variants of the molecule, 4K–PTTH–I, II and III. (There is also known to be a PTTH molecule of higher molecular weight, the so-called 22K–PTTH).

The remarkable outcome of this study is the discovery that the 4K–PTTH variants have a striking amino acid sequence homology with insulin and, indeed, with a related member of the insulin family of peptides, the insulin-like growth factor 1 (IGF 1). Almost 50% of the 4K–PTTH–II sequence is identical to that of the human insulin A chain and to IGF 1 and an even greater homology (58%) is seen when the sequence is compared with the insulin A chain of the codfish (Fig. 3). Furthermore, if we consider the invariant residues of the known A chains of insulin (cf. Cutfield *et al.* 1979), it is seen that Gly (1), Ile (2), Val (3), Cys (6), Cys (7), Cys (11), Leu (16) and Tyr (19) are all present within the *Bombyx* PTTH molecule. The only residue change at a normally invariant insulin A chain position is Glu for Gln at residue 5 and such a substitution requires only a single nucleotide base change.

We do not yet know whether the other invariant residues of the insulin A chain at positions 20 (Cys) and 21 (Asn) exist in the *Bombyx* PTTH molecule, and, of great significance we do not know if there is a second chain to the insect peptide, which might be the equivalent of the insulin B chain. We await, with great interest the further results of Nagasawa and his colleagues in the hope that they will clarify these important issues (see 'Note added in proof at end of Chapter), but meanwhile it is difficult to avoid the conclusion that the *Bombyx* PTTH molecule is related to the insulin family of peptides and that it most probably represents an early ancestral form. Additional evidence in support of the close relationship between insulin and PTTH in *Bombyx* has come from the immunocytochemical studies of Ishizaki *et al.* (1985) where it was shown that four pairs of cells within the MNC group react positively to

	10
AKH 1 (Locusta)	pGlu - Leu - Asn - Phe - Thr - Pro - Asn - Trp - Gly - Thr - NH_2
CC - 1 (Periplaneta)	pGlu - Val - Asn - Phe - Ser - Pro - Asn - Trp - NH_2
CC - 2	pGlu - Leu - Thr - Phe - Thr - Pro - Asn - Trp - NH_2
Glucagon (1 - 12)	His - Ser - Gln - Gly - Thr - Phe - Thr - Ser - Asp - Tyr - Ser - Lys -
AKH (Manduca)	pGlu - Leu - Thr - Phe - Thr - Ser - Ser - Trp - Gly - NH_2

Fig. 4. Amino acid sequences of the adipokinetic hormones of *Locusta* and *Manduca* and the cardioacceleratory peptides CC–1 and CC–2 from *Periplaneta* compared with the amino terminal (1–12) sequence of glucagon. (For references see text.)

both insulin antibodies and to a monoclonal antibody directed against the *synthetic Bombyx* PTTH sequence 1–10.

A final question, of general interest, is whether the peptidergic material of other insects, identified in separate laboratories as being insulin-like, is a homologue of the PTTH of *Bombyx*. It is tempting to speculate that it is so, although further work is required before this can be verified.

(b) *Glucagon:* Certain insects have been known for many years to have one or more hyperglycaemic factors responsible for elevating blood sugar levels under different physiological conditions (Steele 1961, 1963). Knowledge of such factors has initiated the search for an insect glucagon homologue and although the evidence for such a molecule is not as well documented as that for insulin, there are several reports suggesting the presence of a glucagon-like substance based mainly upon the radioimmunoassay of extracts of insect nervous tissues (Tager *et al.* 1975, 1976, 1980; Kramer *et al.* 1980; Kramer 1979, 1985; Duve & Thorpe 1984a; Thorpe & Duve 1984). However, as yet, no amino acid sequence data is available from these studies to confirm the degree to which such molecules are related to glucagon.

Surprisingly, as in the case with insulin, evidence for glucagon-related insect molecules has recently come indirectly from studies of insect peptide hormones produced in the corpus cardiacum. Thus, Scarborough *et al.* (1984) have drawn attention to a sequence homology of part of the amino terminus of glucagon with one of the two recently characterised cardioacceleratory and hyperglycaemic peptides of the corpus cardiacum of the cockroach *Periplaneta*, the so-called periplanetins, CC_1 and CC_2. The amino acid sequence analysis of these peptides has shown that they are undoubtedly members of the adipokinetic hormone (AKH) family, all of which appear to have some degree of hyperglycaemic activity in addition to their major roles in either lipid mobilisation and/or control of heart rate. Another member of this family, also showing a partial structural similarity to glucagon is the recently identified AKH of *Manduca sexta* (Ziegler *et al.* 1985) (Fig. 4).

It remains to be seen whether the various insect peptides described by earlier workers as glucagon-like on the basis of immunological studies (mainly RIA) are the self-same peptides as those of the AKH family currently being characterised or whether they belong to an entirely different peptide series.

(c) *Pancreatic polypeptide:* We have earlier isolated from *Calliphora* a peptide apparently homologous with pancreatic polypeptide (PP) (Duve *et al.* 1981, 1982). The substance was purified during the processing of the insulin-like peptide described previously and was monitored by a bovine PP RIA (Fig. 2). After a 6-stage column chromatography purification method, the peptide was analysed for amino acid composition with the following result (bovine PP in parenthesis): Asx 3 (3); Thr 2 (2); Ser 1 (0); Glx 6 (6); Pro 4 (5); Gly 2 (1); Ala 4 (5); Cys not determined (0); Val 1

(0); Met 1 (2); Ile 2 (1); Leu 3 (3); Tyr 4 (4); Phe 0 (0); Lys 0 (0); Arg 3 (4); Trp not determined (0). The amino acid composition comprises 36 residues, the same number as for all mammalian, reptilian and avian PP molecules so far identified. Sequence data obtained so far indicate an extremely close, if not identical relationship with bovine PP, but this needs to be confirmed. The available evidence, however, does support the concept of a well conserved PP homologous molecule.

(d) *Gastrin/cholecystokinin (CCK):* Gastrin and CCK are believed to have been derived from a common ancestral peptide in that they share a common COOH terminus pentapeptide sequence (Larsson & Rehfeld 1977). Great interest in these peptides has been generated by the finding within the last decade that as well as occurring in the nervous system they also occur in the gut. Gastrin is believed to have only a limited presence in hypothalamo-hypophysial neurons (Rehfeld *et al.* 1984) whereas CCK is now known to be a widespread neuropeptide (Rehfeld 1978; Marley *et al.* 1984).

Interest in the possible status of CCK as a neuropeptide of invertebrates has initiated studies within several phyla (see other Chapters, this Volume). In *Calliphora* a series of antisera raised against the common C–terminus has been used to map the distribution of gastrin/CCK-like immunoreactivity in neurones and fibres of the nervous system (Duve & Thorpe 1981, 1984b). (For other insect immunocytochemical studies see Chapter 3). In order to determine the degree to which the peptide identified immunocytochemically resembles either gastrin or CCK we have carried out isolation and purification studies.

Published methods for the extraction of mammalian CCK and gastrin vary somewhat, and we have found that different extraction media used on *Calliphora* (acidic or neutral, with or without boiling) result in the isolation of apparently different molecular forms, as evidenced by the results of gastrin/CCK RIAs on the eluates from Sephadex G–50 gel-filtration. After initial experiments with immunoaffinity chromatography (Dockray *et al.* 1981) we have recently concentrated our efforts on the gastrin/CCK-like immunoreactive material, with a K_{av} of 0.2, that appears following extraction in 0.2 M HCl and gel-filtration on Sephadex G–50 (see Duve *et al.* 1985 for details).

A disturbing feature of the purification process is the great loss (up to 90%) of immunoreactive material experienced after the gel-filtration step. The peptide appears to be extremely hydrophobic in that on rp–hplc, it requires a high concentration of organic solvent (either ethanol or acetonitrile) to elute it from the column. Furthermore, on freeze drying, or on collection after hplc, losses due to adsorption are extremely high. This phenomenon can be prevented to some extent by the addition of bovine serum albumin to the collecting tubes, although in later stages of the purification prior to sequencing, such a precaution is impracticable. Despite these difficulties, the partially purified material has been tested in RIA with a series of region- or sequence-specific CCK and gastrin antisera. The results indicate that the molecule bears a closer relationship to CCK than to gastrin (Fig. 5).

(e) *Endorphins and enkephalins:* The endorphins and enkephalins were discovered some 10 years ago as endogenous peptides within the brain of mammals having similar effects to those of plant alkaloid-derived drugs such as morphine. Hence the generic term 'end-orphins'. Subsequently, much has been learned about various aspects of these opiod peptides: cytological localisation; processing from precursor molecules such as proopiomelanocortin from which β–endorphin is derived, and preproenkephalin A and B from which both leu– and met–enkephalin are derived; and something of their physiological roles. Nevertheless, the great complexity of these systems of peptides is still unfolding.

Immediately following their discovery in mammals it was suggested that invertebrates do not possess any of the opioid peptides (Simantov *et al.* 1979). This has now been shown to be incorrect, however, and studies based on the use of immunocytochemistry, receptor binding and the isolation and purification of peptides suggest that endorphin-related substances are present within a wide range of invertebrates. Indeed, the presence of enkephalin receptors has been demonstrated in the cockroach, *Leucophaea maderae* (Stefano & Scharrer 1984).

Our own studies on the opioid-like peptides of *Calliphora* were initiated by the finding that α–endorphin-immunoreactive neurones occurred in the MNC, with pronounced axonal transport of the material along nerve fibres within the median bundle leading to the corpus cardiacum (Duve & Thorpe 1983). Since then, we have studied the presence of other opioid-like peptides (leu– and met–enkephalin) by immunocytochemistry and we have also surveyed the endorphin-related peptides using a range of RIAs for α–endorphin, β–endorphin, the enkephalins, and ACTH. The preliminary report of these findings has been published (Duve *et al.* 1986).

Currently, in our laboratory, we are continuing our studies of these peptides extracted in 0.2 M HCl from whole flies. The RIA profiles of the eluate of a Sephadex G–50 F gel filtration column are shown in Figure 6a. The α–endorphin-like peptide has been purified through ion exchange and 6 rp–hplc stages (see Fig 6b,c) to the point where it will be possible to obtain the amino acid sequence of the molecule. At the present time, one co-eluting peptide, present in greater quantity than the α–endorphin-related peptide has been sequenced.

Discussion

This chapter has summarised briefly the evidence for the presence of vertebrate-like peptides in insects. Despite the fact that very few species have been studied and although there are still many uncertainties as to the exact degree of relatedness of the peptides to their mammalian counterparts, one can no longer be in

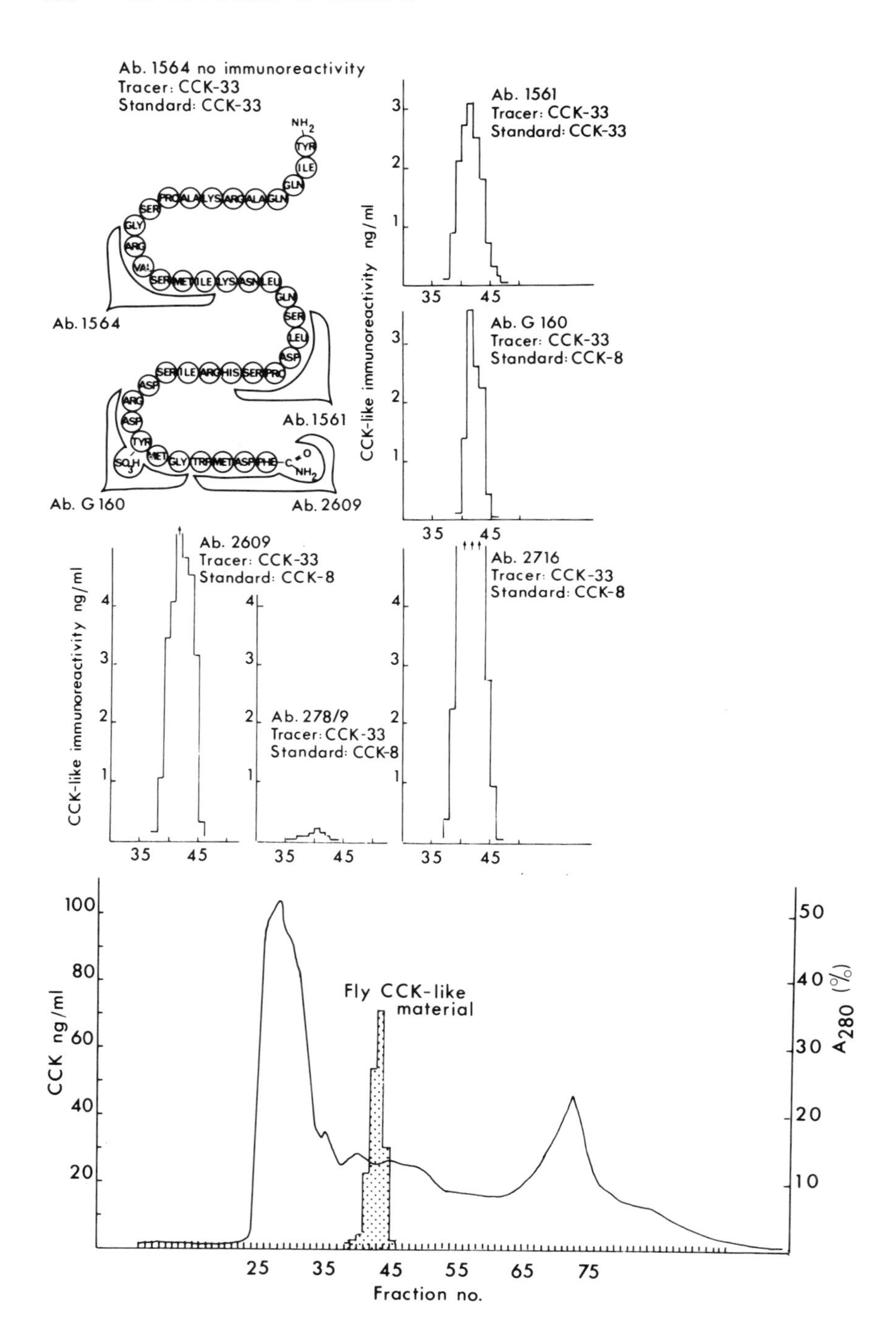
Ab. 1564 no immunoreactivity
Tracer: CCK-33
Standard: CCK-33
Ab. 1564
Ab. 1561
Ab. G 160
Ab. 2609
Ab. 1561
Tracer: CCK-33
Standard: CCK-33
Ab. G 160
Tracer: CCK-33
Standard: CCK-8
CCK-like immunoreactivity ng/ml
Ab. 2609
Tracer: CCK-33
Standard: CCK-8
Ab. 278/9
Tracer: CCK-33
Standard: CCK-8
Ab. 2716
Tracer: CCK-33
Standard: CCK-8
Fly CCK-like material
CCK ng/ml
A_{280} (%)
Fraction no.

doubt that at least some of the regulatory peptides discovered in vertebrates, have their origins in lower levels of organisation.

What is the significance of these findings? How do they affect our thinking of biochemical evolution in general? What do they mean in terms of past, present and future studies in insect physiology and biochemistry? How might the results be applied to the benefit of mankind?

In terms of biochemical evolution, the results demonstrate a conservation of particular amino acid sequences over long periods of time. The origins of both insects and vertebrates are difficult, if not impossible, to pinpoint accurately within the geological record, but the Carboniferous (250 m years) certainly saw many winged insects (some of which have been preserved perfectly in amber and which show incredible morphological similarity to present-day animals) and the Ordovician (400 m years) is generally accepted as the first era in which ostracoderms (ancestors of the cyclostomes and thus all vertebrates, are found). The important point, however, is not the precise evolutionary starting point of these two groups of animals, but the fact that both must have originated from a common ancestor at an even more distant point in time, perhaps as far back in the archeozoic as 1.5×10^9 years.

We know that certain molecular species have had a continuous and little-changing existence. Cytochrome C, haemoglobin, and calmodulin are three obvious examples of protein products that have been retained throughout evolution relatively unchanged, presumably because of their key role in cellular and metabolic processes.

The question we have to consider, therefore, is not whether it is possible for continuous conservation of protein structure to have been maintained throughout evolution, but whether regulatory peptides of the type under discussion had such a key role in the life of organisms, that they were retained. It is at this point that we need to avoid fixed concepts about *particular* roles of *particular* proteins or peptides in *particular* types of animals.

Consider the finding of the Japanese group of Nagasawa and his co-workers (1984), which shows that the prothoracicotropic hormone of *Bombyx* has a close degree of homology to the insulin A chain (see 'Note added in proof at end of Chapter). If we remove the specific name-tag by which the peptide has become known, and simply consider its chemical structure, then we are left with an insect

Fig. 5. (Bottom panel) Sephadex G–50 (SF) gel-filtration profile of 0.2 M HCl extracts of *Calliphora*. Eluent 1% formic acid. Immunoreactive fractions monitored with a C–terminal gastrin/cholecystokinin (CCK) antibody. No. 2609/8, CCK_8 standard and CCK_{33} tracer.

(Top panel) Sequence-specific radioimmunoassays over the peak of immunoreactive fractions from Sephadex G–50 (SF) gel-filtration. Note decreasing immunoreactivity to antibodies increasingly specific towards the N–terminus of porcine CCK.

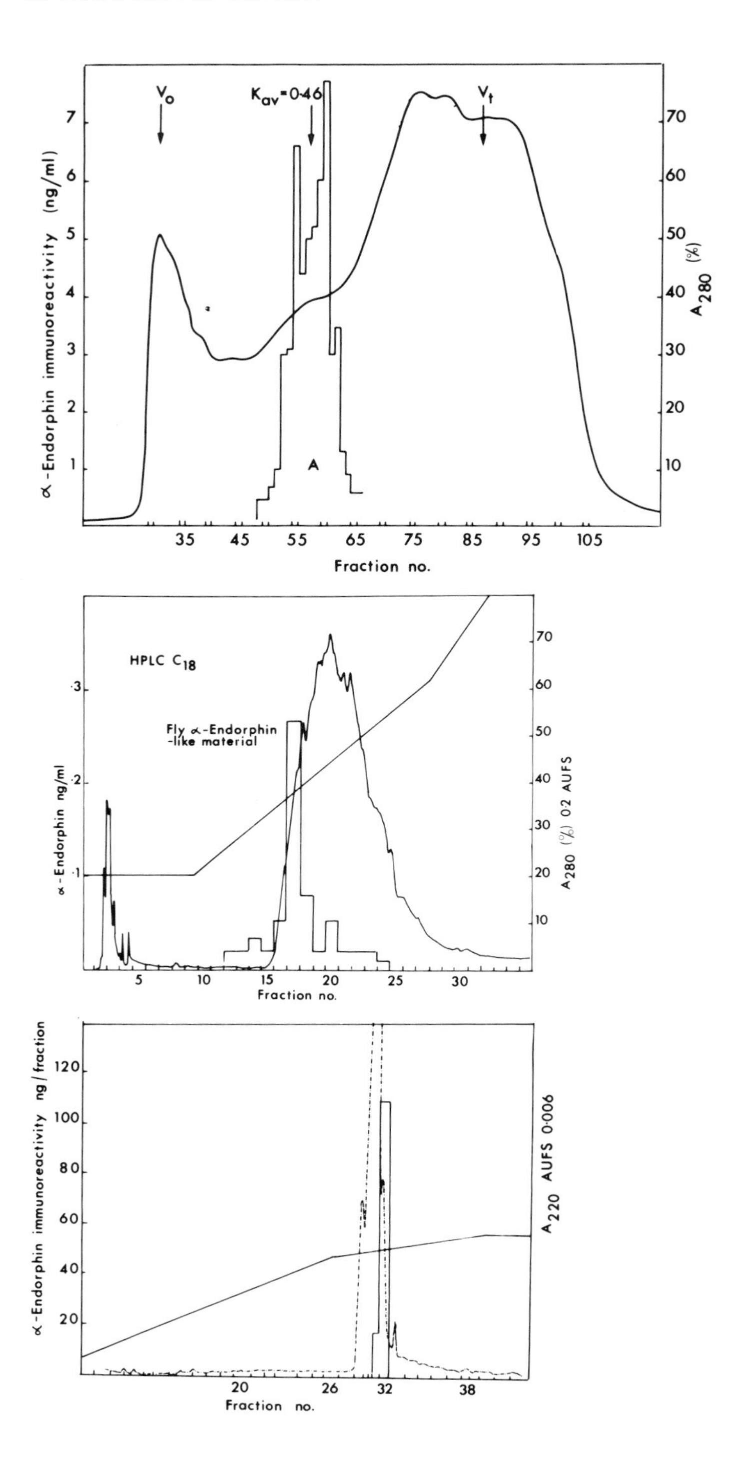

V_o
K_{av}=0·46
V_t
α-Endorphin immunoreactivity (ng/ml)
A_{280} (%)
A
Fraction no.
HPLC C_{18}
Fly α-Endorphin -like material
α-Endorphin ng/ml
A_{280} (%) 0·2 AUFS
Fraction no.
α-Endorphin immunoreactivity ng/fraction
A_{220} AUFS 0·006
Fraction no.

peptide which, in reality, is a possible 'fore-runner' of a molecule recognised and named as insulin (or insulin-like growth factor) in vertebrates. There can be no doubt that this conserved amino acid sequence has been retained because of its ability to convey an important message from one cell to another. In the insect, it is recognised mainly because of the signal it provides for the production and release of a cell product (ecdysone from the prothoracic glands) whereas, in vertebrates, its general role is linked with many aspects of metabolism through the variety of different ways it is received by cells.

It is certainly not mandatory when studying vertebrate-type peptides in invertebrates to anticipate that the role should be the same as that known for vertebrates. Indeed, even within vertebrate phyla, there are extremely good examples of the changing physiological roles of certain peptides as evolution progresses. The wide spectrum of activities of prolactin in vertebrates is evidence of this, although, of course, it must not be forgotten that the peptide/cell interactions will be similar, if not identical in that they depend upon the prolactin receptor. Consequently, it is of interest that the insulin-like peptide of at least certain types of insects may have broadly vertebrate-type roles in the control of metabolism.

One point which requires comment is the fact that the regulatory peptides of vertebrate type found in the invertebrates are, for the most part, the products of nerve cells, whereas in the vertebrates themselves, they occur in both nerve cells and in endocrine cells often at a distance from the CNS. One explanation of this is that the nerve cell represents the most primitive signalling or coordination cell type. In the coelenterates, the nerve cell arrangement (without a coordinating centre) represents the most primitive state of organisation and coordination there is by direct neurone/neurone or neurone/cell effector contact. Chemical messages produced by the

Fig. 6 (Top panel) Sephadex G–50(F) gel filtration profile of 0.2 M HCl extract of *Calliphora*. Eluent 1% formic acid. Histogram indicates α–endorphin-immunoreactive material monitored with RIA using an antibody to synthetic α–endorphin.

(Middle panel) HPLC profile of immunoreactive material from Sephadex G–50 (F) gel filtration, column, semi-preparative ODS; gradient, 16–50% acetonitrile/water/0.1% TFA.

(Lower panel) HPLC profile of immunoreactive material following several chromatographic steps including HPLC (Alkyl Phenyl column) and anion exchange MONO Q chromatography. Column, analytical ODS; gradient, 6–24–29% acetonitrile/water/0.1% TFA. Fraction 32 was rechromatographed on analytical ODS and the α–endorphin immunoreactive fraction used for sequence analysis.

neurones allow for all the varied activities of the organism (cf. Grimmelikhuijzen & Spencer for accounts of vertebrate-type peptides in Coelenterates; Chapter 10).

As the complexity of activities increased and perhaps as greater quantities of a specific chemical became necessary to organise the activities, then it became "worthwhile" to collect a particular cell-secreting type in a discrete organ outside the brain. The insects represent an interesting mid-way point in this development because although they have nerve cells producing peptides within the brain and other parts of the central nervous sytem and obviously releasing them into the CNS itself, in addition, they have neurosecretory peptidergic cells such as those of the MNC, that have their perikarya within the brain, but whose axons lead to storage areas (such as the corpus cardiacum) for release into the blood system. In this respect they are performing a more 'endocrine' type of function than a purely nervous one, in that the neuronal product has the potential to affect all parts of the body by virtue of its general relase into the haemolymph.

Perhaps in the past, there has been too great an inclination to regard insects as highly specialised animals, quite separate and different from all other groups. While this is in one sense self-evident we must remember that taxonomic groupings are an 'invention' of man, for man's benefit and that all animals are related to some extent through their common evolutionary origins. The presence in insects of regulatory peptides similar to those possessed by vertebrates ought now to be considered when fundamental physiological and biochemical processes are studied. Perhaps our approach to insect metabolic processes has to some extent been prejudiced by too great an emphasis upon 'insect' hormones such as AKH, eclosion hormone and PTTH. A first priority, however, before such a reappraisal is the precise chemical identification of the vertebrate-type peptides. Only then can direct comparisons be made and, with synthesised peptides of the same type, accurate assessment of their role in the insect be determined.

The methods of peptide isolation described in this Chapter and elsewhere, are difficult and laborious, and in the future, recombinant DNA techniques will certainly be applied in the insect peptide field. One problem with this approach, however, is that at least a part of the insect peptide sequence must be known in order to construct an oligonucleotide probe complementary to the messenger RNA for that paticular peptide. This oligonucleotide probe could then be used directly to screen a cDNA or genomic library (for *Calliphora* in our case) or it could be used as a primer for reverse transcriptase in the synthesis of a cDNA highly enriched for the specific amino acid sequence being studied. Once this has been achieved, the gene can be cloned and, hopefully, the *complete* peptide could be obtained in quantity and sequenced. This approach would undoubtedly help in the identification of insect peptides of vertebrate type. The fact that at least a part of the insect sequence must be identified prevents the direct use of probes based on known mammalian peptide sequences: any variation

A chains

```
                         1           5              10              15              20
4K-PTTH-II               H-Gly-Ile-Val-Asp-Glu-Cys-Cys-Leu-Arg-Pro-Cys-Ser-Val-Asp-Val-Leu-Leu-Ser-Tyr-Cys-OH
Human Insulin            H-Gly-Ile-Val-Glu-Gln-Cys-Cys-Thr-Ser-Ile-Cys-Ser-Leu-Tyr-Glu-Leu-Glu-Asn-Tyr-Cys-Asn-OH
Porcine relaxin    H-Arg-Met-Thr-Leu-Ser-Glu-Lys-Cys-Cys-Glu-Val-Gly-Cys-Ile-Arg-Lys-Asp-Ile-Ala-Arg-Leu-Cys-OH
```

B chains

```
                   1           5               10              15              20
4K-PTTH-II         pGlu-Gln-Pro-Gln-Ala-Val-His-Thr-Tyr-Cys-Gly-Arg-His-Leu-Ala-Arg-Thr-Leu-Ala-Asp-Leu-Cys-
Human Insulin                  H-Phe-Val-Asn-Gln-His-Leu-Cys-Gly-Ser-His-Leu-Val-Glu-Ala-Leu-Tyr-Leu-Val-Cys-
Porcine relaxin    pGlu-Ser-Thr-Asn-Asp-Phe-Ile-Lys-Ala-Cys-Gly-Arg-Glu-Leu-Val-Arg-Leu-Trp-Val-Glu-Ile-Cys-

                   23  25
4K-PTTH-II         Trp-Glu-Ala-Gly-Val-Asp-OH
Human Insulin      Gly-Glu-Arg-Gly-Phe-Phe-Tyr-Thr-Pro-Lys-Thr-OH
Porcine relaxin    Gly-Val-Trp-Ser-OH
```

Identical residues of PTTH and insulin underlined, ___; and with porcine relaxin, ---.

Fig. 7. Amino acid sequences of 4K–PTTH–II from *Bombyx mori* compared with human insulin and porcine relaxin. (From Nagasawa *et al.* 1986.)

from the vertebrate amino acid sequence would mean that the vertebrate oligonucleotide probe would not complement the particular insect mRNA.

A final point concerns the application of this work as a tool in the production of novel insecticides. Knowledge of the neurochemistry of insects could be the means to discovering new ways of killing insects. If, as the present authors believe, the vertebrate-type peptides are important regulatory agents in metabolism, then it ought to be possible to devise methods of breaking particular links in the synthesis of these key factors. Research directed along these lines could provide Mankind with much needed novel insecticides in the not too distant future.

Acknowledgements

The original studies cited in this chapter were supported by the Science & Engineering Research Council of Great Britain (Grant No. GR/D.0338.3) and in part by a NATO Collaborative Research Grant (No. 265/83/D1).

Note added in proof

Since the preparation of this Chapter, Nagasawa *et al.* (1986) have reported the complete amino acid sequence of the 4K–PTTH–II molecule of *Bombyx mori.* It has now been shown that contrary to the earlier predictions of the group, the molecule does indeed consist of two non-identical peptide chains (A and B chains). The A chain has 20 amino acid residues and the B chain is a mixture of four microheterogenous peptides, two of which consist of 28 residues and the other two of 26 residues. The sequences of the A and B chains are shown in Figure 7 where they are compared with human insulin and porcine relaxin. The molecule shows considerable sequence homology with human insulin (eighteen sequence identities) and a lesser degree of homology to porcine relaxin (11 sequence identities). More important, however, is the fact that the majority of the hydrophobic core residues, including A2 Ile, A6 Cys, A11 Cys, A16 Leu, A20 Cys, B11 Leu, B15 Leu and B19 Cys are identical in PTTH–II and insulin from any source.

There seems little doubt that PTHH–II is a member of the insulin family of peptides and that it represents an early derivative of the ancestral form of the molecule that has subsequently been strongly conserved during evolution.

Nagasawa, H., Kataoka, H., Isogai, A., Tamura, S., Susuki, A., Mizoguchi, A., Fujiwara, Y., Suzuki, A., Takahashi, S.Y. & Ishizaki, H. (1986) Amino acid sequence of a prothoracicotropic hormone of the silkworm, *Bombyx mori. Proc. Natl. Acad. Sci. USA*, **83**, 5840–5843..

References

Cutfield, J.F., Cutfield, A.M., Dodson, E.J., Dodson, G., Edmin, S.F. & Reynolds, C.D. (1979) Structure and biological activity of hagfish insulin. *J. Mol. Biol.* **132**, 85–100.

Dockray, G.J., Duve, H. & Thorpe, A. (1981) Immunochemical characterization of gastrin/cholecystokinin-like peptides in the brain of the blowfly *Calliphora vomitoria. Gen. & Comp.Endocrinol.*, **45**, 491–496.

Duve, H. & Thorpe, A. (1979) Immunofluorescent localization of insulin-like material in the median neurosecretory cells of the blowfly, *Calliphora vomitoria* (Diptera). *Cell & Tiss. Res.* **200**, 187–191.

Duve, H. & Thorpe, A. (1981) Gastrin/cholecystokinin (CCK)-like immunoreactive neurones in the brain of the blowfly, *Calliphora erythrocephala* (Diptera). *Gen. Comp. Endocrinol.*,**43**, 381–391.

Duve, H. & Thorpe, A. (1983) Immunocytochemical identification of α–endorphin-like material in neurones of the brain and corpus cardiacum of the blowfly, *Calliphora vomitoria* (Diptera). *Cell & Tiss. Res.* **233**, 415–426.

Duve, H. & Thorpe, A. (1984a) Comparative aspects of insect–vertebrate neurohormones. In: *Insect Neurochemistry and Neurophysiology*, eds. A.A. Borkovec & T.J. Kelly, pp. 171–196. New York & London: Plenum Press.

Duve, H. & Thorpe, A. (1984b) Immunocytochemical mapping of gastrin/CCK-like peptides in the neuroendocrine system of the blowfly, *Calliphora vomitoria* (Diptera). *Cell & Tiss. Res.* **237**, 309–320.

Duve, H., Thorpe, A. & Lazarus, N.R. (1979) Isolation of material displaying insulin-like immunological and biological activity from the brain of the blowfly *Calliphora vomitoria. Biochem. J.* **184**, 221–227.

Duve, H., Thorpe, A., Lazarus, N.R. & Lowry, P.J. (1982) A neuropeptide of the blowfly *Calliphora vomitoria* with an amino acid composition homologous with vertebrate pancreatic polypeptide. *Biochem. J.* **201**, 429–432.

Duve, H., Thorpe, A., Neville, R. & Lazarus, N.R. (1981) Isolation and partial characterization of pancreatic polypeptide-like material in the brain of the blowfly *Calliphora vomitoria. Biochem. J.* **197**, 767–770.

Duve, H., Thorpe, A. & Rehfeld, J.F. (1985) Localisation and characterisation of cholecystokinin (CCK)-like peptides in the brain of the blowfly, *Calliphora vomitoria*. In: *Neurosecretion and the Biology of Neuropeptides*, eds. H. Kobayashi, H.A. Bern & A. Urano. pp. 401–409. Tokyo: Japan Scientific Societies Press, Berlin: Springer–Verlag.

Duve, H., Thorpe, A. & Scott, A.G. (1986) Localisation and characterisation of opioid-like peptides in the nervous system of the blowfly, *Calliphora vomitoria*. In: *CRC Handbook of Comparative Aspects of Opioid and Related Neuropeptide Mechanisms*, Vol. 1, ed. G.B. Stefano, pp. 197–211. CRC Press, Inc. Boca Raton, Florida.

Hökfelt, T., Johansson, O., Ljungdahl, Å., Lundberg, J.M. & Schultzberg, M. (1980) Peptidergic neurones. *Nature* **284**, 515–521.

Ishizaki, H., Nagasawa, H. & Suzuki, A. (1985) Prothoracicotropic hormone of the silkmoth *Bombyx mori*: Amino acid sequence homology with insulin. *Diabetes Res. & Clin. Pract. Suppl.***1**, S262.

Kramer, K.J. (1979) Insulin-like and glucagon-like hormones in insects. In: *Experimental Entomology: Insect Neurohormones*, ed. T.A. Miller, pp. 116–136. Berlin, Heidelberg, New York: Springer–Verlag.

Kramer, K.J.. (1985) Vertebrate hormones in insects. In: *Comprehensive Insect Physiology, Biochemistry and Pharmacology*, Vol. 7, Chapter 15, pp. 511–536. Pergamon Press, New York.

Kramer, K.J., Tager, H.S. & Childs, C.N. (1980) Insulin-like and glucagon-like peptides in insect hemolymph. *Insect Biochem.* **10**, 179–182.

Larsson, L.–I. (1980) Problems and pitfalls in immunocytochemistry of gut peptides. In: *Gastrointestinal Hormones*, ed. G.B. Jerzy Glass, pp. 53–70. New York: Raven Press.

Larsson, L.–I. (1983) Methods for immunocytochemistry of neurohormonal peptides. In: *Handbook of Chemical Neuroanatomy*, Vol. 1, Chapter IV, eds. A. Björklund & T. Hökfelt, pp. 147–209. Elsevier Science Publishers B.V.

Larsson, L.–I. & Rehfeld, J.F. (1977) Characterization of antral gastrin cells with region-specific antisera. *J. Histochem. & Cytochem.* **25**, 1317–1321.

LeRoith, D., Lesniak, M.A. & Roth, J. (1981) Insulin in insects and annelids. *Diabetes* **30**, 70–76.

Maier, V., Mezger, S., Herlitzius, H., Fuchs, J. & Pfeiffer, E.F. (1985) Insulin and experimental diabetes in the honeybee (*Apis mellifera*). *Diabetes Res. & Clin. Pract. Suppl.* **I**, S353.

Maier, V., Witznick, G., Keller, R. & Pfeiffer, E.F. (1978) Insulin-like and glucagon-like immunoreactivities in the honeybee (*Apis mellifera*). *Acta Endocrinologica* **87**, Suppl. 215, 69–70.

Marley, P.D., Rehfeld, J.F. & Emson, P.C. (1984) Distribution and chromatographic characterisation of gastrin and cholecystokinin in the rat nervous system. *J. Neurochem.* **42**, 1523–1535.

Motta, W. (1980) *The endocrine functions of the brain.* New York: Raven Press.

Nagasawa, H., Kataoka, H., Isogai, A., Tamura, S., Suzuki, A., Ishizaki, H., Mizoguchi, A., Fujiwara, Y. & Suzuki, A. (1984) Amino-terminal amino acid sequence of the silkworm prothoracicotropic hormone: homology with insulin. *Science* **226**, 1344–1345.

Normann, T.C. & Duve, H. (1969) Experimentally induced release of neurohormone influencing haemolymph trehalose level in *Calliphora erythrocephala* (Diptera). *Gen. & Comp. Endocrinol.* **12**, 449–459.

O'Connor, K. & Baxter, D. (1985) The demonstration of insulin-like material in the honeybee, *Apis Mellifera. Comp. Biochem. Physiol.* **3**, 755–760.

Rehfeld, J.F. (1978) Immunochemical studies on cholecystokinin: II Distribution and molecular heterogeneity in the central nervous system and small intestine in man and hog. *J. Biol. Chem.* **253**, 4022–4030.

Rehfeld, J.F., Hansen, H.F., Larsson, L.–I., Stengaard–Pedersen, K. & Thorn, N.A. (1984) Gastrin and cholecystokinin in pituitary neurons. *Proc. Nat. Acad. Sci., USA*, **81**, 1902–1905.

Scarborough, R.M., Jamieson, G.C., Kalish, F., Kramer, S.J., McEnroe, G.A., Miller, C.A. & Schooley, D.A. (1984) Isolation and primary structure of two peptides with cardioacceleratory and hyperglycemic activity from the corpora cardiaca of *Periplaneta americana. Proc. Nat. Acad. Sci., USA* **81**, 5575–5579.

Simantov, R., Goodman, R., Aposhian, D. & Snyder, S.H. (1979) Phylogenetic distribution of a morphine-like peptide 'enkephalin'. *Brain Res.* **111**, 204.

Stefano, G.B. & Scharrer, B. (1981) High affinity binding of an enkephalin analog in the cerebral ganglion of the insect *Leucophaea maderae* (Blattaria). *Brain Res.* **225**, 107–114.

Steele, J.E. (1961) Occurrence of a hyperglycemic factor in the corpus cardiacum of an insect. *Nature* **192**, 680–681.

Steele, J.E. (1963) The site of action of insect hyperglycemic hormone.*Gen. & Comp. Endocrinol.* **3**, 46–52.

Tager, H.S. & Kramer, K.J. (1980) Insect glucagon-like peptides: evidence for high molecular weight form in midgut from *Manduca sexta* (L.). *Insect Biochem.* **10**, 617–619.

Tager, H.S., Markese, J., Kramer, K.J., Spiers, R.D. & Childs, C.N. (1976) Glucagon-like and insulin-like hormones of the insect neurosecretory system. *Biochem. J.* **156**, 515–520.

Tager, H.S., Markese, J., Spiers, R.D. & Kramer, K.J. (1975) Glucagon-like immunoreactivity in insect corpus cardiacum. *Nature* (London) **254**, 707–708.

Thorpe, A. & Duve, H. (1984) Insulin- and glucagon-like peptides in insects and molluscs. *Mol. Physiol.* **5**, 235–260.

Ziegler, R., Eckart, K., Schwarz, H. & Keller, R. .(1985).. Amino acid sequence of *Manduca sexta* adipokinetic hormone elucidated by combined fast atom bombardment (FAB)/tandem mass spectrometry. *Biochem. Biophys. Res. Comm.*, **133**, 337–342.

C. H. WHEELER, G. GÄDE &
G. J. GOLDSWORTHY

Humoral functions of insect neuropeptides

Introduction: Hormones and development

Since Kopec's demonstration of a humoral role of the brain in insect development, peptide hormones have had a central historic importance in the study of insect endocrinology: the brain produces prothoracicotropic hormones (PTTH's; see Bollenbacher & Granger 1985), thus stimulating moulting by initiating ecdysone synthesis and release by the prothoracic glands. A second neuropeptide, eclosion hormone (see Reynolds & Truman 1983), initiates the necessary behaviour patterns associated with ecdysis and its timing. Subsequently, a third neuropeptide, bursicon (see Reynolds 1983; 1985), controls the tanning of the new cuticle and stimulates endocuticle deposition. Neuropeptides are also implicated in the control of corpus allatum activity; both allatohibins and allatotropins being identified in various insects (see Tobe & Stay 1985). Unfortunately, chemical characterisation of these peptides has progressed slowly. The physiological actions of most of these peptides concerned with development have been the subject of numerous and extensive reviews (see for example Downer & Laufer 1983; Kerkut & Gilbert 1985) and will not be dealt with here.

Recently considerable progress has been made in the characterisation of PTTH's from *Bombyx mori* (Nagasawa *et al.* 1984, 1986). Characterisation of these polypeptides has been hampered by their heterogeneity (in several species high and low molecular weight forms exist) but also, no doubt, due to the impurity of the starting material used. The sequence of 4K–PTTH–II one of the small *Bombyx* hormones, has now been determined (Nagasawa *et al.* 1986). Interestingly, these molecules are isolated from adults in which they have as yet no known function, and the 4K–PTTH's are inactive in *Bombyx* pupae but induce ecdysis in dauer *Samia* pupae! This is however the only clear-cut example to date where a physiological function in an insect has been shown to be under the control of peptides with close similarities to vertebrate peptides, namely insulin and porcine relaxin with which they show remarkable homologies. Like the vertebrate hormones, the 4K–PTTH–II's have A and B chains linked by disulphide bridges but the molecules also show micro-heterogeneity at the N–termini (Nagasawa *et al.* 1986).

In the limited space available for this review we will describe what is known of the actions of some other insect peptides which have been studied and characterised on the basis of well-defined biological activities in insects, although we shall see that for some of these peptides their actions in various bioassays are not clearly established as physiologically significant roles within the insect.

Peptides and ion and water balance

Diuretic peptides. Insect diuretic hormones (Phillips 1983; Mordue & Morgan 1985) increase the rate of fluid secretion by the Malpighian tubules to void excess water (after feeding on blood, or vegetation containing a high water-content or during flight; see Maddrell 1986). By a potassium-driven 'filtration' process, the distal ends of the tubules produce a primary urine, and selective reabsorption proximally in the tubules and in the rectum determines the final composition of the excreta. In locusts, there are at least two distinct diuretic peptides which stimulate primary urine production and reduce reabsorption in the rectum. These peptides are present in varying amounts in the brain and storage parts of the corpora cardiaca, and in the ventral nerve cord ganglia.

Proux and his colleagues (1982) have isolated (mainly from sub-oesophageal and thoracic ganglia) a peptide which is immunochemically related to vertebrate vasopressin, but the diuretic peptide studied by Mordue and his co-workers (see Mordue & Morgan 1985; Morgan & Mordue 1985), mainly from the storage lobe of the corpora cardiaca, is chemically distinct from the vasopressin-like peptide. The instability of these peptides has hindered full characterisation and easy purification. Why should diuretic peptides be distributed widely in the nervous system, and why more than one? It remains to be shown whether they are released from all the sites at which they are found. Perhaps at some locations they may have neuro-transmitter/neuromodulatory functions. Although diuretic peptides have been identified in several other insects, there has been little progress in the characterisation of these molecules. For a discussion of insect diuretic peptides see Phillips (1983) and Mordue & Morgan (1985).

Chloride transport stimulating hormone. The reabsorption of chloride has been studied *in vitro* using voltage-clamped preparations of locust recta. In *Schistocerca gregaria* (see Phillips 1983) and *Romalea microptera* (Spring 1986) the reabsorption of chloride from the rectum is stimulated by a neuropeptide hormone released from the corpora cardiaca; its action is probably mediated via cAMP. Preliminary studies suggest that the active peptide has a molecular weight of c. 8000 and carries a net negative charge at neutral or alkaline pH; it is chemically distinct from AKH–I, diuretic and anti-diuretic hormones (see Phillips *et al.* 1980; Mordue & Morgan 1985).

The locust rectum is also important in acid-base balance and in nitrogenous excretion (NH_4^+, originating from the oxidation of amino acids). Apparently the rectum derives its metabolic energy largely from the oxidation of proline. The chloride transport stimulating hormone is probably involved in the control of all these processes (Phillips *et al.* 1986).

Hormones and energy metabolism

Flight is the most metabolically-demanding activity that insects undertake. Considering the diversity of insects, it is not surprising that a variety of strategies have evolved to satisfy and control this demand for energy: some insects use both carbohydrates and lipids, with one fuel predominating during different phases of flight, others use essentially only lipid or only carbohydrate throughout flight, while yet others use amino acids. Although too few insect species have been studied in detail, it is likely that peptide hormones are involved in maintaining adequate supplies of fuels to the muscles in most flying insects (see Fig. 1). Clearly, however, not all adult insects (and no pre-imaginal stages) fly, but even in non-fliers the supply of energy to the tissues may be restricted during non-feeding periods or as a result of seasonal food deprivation. At such times, hormones may regulate the availability and utilisation of stored fuels; any shortfall between the demand for energy substrates by the tissues and their availability in the haemolymph must be made good by mobilising reserves.

Peptides affecting carbohydrate metabolism

Neurohormone D. Neurohormone D from the brain/corpus cardiacum complex of *Periplaneta* was the first of several cockroach peptides discovered that increase the frequency and amplitude of heartbeat in semi-isolated preparations (Gersch 1983). Other properties are claimed for this peptide, such as the stimulation of colour change and increased dye excretion by Malpighian tubules in the stick insect, *Carausius morosus*, as well as increased electrical activity of the phallus nerves of the cockroach (Gersch 1983). However, it is uncertain if these effects are all due to a single peptide. The structure of Neurohormone D is identical to that of the myoactive factor I (O'Shea *et al.* 1984) of the cockroach corpus cardiacum (see Goldsworthy *et al.* 1986a).

Hypertrehalosaemic peptides. Aqueous extracts of corpora cardiaca from *Periplaneta americana* possess potent hypertrehalosaemic activity: concentrations of haemolymph trehalose (rather than glucose) increase in response to such extracts (see Goldsworthy and Gäde 1983). The presence of 'hypertrehalosaemic' factors is inferred in a wide range of insect species (see Steele 1983; Goldsworthy & Gäde 1983) in which glycogenolysis, phosphorylase activation, or hypertrehalosaemia occur in response to injections of extracts of their own corpora cardiaca, or in which extracts of their corpora cardiaca are hypertrehalosaemic in cockroaches (see Goldsworthy & Gäde 1983; Gäde 1984; Gäde & Scheid 1986). However, with the exception of locust AKH–I and AKH–II (the locust 'hypertrehalosaemic' or, more strictly, phosphorylase-activating peptides) and two further examples discussed below, clear-cut roles or strong evidence of release under physiological conditions, have not generally been described.

In locusts, adipokinetic hormones released during flight stimulate glycogen breakdown (Van Marrewijk *et al.* 1980; Gäde 1981), although this is probably secondary to lipid mobilisation (see Goldsworthy & Gäde 1983; Goldsworthy & Wheeler 1984). Van Marrewijk *et al.* (1986) have shown recently that phosphorylase activation occurs within only 5 min of the start of flight but this is prevented by

Fig. 1. A schematic diagram to show different forms of lipid mobilisation in locusts (upper part) and *Leptinotarsa* (lower part) during flight. In locusts, AKH, which is thought to act via adenylate cyclase, stimulates the conversion of stored triacylglycerols (TGL) to diacylglycerols (DGL). These are loaded onto the existing haemolymph lipoprotein Ayellow (Ay) which binds non-lipid carrying C_L–proteins to become lipoprotein A^+, which transports lipids to the flight muscles. At the flight muscles lipids are hydrolysed by the membrane-bound lipoprotein lipase (LPL) and the free fatty acids released provide the energy for long-term flight. In *Leptinotarsa*, the action of an 'adipokinetic hormone' stimulates the conversion of triacylglycerols to acetyl CoA which together with alanine is used to synthesize proline. Proline is released into the haemolymph and in the flight muscles is broken down to yield C2 units (alanine), which return to the fat body, and Acetyl CoA which is used for energy production.

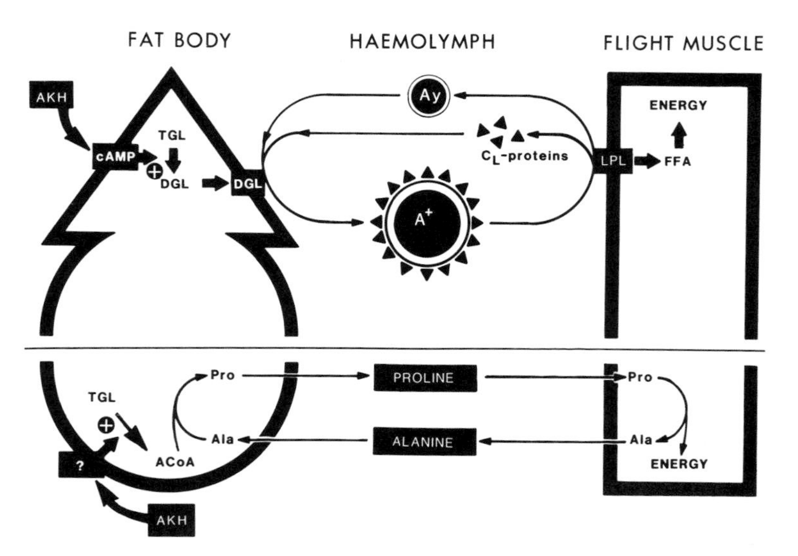

cardiacectomy or injection of trehalose (which prevents release of adipokinetic hormones), providing further evidence for the rapid release of AKH during flight (see Goldsworthy & Wheeler 1984).

In *Calliphora erythrocephala*, a hormonal role is also proposed for a hypertrehalosaemic factor thought to maintain trehalose homeostasis during flight (Vejberg & Normann 1974). After cardiacectomy or denervation of the corpora cardiaca, sustained flight activity is reduced, apparently because haemolymph trehalose levels cannot be maintained. However, injections of extracts of corpora cardiaca do not induce hypertrehalosaemia in flight-exhausted flies, and release of the factor into the haemolymph has not been demonstrated directly, but physical disruption of the denervated glands *in situ* in such exhausted flies allows the resumption of flight (Vejberg & Normann 1974). Despite recent interest in the endocrinology of blowflies (see Chapter 6), these uncertainties in our understanding of the maintenance of haemolymph carbohydrate homeostasis in blowflies have not been resolved.

There is strong evidence for a physiological role for a 'hypertrehalosaemic' peptide in larvae of *Manduca sexta*. During the moult from the IVth to Vth stadium, larvae do not feed for c. 25h. During this period of 'fasting', the activity of glycogen phosphorylase in the fat body increases dramatically; an effect which can be induced experimentally in mid-stadium by starvation. Activation of the enzyme is prevented by cardiacectomy, but is stimulated by injections of extracts of these glands (Siegert & Ziegler 1983). Because haemolymph trehalose levels do not alter appreciably after the activation of phosphorylase (see also Goldsworthy & Gäde 1983), the active peptide is termed a 'glycogen phosphorylase activating hormone'. There is little doubt that it is important in providing haemolymph carbohydrate for use during non-feeding periods, but release of the larval peptide into the haemolymph has not yet been demonstrated directly.

What do we know about the chemistry of the hypertrehalosaemic peptides? Despite many attempts of various groups to isolate and characterise these peptides, especially from the American cockroach, their chemical identities proved elusive (see Goldsworthy & Gäde 1983) until recently when the myoactive factors, M–I and M–II, from the cockroach corpora cardiaca were isolated (O'Shea *et al.* 1984) and sequenced using Fast Atom Bombardment Mass Spectrometry (FABMS; Witten *et al.* 1984). They are both octapeptides belonging to the AKH/RPCH family (see Table 1) and the hypertrehalosaemic factors isolated from the American cockroach proved to be identical (Gäde 1985a).

Another peptide, the hypertrehalosaemic factor II, which increases trehalose levels in the blood of ligated stick insects, *Carausius morosus* (Gäde & Lohr 1982), is a novel decapeptide also belonging to the AKH/RPCH family (see Table 1; Gäde & Rinehart 1986). Its physiological significance in *Carausius* has yet to be determined. Another hypertrehalosaemic peptide was isolated from the corpora cardiaca of the

Table 1. *The AKH/RPCH family of peptides*

Amino acid sequence	ED_{50} pmol	EDmax pmol	Maximum response %	Name of peptide
pGlu-Leu-Asn-Phe-Thr-Pro-Asn-Trp-Gly-Thr-NH_2	1	3	100	Locust AKH-I
pGlu-Leu-Thr-Phe-Thr-Pro-Asn-Trp-Gly-Thr-NH_2	2	8	100	***Carausius***-2
pGlu-Val-Asn-Phe-Ser-Pro-Gly-Trp-Gly-Thr-NH_2	?	?	?	***Nauphoeta & Blaberus*** HTH
pGlu-Leu-Thr-Phe-Thr-Ser-Ser-Trp-Gly-NH_2	10	40	45	***Manduca & Heliothis*** AKH
pGlu-Leu-Asn-Phe-Ser-Pro-Gly-Trp-NH_2	4.6	20	100	Crustacean RPCH
pGlu-Leu-Asn-Phe-Ser-Ala-Gly-Trp-NH_2	2	3	50–60	***Locusta*** AKH-II
pGlu-Leu-Asn-Phe-Ser-Thr-Gly-Trp-NH_2	12* 2	25* 5	50–60* 90	***Schistocerca*** AKH-II
pGlu-Leu-Thr-Phe-Thr-Pro-Asn-Trp-NH_2	5	30	~90	***Periplaneta*** M-II
pGlu-Val-Asn-Phe-Ser-Pro-Asn-Trp-NH_2	5	20	~90	***Periplaneta*** M-I Neurohormone D

*Assayed in ***Schistocerca***

cockroach *Nauphoeta cinerea* and the structure as a decapeptide elucidated (see Table 1; Gäde & Rinehart 1986). The same compound has been sequenced as a hypertrehalosaemic hormone from the cockroach *Blaberus discoidalis* (Hayes & Keeley 1986).

The mode of action of the hypertrehalosaemic peptides

Most of our information on the mechanism of action of hypertrehalosaemic peptides comes from studies using the American cockroach. Early investigations using crude extracts of whole corpora cardiaca, or partially purified material, suggested that the hypertrehalosaemic response is mediated by cAMP (see Steele 1983; Goldsworthy & Gäde 1983). Calcium ions have, however, also been proposed as a second messenger (McClure & Steele 1981): activation of phosphorylase by a highly purified hypertrehalosaemic factor was dependent upon extracellular Ca^{2+} and could be mimicked by a calcium ionophore. Membrane permeability to Ca^{2+} may be affected by an increase in intracellular cAMP concentration induced by the peptide. Recent studies using synthetic hypertrehalosaemic peptides demonstrated a slight increase in cAMP concentration in the fat body (Gäde 1985b), although Orr *et al.* (1985) found that synthetic M–I and M–II did not have such an effect. The exact mechanism of action of these peptides therefore remains problematic.

There is circumstantial evidence to suggest that hypertrehalosaemic hormones act at sites in addition to fat body phosphorylase in order to effect trehalose synthesis. In response to crude extracts of corpora cardiaca, fat bodies of *Leucophaea maderae* exhibit increased citric acid cycle activity, and oxidation of palmitate and acetate (Wiens & Gilbert 1965), and in *Periplaneta americana,* glucose–6–phosphate, fructose–6–phosphate, and citrate increase in concentration (Steele *et al.* 1984). Glycolysis in locust flight muscles appears to be regulated by an effect of citrate on aldolase (Ford & Candy 1972; Storey 1980) and it is possible that factors in the corpora cardiaca stimulate fatty acid oxidation in the fat body (to provide the energy for trehalose synthesis) thus causing an increase in fat body citrate which could block glycolysis. Glucose–1–phosphate formed by the action of phosphorylase on glycogen would therefore be directed into trehalose synthesis. Citrate does, indeed, accumulate in the fat body of *Periplaneta* when synthetic hypertrehalosaemic hormones (M–I and M–II; 5pmol) are injected *in vivo* (Gäde 1986b).

Peptides affecting lipid metabolism

Locust corpora cardiaca contain 2 peptides which are released during flight and effect lipid mobilisation from the fat body. The actions of these adipokinetic hormones (AKH) on fat body and flight muscle metabolism in locusts are well-documented and have been reviewed recently (Goldsworthy 1983; Goldsworthy & Wheeler 1984; Wheeler & Goldsworthy 1985a). We will deal here with these topics only briefly, and concentrate on recent developments in locusts and other insects.

Adipokinetic hormones in locusts. The 'classical' action of AKH in locusts is to increase the concentration of diacylglycerols in the haemolymph; both AKH–I and

Fig. 2. Typical dose-response curves for members of the AKH/RPCH family assayed for hyperlipaemic activity in *Locusta* (Goldsworthy *et al.* 1986a). Note the different scales for the abscissae.

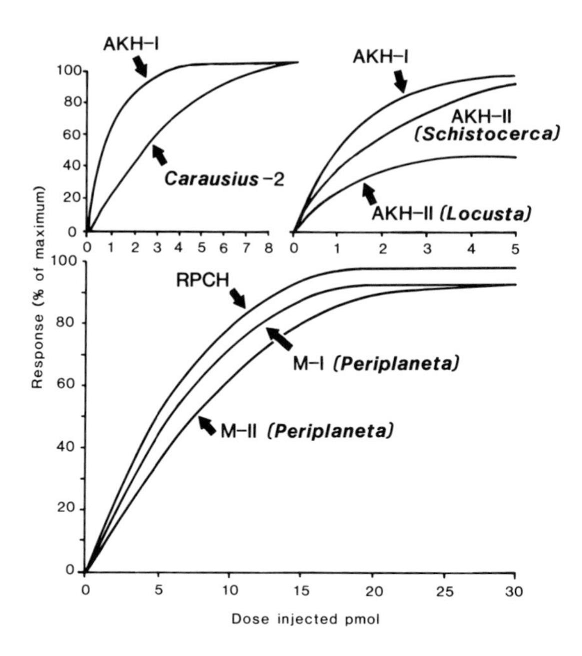

AKH–II elicit this response although AKH–I is more effective (see Fig. 2). There are 3 adipokinetic peptides known in locusts (Table 1); the decapeptide AKH–I the major peptide in all three locust species investigated (*Locusta migratoria, Schistocerca gregaria* and *S. nitans*) and the octapeptides LAKH–II from *Locusta* and SAKH–II from *Schistocerca* species. LAKH–II and SAKH–II differ by only a single amino acid residue (Siegert *et al.* 1985; Gäde *et al.* 1984, 1986). Recently, Hekimi and O'Shea (1985) have identified precursors for locust adipokinetic hormones, and these proteins are apparently being sequenced. The action of AKH at the fat body is probably mediated by cAMP (see Goldsworthy 1983); again, AKH–I and LAKH–II are not equipotent but, interestingly, in the cAMP bioassay LAKH–II is more effective than AKH–I (Goldsworthy *et al.* 1986b). The actions of AKH at the flight muscle (see Goldsworthy 1983) do not appear to be mediated by cyclic nucleotides.

The actions of AKH in locusts mostly concern metabolic adaptations for long-term (migratory) flight, and it is unfortunate that most studies have involved relatively short periods of tethered flight in the laboratory, or injection of AKH into resting locusts. Nevertheless, AKH's are released during tethered flight (see Goldsworthy & Wheeler 1984) but the suggestion by Orchard & Lange (1983) of separate roles for the two peptides, in mobilising lipids (AKH–I) and in maintaining haemolymph trehalose levels (AKH–II), is not supported by recent evidence; for both lipid mobilisation and phosphorylase activation, AKH–I is markedly more effective than AKH–II (Goldsworthy & Wheeler 1984; Goldsworthy *et al.* 1986a,b). One area which has received little attention is the effect of AKH on behaviour; AKH influences flight speed and 'willingness' to fly (see Goldsworthy 1983), but while endocrine influences on flight-associated behaviour have been studied in other migratory insects, such studies have concentrated on the duration of tethered flight and the influence of Juvenile Hormone, rather than on neuropeptides (see Pener 1985).

Adipokinetic hormones in other insects

Many insects other than locusts possess peptides which induce hyperlipaemia in locusts: although these effects are pharmacological, they can be useful tools in isolating and characterising peptides and their receptors (see structure/activity relationships below). Nevertheless, some of these insects possess adipokinetic peptides which are active both contra- and intra-specifically. For example an AKH-like peptide from the cricket *Gryllus bimaculatus* raises haemolymph lipid levels when injected into *Gryllus* or into *Locusta* (Gäde & Scheid 1986).

During flight in *Danaus plexipus*, a migratory butterfly, the concentration of haemolymph carbohydrate decreases while that of lipid increases (Dallmann & Hermann 1978), and injections of extracts of heads, of corpora cardiaca, or of thoracic nerves, cause lipid mobilisation in resting butterflies. The active factor from the corpora cardiaca appears to be a peptide. Another butterfly, *Vanessa cardui*, possesses an adipokinetic peptide (Herman & Dallman 1981), but as yet its biology and chemistry have not been determined.

Moths also possess adipokinetic peptides, and the most comprehensive studies relate to an adipokinetic peptide released during flight in *Manduca sexta*. In this moth, lipids appear to be the major fuel throughout flight (Ziegler & Schulz 1986), and high levels are present in the haemolymph even during rest. Flight leads initially to a decrease in the levels of haemolymph lipid, with a new steady state equilibrium between utilisation and mobilisation of lipid only being established after 30min of flight. When moths are flown for different times, the changes in haemolymph lipid concentrations measured during a subsequent rest suggest that mobilisation of lipids from the fat body increases during flight; it is assumed that this is caused by the release of the adipokinetic peptide, because cardiacectomised moths fly poorly and do not mobilise lipids during rest, after flight (Ziegler & Schulz 1986). The adipokinetic hormone from adult *Manduca* is a nonapeptide member of the AKH/RPCH family (Ziegler *et al.* 1984, 1985: see Table 1). Interestingly, the same peptide is found in the corpora cardiaca of the corn ear worm moth, *Heliothis* (Jaffe *et al.* 1986).

In the beetle, *Leptinotarsa decemlineata*, an 'adipokinetic hormone' has an important function in flight metabolism; not to stimulate transport of neutral lipids in the haemolymph, but to make available (from triacylyglycerol, presumably by activation of a lipase) acetyl CoA for the synthesis of proline in the fat body (Weeda 1981). In this beetle (Mordue & De Kort 1978), and in the tsetse fly, *Glossina morsitans* (Bursell 1963), proline is the major fuel oxidised by the flight muscle: this has the advantage that it is water-soluble, and thus obviates the need for a well-developed and metabolically-expensive lipoprotein carrier mechanism as found in insects which mobilise diacylglycerols as a fuel for flight (see Wheeler & Goldsworthy 1985a). Alanine acts as a 'carrier' of 2-carbon skeletons (from the acetyl CoA): the proline thus formed is oxidised in the flight muscles, liberating alanine which functions as a re-usable 'metabolic shuttle' by returning to the fat body for further proline synthesis (Fig. 1). In *Leptinotarsa*, injections of extracts of corpora cardiaca from *Leptinotarsa, Locusta,* or *Periplaneta*, or of synthetic AKH–I decrease the haemolymph concentration of alanine, which is indicative of proline synthesis: the same materials increase fat body proline synthesis *in vitro* (Weeda 1981). The activity in the corpora cardiaca of *Leptinotarsa* appears to be due to two peptides and these may be the same as M–I and M–II (Gäde & Scheid 1986). A corpora cardiaca factor which stimulates proline synthesis in *Glossina* fat cells (Pimley & Langley 1982) appears to be a small peptide but it is not well characterised (Pimley 1984). Considering that most of the proline used by tsetse flies during flight is derived from the gut, it is unclear whether this peptide actually functions in increasing proline synthesis *in vivo* but other possible actions such as stimulating synthesis/release of proline from the gut have not apparently been studied.

Indirect effects of adipokinetic hormone

Most insects transport lipids in the form of diacylglycerols which are carried as part of specific lipoproteins (Goldsworthy 1983; Chino 1985; Wheeler & Goldsworthy 1985; Beenakkers *et al.* 1986). In those species which also mobilise extra lipids in response to AKH's, there are specific 'rearrangements' in lipoproteins which occur during flight. Orthoptera (Goldsworthy 1983) and Lepidoptera (Shapiro & Law 1983) commonly possess a single major lipoprotein species which carries lipids for non-flight specific purposes but during flight (or starvation) this, by combination with non-lipid carrying haemolymph apoproteins and lipids from the fat body, transforms to a lower density lipoprotein which specifically carries lipids to the flight muscles (see Goldsworthy 1983). In *Locusta* the binding of apoproteins during low density lipoprotein formation is important in controlling the hydrolysis of lipids by a lipoprotein lipase at the flight muscles (Wheeler & Goldsworthy 1985a,b, 1986; Wheeler *et al.* 1986), and this may be so for other insects. By stimulating the formation of low density lipoproteins, AKH indirectly regulates the supply and uptake of lipids at the flight muscles: only when sufficient lipoprotein is available will the flight muscle switch from carbohydrate- to lipid-based metabolism.

Structure/activity relationships in the AKH/RPCH family

Hormonal activity depends on the binding of hormone molecules by specific membrane-bound receptor proteins in target cells. Thus, studies on the structure/activity relationships of closely related peptides should help to elucidate the special structural features of the peptide molecule requisite for such recognition.

In their pioneering work on the structure/activity relationships of adipokinetic peptides, Stone *et al.* (1978) identified some key requirements for activity in the locust hyperlipaemic assay: at least 8 amino acid residues from the N–terminus; a C–terminal threonine amide; and an N–terminal L–enantiomer of pyroglutamic acid. In addition, the sequence Pro–Asn–Trp (positions 6 to 8) is important for activity of the decapeptide AKH–I, and this is evidence in support of there being a β–bend in the molecule at this position.

Our previous studies (Goldsworthy *et al.* 1986a,b; Gäde 1986) show that the naturally occurring members of the AKH/RPCH family appeared to fall into two categories: peptides (AKH–I, *Carausius*–2, RPCH, M–I, M–II) which, although varying in potency, elicit at sufficiently high doses the maximum possible adipokinetic response in locusts (Fig. 2); and peptides (the AKH–II's and *Manduca* AKH) which appeared to induce a truncated response (50–60% and 40–50% of the maximum respectively), even at very high doses (Fig. 2, Table 1). However, with the availability of synthetic SAKH–II we have re-investigated the effect of this peptide in *Schistocerca* and *Locusta* species. We find that SAKH–II although

Table 2. *Structure/activity relationships for members of the AKH/RPCH family assayed for hypertrehalosaemic activity in* Periplaneta americana *(see Gäde 1986a)*

Name of Peptide	ED_{50}pmol	Approximate EDmax pmol	Maximum response mg/ml %	Maximum response
Carausius–2	0.9	4	35	100
Periplaneta M–I	1.9	5	28	80
Periplaneta M–II	1.8	5	28	80
Crustacean RPCH	4.5	6*	21	60
Locust AKH–I	5.9	8*	21	60
Locusta AKH–II	?	?	16	46
Manduca AKH	No activity	–	–	–

* Denotes EDmax of the truncated response

producing only a truncated response in *Schistocerca* (with doses up to 50pmol), a near maximal adipokinetic response can be elicited, in *Locusta,* by as little as 5pmol (Goldsworthy & Wheeler 1986). Our inability to show such a response previously was probably due to losses of natural hormone during purification and preparation; this is borne out by recent studies in which maximal lipid mobilisation in *Locusta* was obtained in response to injection of 5pmol of natural SAKH–II prepared by a different extraction procedure. Our data suggest that the naturally occurring amino acid substitutions in those members which contain proline (always at position 6, so far), are relatively unimportant in determining the maximum response, and in influencing the overall potency of the octapeptides. The two members eliciting a truncated response, lack proline and this may be important in determining the maximum activity, especially in nona- and decapeptides, but the good response elicited by SAKH–II in *Locusta* (but not in *Schistocerca*) suggests that proline (and a β-bend) may not be of absolute importance in determining the activity of the octapeptides.

In the cockroach hypertrehalosaemic assay, only the two octapeptides, M–I and M–II, and the decapeptide stick insect hypertrehalosaemic factor-2 induce a maximum response (Gäde 1986a). AKH–I and RPCH are markedly less potent and fail to elicit a 'full' response even at very high doses (Table 2). These differences cannot yet be explained in terms of the structures of the peptides (see Gäde 1986a), and an understanding of their significance awaits studies of the receptors. Again, however, one feature may again be the apparent importance of the proline residue; significantly AKH–II elicits an extremely poor hypertrehalosaemic response (Gäde 1986a), whereas the *Manduca*–AKH has no activity at all in this cockroach assay, even at doses up to 75pmol of synthetic peptide (Goldsworthy, unpublished observations).

General conclusions

Until recently studies on insect peptides were usually related directly to their roles in controlling specific biological functions such as ecdysis, tanning, diuresis, and the mobilisation of energy metabolites, but in recent years with the development of new technologies it has become possible, even in vogue, to identify, purify, and sequence peptides long before their physiological roles are understood, or even studied! The ability to isolate and sequence peptides is outpacing efforts to assess their roles as hormones, neurotransmitters or neuromodulators.

The reasons for such a revolutionary change in approach are two-fold: first, developments in analytical techniques such as HPLC (High Performance Liquid Chromatography) and FABMS, which provide powerful tools in studies of peptides, have allowed easier and faster progress in the isolation and characterisation of some biologically active peptides; especially those which are chemically rather stable. Second, the ease with which immunochemical techniques can be used to identify short sequences of amino acids in tissue sections, in body fluids, and in extracts of tissues has enabled an 'explosion' of literature cataloguing such (often indeterminate) sequences in many invertebrates.

Thus, antisera raised to vertebrate peptides have 'identified' (by RIA or most often, by immunocytochemistry; see Chapter 3) vertebrate-like peptides in insects. This immunochemical data should be interpreted with care, because even monoclonal antibodies recognise only short amino acid sequences, and these may occur in completely unrelated and functionally different molecules. Even when such immunochemically 'defined' materials are isolated, it is problematic to ascribe precise biological functions to the peptides. Nevertheless, such studies backed by rigorous chemical purification and characterisation have demonstrated the presence in insects of peptides with striking similarities to vertebrate peptides (see Chapter 6). The question remains, however, what do such molecules do within the insect?

In those instances where the control of physiological processes has been investigated and regulatory peptides subsequently isolated and sequenced, the peptides usually proved to be totally dissimilar to any known vertebrate regulatory peptide (they may, however, have common antigenic determinants with unknown vertebrate peptides: see Sasek *et al.* 1985). Certain peptides have been conserved within both vertebrate and invertebrate tissues during evolution but the presence of 'vertebrate-like' peptides in insects and their localisation within single cells, or small groups of cells, may suggest that they function mainly as neurotransmitters/neuromodulators. On the other hand, insect regulatory peptides acting at sites peripheral to the nervous system appear mostly to be chemically unrelated to humorally active regulatory peptides of vertebrates, the major exception being PTTH–II. If this is true, then the determination of the biological functions of vertebrate-like peptides in insects will in the main be the preserve of the neurobiologist (a formidable task considering the relatively slow progress in understanding the actions of 'conventional' neurotransmitters), while investigation of the humorally-active 'insect' peptides will remain a challenge to the experimental endocrinologist.

Acknowledgements

Original research described in this paper was supported by grants from The Royal Society and SERC to GJG, and from the Deutsche Forschungsgemeinschaft to GG. GG was supported by a Heisenberg Fellowship from the Deutsche Forschungsgemeinschaft.

References

Beenakkers, A.M.Th., Van Der Horst, D.J. & Van Marrewijk, W.J.A. (1986) Insect lipids and lipoproteins and their role in physiological processes. *Prog. Lipid Res.* **24**, 19–67.

Bollenbacher, W.E. & Granger, N.A. (1985) Endocrinology of prothoracicotropic hormones. In: Comprehensive insect physiology, biochemistry and pharmacology (eds. G.A. Kerkut & L.I. Gilbert). pp.109–183. Pergamon Press, Oxford.

Bursell, E. (1963) Aspects of the metabolism of amino acids in the Tsetse fly, *Glossina* (Diptera). *J. Insect. Physiol.* **9**, 439–452.

Chino, H. (1985) Lipid transport: Biochemistry of haemolymph lipophorin. In: Comprehensive insect physiology, biochemistry and pharmacology (eds. G.A. Kerkut & L.I. Gilbert), pp. 115–135. Pergamon Press, Oxford.

Dallmann, S.H. & Herman, W.S. (1978) Hormonal regulation of haemolymph lipid concentration in the Monarch Butterfly, *Danaus plexippus. Gen. Comp. Endocr.* **36**, 142–150.

Downer, R.G.H. & Laufer, H. (1983) *Endocrinology of insects,* **1**. Alan Liss, New York.

Ford, W.C.L. & Candy, D.J. (1972) The regulation of glycolysis in perfused locust flight muscle. *Biochem. J.* **130**, 1101–1112.

Gäde, G. (1981) Activation of fat body glycogen phosphorylase in *Locusta migratoria* by corpus cardiacum extract and synthetic adipokinetic hormone. *J. Insect Physiol.* **27**, 155–161.

Gäde, G. (1984) Adipokinetic and hyperglycaemic factors of different insect species: separation with high performance liquid chromatography. *J. Insect Physiol.* **30**, 729–736.

Gäde, G. (1985a) Hypertrehalosaemic hormones and myoactive factors from cockroach corpus cardiacum are very likely identical. *Naturwissenschaften* **72**, 95–96.

Gäde, G. (1985b) Mode of action of the hypertrehalosaemic peptides from the American cockroach. *Z. Naturforsch.* **40C**, 670–676.

Gäde, G. (1986a) Relative hypertrehalosaemic activities of naturally occurring neuropeptides from the AKH/RPCH family. *Z. Naturforsch.***41C**, 315–320.

Gäde, G. (1986b) Studies on the effect of cockroach hypertrehalosaemic hormones on adenylate cyclase, phosphorylase and glycolysis. *Proc. Int. Union Physiol. Sci.*, XXXth Congress, Vancouver. Abstract p269.03.

Gäde, G. & Lohr, P. (1982) Restricted specificity of a hyperglycaemic factor from the corpus cardiacum of the stick insect *Carausius morosus. J. Insect. Physiol.* **28**, 805–811.

Gäde, G. & Rinehart, K.L. Jr. (1986) Amino acid sequence of a hypertrehalosaemic neuropeptide from the corpus cardiacum of the cockroach, *Nauphoeta cinerea. Biochem. Biophys. Res. Comm.* , **141**, 714–781..

Gäde, G. & Rinehart, K.L. Jr. (1987) Primary structure of the hypertrehalaosaemic factor II from the corpus cardiacum of the Indian stick insect, *Carausius morosus*, determined by fast atom bombardment mass spectrometry. *Biol. Chem.* Hoppe–Seyler **368**, 67–75.

Gäde, G. & Scheid, M. (1986) A comparative study on the isolation of adipokinetic and hypertrehalosaemic factors from insect corpora cardiaca. *Physiol. Entomol.* **11**, 145–157.

Gäde, G., Goldsworthy, G.J., Kegel, G. & Keller, R. (1984) Single step purification of locust adipokinetic hormones I and II by reversed-phase high-performance liquid chromatography, and amino acid composition of the hormone II. Hoppe–Seyer's *Z. Physiol. Chem.* **365**, 393–398.

Gäde, G., Goldsworthy, G.J., Schaffer, M.H., Cook, J.C. & Rinehart, K.L. Jr. (1986) Sequence analyses of adipokinetic hormones II from corpora cardiaca of *Schistocerca nitans*, *Schistocerca gregaria*, and *Locusta migratoria* by fast atom bombardment mass spectrometry. *Biochem. Biophys. Res. Comm.* **134**, 723–730.

Gersch, M. (1983) Chemistry of other neurohormonal factors. In: Endocrinology of Insects (eds. R.G.H. Downer & H. Laufer), pp. 121–129. Alan Liss, New York.

Goldsworthy, G.J. (1983) The endocrine control of flight metabolism in locusts. In: Advances in insect endocrinology, **17** (eds. M.J. Berridge, J.E. Treherne & V.B. Wigglesworth), pp. 149–204. Academic Press, New York.

Goldsworthy, G.J. & Gäde, G. (1983) The chemistry of hypertrehalosemic factors. In: Endocrinology of Insects, **1** (eds. R.G.H. Downer & H. Laufer), pp. 109–119. Alan Liss, New York.

Goldsworthy, G.J. & Wheeler, C.H. (1984) Adipokinetic hormones in locusts. In: Biosynthesis, metabolism and mode of action of invertebrate hormones (eds. J.A. Hoffmann & M. Porchet), pp. 126–135. Springer–Verlag, Heidelberg.

Goldsworthy, G.J. & Wheeler, C.H. (1986) Structure/activity relationships in the adipokinetic hormone/red pigment concentrating hormone family. In: Insect neurochemistry and neurophysiology 1986 (eds. A.B. Borkovec & A.B. Gelman), pp.183–186. Humana Press, New Jersey.

Goldsworthy, G.J., Mallison, K., Wheeler, C.H. & Gäde, G. (1986a) Relative adipokinetic activities of members of the AKH/RPCH family. *J. Insect Physiol.* **32**, 433–438.

Goldsworthy, G.J., Mallison, K. & Wheeler, C.H. (1986b) The relative potencies of two known locust adipokinetic hormones. *J. Insect Physiol.* **32**, 95–101.

Hayes, T.K. & Keeley, L.L. (1986) Isolation and structure of the hypertrehalosaemic hormone from *Blaberus discoidalis* cockroaches. In: Insect neurochemistry and neurophysiology 1986 (eds. A.B. Borkovec & D.B. Gelman), Humana Press, New Jersey.

Hekimi, S. & O'Shea, M. (1985) Adipokinetic hormones in locusts: synthesis and developmental regulation. *Soc. Neurosci. Abstr.* **11**, 959.

Herman, W.S. & Dallmann, S.H. (1981) Endocrine biology of the Painted Lady butterfly, *Vanessa cardui*. *J. Insect Physiol.* **18**, 1265–1285.

Jaffe, H., Raina, A.K., Riley, C.T., Fraser, B.A., Holman, G.M., Wagner, R.M., Ridgway, R.L. & Hayes, D.K. (1986) Isolation and primary structure of a peptide from the corpora caridaca of *Heliothis zea* with adipokinetic activity. *Biochem. Biophys. Res. Comm.* **135**, 622–628.

Kerkut, G.A. & Gilbert, L.I. (1985) *Comprehensive insect physiology, biochemistry and pharmacology*, **7** & **8**. Pergamon Press, Oxford.

Maddrell, S.H.P. (1986) Hormonal control of diuresis and metabolism in insects. In: Insect neurochemistry and neurophysiology 1986 (eds. A.B. Borkovec & D.B. Gelman), Humana Press, New Jersey.

McClure, J.B. & Steele, J.E. (1981) The role of extracellular calcium in hormonal activation of glycogen phosphorylase in cockroach fat body. *Insect Biochem.* **11**, 605–613.

Mordue, W. & De Kort, C.A.D. (1978) Energy substrates for flight in the Colorado beetle, *Leptinotarsa decemlineata. Say. J. Insect. Physiol.* **24**, 221–224.

Mordue, W. & Morgan, P.J. (1985) Chemistry of peptide hormones: In: Comprehensive insect physiology, biochemistry and pharmacology **7** (eds. G.A. Kerkut & L.I. Gilbert), pp. 153–183. Pergamon Press, Oxford.

Morgan, P.J. & Mordue, W. (1985) Cyclic AMP and locust diuretic hormone action. Hormone induced changes in cAMP levels offers a novel method for detecting biological activity of uncharacterized peptide. *Insect Biochem.* **15**, 247–257.

Nagasawa, H., Kataoka, H., Isogai, A., Tamura, S., Suzuki, A., Ishizaki, H., Mizoguchi, A., Fujiwara, Y. & Suzuki, A. (1984) Amino-terminal amino acid sequence of the silkworm prothoracicotropic hormone: Homology with insulin. *Science,* **226**, 1344–1345.

Nagasawa, H., Kataoka, H., Isogai, A., Tamura, S., Suzuki, A., Mizoguchi, A., Fujiwara, Y., Suzuki, A., Takahashi. S.Y. & Ishizaki, H. (1986) Amino acid sequence of a prothoracicotropic hormone of the silkworm *Bombyx mori. Proc. Natl. Acad. Sci.* **83**, 5840–5843.

Orchard, I. & Lange, A.B. . (1983) Release of identified adipokinetic hormones during flight and following neural stimulation in *Locusta migratoria. J. Insect Physiol.* **29**, 425–429.

Orr, G.L., Gole, J.W.D., Jahagirdar, A.P., Downer, R.G.H. & Steele, J.E. (1985) Cyclic AMP does not mediate the action of synthetic hypertrehalosaemic peptides from the corpus cardiacum of *Periplaneta americana. Insect Biochem.* **15**, 703–709.

O'Shea, M., Witten, J. & Schaffer, M. (1984) Isolation and characterization of two myoactive neuropeptides: further evidence of an invertebrate peptide family. *J. Neurosci.* **4**, 521–529.

Pener, M.P. (1985) Hormonal effects on flight and migration. In: Comprehensive insect physiology, biochemistry and pharmacology, **8** (eds. G.A. Kerkut & L.I. Gilbert), pp. 491–550. Pergamon Press, Oxford.

Phillips, J.E. (1983) Endocrine control of salt and water balance: Excretion. In: Endocrinology of insects, **1** (eds. R.G.H.Downer & H. Laufer), pp. 411–425. Alan Liss, New York.

Phillips, J.E., Mordue, W., Meredith, J. & Spring, J. (1980) Purification and characteristics of the chloride stimulating factor from locust corpora cardiaca: A new peptide. *Can. J. Zool.* **58**, 1851–1860.

Phillips, J.E., Thomson, B., Spring, J., Proux, B. & Irvine, B. (1986) Recent observations on control of locust hindgut activities by cAMP and CTSH.

Pimley, R.W. (1984) Chromatographic separation of some corpora cardiaca peptides that influence fat cell activity in female *Glossina morsitans.Insect Biochem.* **14**, 521–525.

Pimley, R.W. & Langley, P.A. (1982) Hormone stimulated lipolysis and proline synthesis in the fat body of the adult tsetse fly. *Glossina morsitans. J. Insect Physiol.* **28**, 781–789.

Proux, J., Rougon–Rapuzzi, G. & Cupo, A. (1982) Enhancement of dye excretion across locust Malpighian tubules by a diuretic vasopressin-like hormone. *Gen. comp. Endocr.* **47**, 449–457.

Reynolds, S.E. (1983) Bursicon. In: Endocrinology of insects, **1** (eds. R.G.H. Downer & H. Laufer), pp. 235–248. Alan Liss, New York.

Reynolds, S.E. (1985) Hormonal control of cutical mechanical properties. In: Comprehensive insect physiology, biochemistry and pharmacology, **8** (eds. G.A. Kerkut & L.I. Gilbert), pp. 335–351. Pergamon Press, Oxford.

Reynolds, S.E. & Truman, J.W. (1983) Eclosion hormone: In: Endocrinology of insects, **1** (eds. R.G.H. Downer & H. Laufer), pp. 217–233. Alan Liss, New York.

Sasek, C.A., Schueler, P.A., Herman, W.S. & Elde, R.P. (1985) An antiserum to locust adipokinetic hormone reveals a novel peptidergic system in the rat central nervous system. *Brain Res.* **343**, 172–175.

Shapiro, J.P. & Law, H.J. (1983) Locust adipokinetic hormone stimulates lipid mobilization in *Manduca sexta. Biochem. Biophys. Res. Comm.* **115**, 924–931.

Siegert, K. & Ziegler, R. (1983) A hormone from the corpora cardiaca controls fat body phosphorylase during starvation in tobacco hornworm larvae. *Nature* **301**, 526–527.

Siegert, K., Morgan, P.J. & Mordue, W. (1985) Primary structures of locust adipokinetic hormones II. *Biol. Chem.* Hoppe–Seyler **366**, 723–727.

Spring, J. (1986) Hormonally-stimulated chloride transport in the Eastern lubber grasshopper *Romalea microptera. J. exp. Zool.* **237**, 3–9.

Steele, J.E. (1983) Endocrine control of carbohydrate metabolism in insects. In: Endocrinology of insects, **1** (eds. R.G.H.Downer & H. Laufer), pp. 427–439. Alan Liss, New York.

Steele, J.E., Coulthart, K.C. & McClure, J.B. (1984) Control of hexose phosphate and citrate in fat body of the cockroach (*Periplaneta americana*) by the corpus cardiacum. *Comp. Biochem. Physiol.* **79B**, 559–563.

Stone, J.V., Mordue, W., Broomfield, C.E. & Hardy, P.M. (1978) Structure-activity relationships for the lipid-mobilizing action of adipokinetic hormone. Synthesis and activity of a series of hormone analogues. *Eur. J. Biochem.* **89**, 195–202.

Storey, K.B. (1980) Kinetic properties of purified aldolase from flight muscles of *Schistocerca americana gregaria.* Role of the enzyme in the transition from carbohydrate to lipid-fuelled flight. *Insect Biochem.* **10**, 647–655.

Tobe, S.S. & Stay, B. (1985) Structure and regulation of the corpus allatum. In: Advances in insect physiology (eds. M.J. Berridge, J.E. Treherne & V.B. Wigglesworth), pp. 305–432. Academic Press, London.

Van Marrewijk, W.J.A., Van Den Broek, A.Th.M. & Beenakkers, A.M.Th. (1980) Regulation of glycogenolysis in the locust fat body during flight. *Insect Biochem.* **10**, 675–679.

Van Marrewijk, W.J.A., Van Den Broek, A.Th.M. & Beenakkers, A.M.Th. (1986) Hormonal control of fat-body glycogen mobilisation for locust flight. *Gen. comp. Endocr.* **54**, 136–142.

Vejberg, K. & Normann, T.C. (1974) Secretion of hyperglycaemic hormone from the corpus cardiacum of flying blowflies, *Calliphora erythrocephala. J. Insect Physiol.* **20**, 1189–1192.

Weeda, E. (1981) Hormonal regulation of proline synthesis and glucose release in the fat body of the Colorado potato beetle, *Leptinotarsa decemlineata. J. Insect Physiol.* **27**, 411–417.

Wheeler, C.H. & Goldsworthy, G.J. (1985a) Lipid transport to the flight muscles in *Locusta*. In: Insect locomotion (eds. M. Gewecke & G. Wendler), pp. 126–135. Paul Parey Press, Berlin.

Wheeler, C.H. & Goldsworthy, G.J. (1985b) C_L–proteins and the regulation of lipoprotein lipase activity in locust flight muscle. *Biol. Chem.* Hoppe–Seyler **367**, 1971–1077.

Wheeler, C.H. & Goldsworthy, G.J. (1986) Lipoprotein/apoprotein interactions during adipokinetic hormone action in *Locusta*. In: Insect neurochemistry neurophysiology 1986 (eds. A.B. Borkovec & D.B. Gelman), pp. 187–190. Humana Press, New Jersey.

Wheeler, C.H., Boothby, K.M. & Goldsworthy, G.J. (1986) C_L–proteins and the regulation of lipoprotein lipase activity in locust flight muscle. *Biol. Chem.* Hoppe–Seyler **367**, 1127–1133.

Wiens, A.W. & Gilbert, L.I. (1965) Regulation of cockroach fat body metabolism by the corpus cardiacum *in vitro*. *Science*, **150**, 614–615.

Witten, J., Schaffer, M.A., O'Shea, M., Cook, J.C., Hemling, M.E. & Rinehart, K.L. (1984) Structures of two cockroach neuropeptides assigned by fast atom bombardment mass spectrometry. *Biochem. Biophys. Res. Comm.* **124**, 350–358.

Ziegler, R. & Schulz, M. (1986) Regulation of lipid metabolism during flight in *Manduca sexta*. *J. Insect Physiol.* **32**, 997–1001.

Ziegler, R., Kegel, G. & Keller, R. (1984) Isolation and amino acid composition of the adipokinetic hormone of *Manduca sexta*. Hoppe–Seyler's *Z. Physiol. Chem.* **365**, 1451–1456.

Ziegler, R., Eckart, K., Schwarz, H. & Keller, R. (1985) Amino acid sequence of *Manduca sexta* adipokinetic hormone elucidated by combined fast atom bombardment (FAB)/tandem mass spectrometry. *Biochem. Biophys. Res. Comm.* **133**, 337–342.

M. O'SHEA, S. HEKIMI, J. WITTEN &
M.K. WORDEN

Functions of aminergic and peptidergic skeletal mononeurones in insects

Introduction: identifiable insect skeletal mononeurones and why they are important

Insect skeletal muscle is innervated by monopolar motoneurones in the segmental ganglia of the CNS which send motor axons to specific muscle targets. The specificity of muscle innervation allows us to identify motoneurones uniquely by their target muscles. Motoneurones innervating particular muscles have a characteristic position and morphology in the central nervous system, and these characteristics can also be used to identify them. An example is provided by the slow coxal depressor or Ds motoneurone of the cockroach *Periplaneta americana*. This neurone, as its name suggests, innvervates coxal depressor muscles and it has an invariant position and morphology. It is one of perhaps 5 motoneurones involved in the innervation of the coxal depressors. In addition to being identifiable, skeletal motoneurones innervating particular muscles in insects are few in number (in contrast to vertebrate muscle innervation for example). Thus the large extensor muscle of the locust tibia, the extensor tibialis muscle, is innervated by only 4 motor cells, each of which has been identified (see below). The identifiability of insect motorneurones and their small number have combined to make insect neuromuscular systems important model preparations.

The simplicity of insect neuromuscular systems has obvious advantages for the study of muscle physiology in these organisms. Insect motor neurones differ from the classical vertebrate model by using multiple transmitters. This feature has allowed us to study fundamental aspects of synaptic physiology, pharmacology and neuronal function which perhaps could not be studied as conveniently in other organisms. Motoneurones identified according to morphology or target muscle may therefore fall into different classes according to transmitter type and physiological action.

We can recognise 5 functional classes of skeletal motoneurones in hemimetabolous insects: the fast excitatory (glutaminergic); the type I slow excitatory (glutaminergic but non-proctolinergic); the type II slow excitatory (glutaminergic and proctolinergic); the inhibitory (gaba-ergic); and the dorsal unpaired median or DUM (octopaminergic) motoneurones. Here, all of these cell types will be referred to as motoneurones, even though some of them may participate only indirectly in the generation of force. Not all agree with this use of the term motoneurone: some workers would, for example, call the DUM cells modulatory neurones (which indeed they are) and restrict the term motoneurone to the excitatory and inhibitory neurones. The DUM cells themselves, like the inhibitory neurones are, however, capable of releasing muscle force (see below). Moreover the distinction between motoneurone and modulator is further complicated by the existence of type II slow excitatory motoneurones. Such neurones use two transmitters (glutamate and proctolin), one of which produces fast muscle contraction while the other has a modulatory role.

In our view the term motoneurone is functionally neutral and can therefore accommodate newly discovered functions. It applies to any non-sensory neurone of the central nervous sytem which innervates skeletal muscle. With such a definition there may be some argument over what is exactly meant by "innervation". Do for example DUM cells actually innervate muscle in the same way as the fast excitatory motoneurones? Probably not. The DUM cells appear not to form classical neuromuscular junctions, but neither are they classical excitatory motoneurones. They do however have specific muscle targets, travel in specific motonerves and have both direct and modulatory effects. In this sense they innervate muscle and represent a class of motoneurones. Excluded from the term motoneurone are muscle effector cells which, while having direct influence on muscle performance, do not actually innervate muscle. An example would be the neurosecretory cells of the locust corpus cardiacum (CC) which release adipokinetic hormone. This hormone, of which there are at least two forms (see Chapters 5 and 6 this volume), can cause muscle contraction and also regulates muscle metabolism. But the cells of the corpus cardiacum do not innervate muscle and are not therefore motoneurones.

In this short essay we will show how in insect neuromuscular systems the combined advantages of physical simplicity on the one hand and pharmacological complexity on the other have been exploited to improve our understanding of amine and peptide transmitter action.

Aminergic and peptidergic motoneurones have been identified in insects. They are small in number, but some are relatively large and innervate specific muscle targets. It is important to realize the advantages conferred by such motoneurones for the study of these transmitter types. Consider, for example, the importance of cholinergic vertebrate motoneurones for our understanding of acetylcholine's synaptic actions. Neuromuscular systems are clearly accessible to physiological study and have been widely exploited in studies of synaptic transmission. Muscle cells not only offer advantage for the study of synaptic physiology, they are also large and relatively homogenous cells and therefore can be used to study biochemical events activated by transmitters. This is particularly important for understanding amine and peptide action because physiological effects are likely to be mediated by intracellular second messengers. Furthermore, because we can identify specific aminergic and peptidergic motoneurones, the actions of these transmitters can be studied directly at the cellular level. While all of the information obtained in the studies described below is clearly relevant to insects, some of it is fundamental and provides insight into the complexities of transmitter action in general.

Identification of an aminergic skeletal neuromuscular system

The only biogenic amine to be specifically associated with an insect skeletal motoneurone is the phenolamine, octopamine. Octopamine was first discovered in the salivary glands of the octopus, but is a ubiquitous amine of the nervous systems of both vertebrates and invertebrates. Little is known about its role in vertebrates, partly because it is present in much lesser amounts than the catecholamine noradrenaline. For reviews of the octopamine literature see Evans (1980, 1985).

A significant advance in our understanding of the function of octopamine in insects was achieved when octopamine-containing neurones were identified in the desert locust *Schistocerca gregaria* (O'Shea & Evans 1977). The first neurone to be identified as octopaminergic is a member of a specialized group of efferent cells called the dorsal unpaired median (DUM) neurones (Hoyle *et al.* 1974). These neurones, which were first described in the locust *Locusta migratoria* (Plotnikova 1969) are morphologically unusual. Their cell bodies reside on the dorsal surface of the segmental ganglia, forming a loose cluster near the midline. Most other neuronal somata, including those of the other motoneurones, are located on the ventral ganglionic surface. The DUM neurones are monopolar, but the primary neurite bifurcates close to the cell body to produce a laterally-directed axon pair. The largest of the DUM neurones project left and right axons symmetrically to peripheral nerve roots. They thus appear to be efferent neurones. Unlike the usual excitatory motoneurone, however, they are bilaterally symmetrical. This is why they were called unpaired. The bifurcated axon distinguishes the DUM cells from most of the neurones of the ganglia, which are paired. Bilaterally paired neurones form contralateral mirror images of one another. In effect the DUM cell embodies its own contralateral homologue in the form of its intrinsic bilateral symmetry. The morphology of the DUM cells is not their only unusual feature. Electrophysiological recordings from the cell bodies reveals their ability to support action potentials that exceed the value of the resting potential – so called "overshooting" action potentials (Crossmann *et al.* 1971). Most insect neuronal somata are electrically inexcitable.

Although the unusual morphology and physiological features of the DUM cells attracted considerable interest, little could be said during the early 1970s of their likely functional role. A major step towards an understanding of DUM cells was taken when Hoyle *et al.* (1974) showed that one of the DUM cells in the locust metathoracic ganglion projected axons to the third thoracic left and right extensor tibialis or ETi muscles. This muscle appeared to be the only target for one particular DUM neurone, and this cell was therefore named DUM–ETi. The implications of this discovery were first, that DUMETi might have some motor function and second, that other DUM neurones might innervate other muscles and be identifiable according to their own specific muscle targets.

What function might be served by the DUM cell of the extensor tibialis muscle? This was a puzzling question because at the time DUMETi was identified it was thought that the ETi muscle received innervation from only 3 neurones – the fast and slow excitatory motoneurones (FETi and SETi) and the common inhibitory (CI) motoneurone. The existence of a fourth neurone suggested that some aspect of the ETi muscle's physiological response to nerve stimulation could not be explained by known actions of the known motoneurones. A partial answer to the question of DUMETi's role was, however, provided by the description of a novel cardiac-like function for the hindleg extensor tibialis muscle (Hoyle & O'Shea 1974). The ETi muscle is shown capable of generating an intrinsic myogenic rhythm of contraction and relaxation which appears to be an intrinsic property of the muscle because it persists when the leg is removed or when the muscle innervation is severed (Hoyle & O'Shea 1974). While the rhythm is myogenic, the CNS nevertheless exercises some control over it; when the muscle innervation is intact the rhythm is far less regular than when the muscle is separated from the CNS (Hoyle & O'Shea 1974). At least part of the control over rhythm frequency appears to be through DUM cells When DUM cells are stimulated the ongoing rhythm is inhibited (Hoyle 1974 and O'Shea, unpublished observations). DUMETi is not the only DUM cell capable of such control over the myogenic rhythm. Although DUMETi has the most potent effect, intracellular stimulation of other DUM cells also slows the rhythm. These observations provided the first evidence for a physiological modulatory role for DUM cells. The myogenic rhythm is now known to be sensitive to octopamine, and the physiological effects of DUM cell stimulation are understood in terms of the octopaminergic nature of the DUM cells (O'Shea & Evans 1977; Evans & O'Shea 1978; O'Shea & Evans 1979).

When the role of DUM cells in regulating the myogenic rhythm was discovered it was suspected that there must be other functions for DUM cells. It was clear that myogenic rhythm regulation was not the whole story because the pro- and mesothoracic extensor tibialis muscles are also innervated by their own DUMETi neurones (from their respective segmental ganglia), but these muscles do not generate a myogenic rhythm. Perhaps the hindleg muscle specialization is related to the fact that the hindleg is greatly elongated, especially in the femoral and tibial segments. It is possible that the myogenic rhythm functions to aid the circulation of hemolymph and air in this greatly elongated appendage. Thus the most likely function for the myogenic rhythm is as an accessory pulsatile organ specifically for the hindleg. What then could be the more general function for DUM cells ?

Although the discovery that DUM cells control a myogenic cardiac-like oscillator was not entirely satisfactory it did provide a way to determine that the transmitter of the DUMETi neurone was octopamine. Two lines of evidence suggested that the DUM cells might contain a biogenic amine. First the DUM cells stain readily with the vital due Neutral Red, and in other systems (notably in leech nervous system) this is a specific stain for monoamine-containing cells (Stuart *et al.* 1974). Second, a number of workers were unable to stain the cells with either the glyoxalic acid or the Falk Hillarp histofluorescence techniques which suggested the presence of an unconventional amine. These methods are used to identify the presence of indole and catacholamines, and should have stained the DUM cells if amines of these classes were responsible for the Neutral Red staining. The first preliminary evidence suggesting that octopamine was the transmitter of the DUM cells was provided by Hoyle (1975) and Hoyle and Barker (1975). Hoyle (1975) showed that application of octopamine to the myogenic fibres of the extensor muscle inhibited the myogenic rhythm. Thus octopamine could mimic the effect of DUM cell stimulation. Hoyle and Barker (1975) showed that preparations of cells taken from the dorsal region of the ganglion, certainly including DUM cells, could synthesise radiolabelled octopamine from radioactive tyrosine. While neither piece of evidence is sufficient to say octopamine is present in DUM cells, they served to focus attention on this amine.

Conclusive evidence for the octopaminergic nature of DUM neurones was provided when biochemical and physiological experiments were performed on the identified DUMETi neurone (O'Shea & Evans 1977; Evans & O'Shea 1977, 1978; O'Shea & Evans 1979). The experimental approach consisted of dissecting the cell body of the DUMETi free of the ganglion. Identification was achieved by electrophysiological methods. The left and right motor nerves to the ETi muscle were monitored with extracellular electrodes. These nerves contain only the 4 axons innervating the ETi muscle. An intracellular recording electrode was then used to penetrate DUM neurone cell bodies in the metathoracic ganglion. The DUMETi neurone was identified by comparing the intracellular soma recording with the extracellular ETi nerve recordings. For DUMETi, but no other DUM neurone, the soma action potentials are 1:1 with extracellular spikes recorded in the ETi nerves. Once identified in this way the DUMETi soma was removed by microdissection. This procedure was repeated many times, and DUMETi somata from different individuals were pooled and assayed for octopamine content using the radioenzymatic assay developed by Molinoff *et al.* (1969). Each DUMETi somata was shown to contain approximately 0.1 pmoles of octopamine. No octopamine was detected in control experiments on the identified fast excitatory motoneurone. Thus, for the first time octopamine was shown to be a constituent of a specifically identified neurone. If octopamine is the DUMETi transmitter, it ought to be released by the DUMETi neurone on the extensor muscle and DUMETi stimulation ought to have the effect of applied octopamine. Thus the inhibitory effect of DUMETi stimulation and octopamine on the myogenic rhythm confirmed the biochemical evidence for a transmitter role for octopamine in DUM cells.

The above data constituted the evidence for assigning octopamine as an amine transmitter of the DUMETi neurone. Further physiological evidence, described below, confirmed this. Now there is accumulating evidence that octopamine is the likely transmitter of all the larger DUM cells in locusts and cockroaches. Indeed, octopamine has also been suggested as the transmitter of DUM type neurones in fireflies, *Photuris versicolor*, which innervate and activate the light organs (Christensen *et al.* 1983). Many questions remain concerning the functions of DUM cells and these will be considered below. The next part of this paper will be concerned with the identification of a peptidergic skeletal motoneuronal system, a discovery which has further complicated the neuropharmacology of the insect neuromuscular system.

Identification of peptidergic skeletal neuromuscular systems

The only neuropeptide to be specifically associated with identified skeletal insect motoneurones is the pentapeptide proctolin.

Arg – Tyr – Leu – Pro – Thr

Proctolin was discovered in the cockroach *Periplaneta americana* by B.E. Brown in 1967 (Brown 1967). At that time it was referred to as "gut-factor" – a biologically active constituent of hindgut that could cause gut muscle contraction. In 1975 this factor was isolated in pure form from approximately 125,000 cockroaches (*Periplaneta americana*) and subjected to chemical analysis (Brown & Starratt 1975; Starratt & Brown 1975). Proctolin is widely distributed among the arthropods and there is some evidence for a proctolin-like peptide in non-arthropod invertebrates and in vertebrates (see O'Shea & Schaffer 1985).

In addition to its presence in the hindgut, proctolin is also a CNS peptide. Brown found proctolin in peripheral nerves of the terminal and thoracic ganglia. It was first suggested as the transmitter of hindgut motoneurones by Brown (1975), who was the first to introduce the concept of the peptidergic motoneurone. The history of establishing proctolin as a transmitter of skeletal motoneurones also has its roots in the early studies of Brown (1967). He found the "gut-factor" in the peripheral skeletal motor nerves of the cockroach thoracic CNS. When synthetic proctolin became available, its activity on insect skeletal muscle was discovered. For example, Piek & Mantel (1977) showed that at nanomolar concentrations proctolin induces and accelerates the myogenic rhythm of contraction and relaxation of the hindleg extensor tibialis muscle of the locust (the same rhythm that is inhibited by octopamine and DUMETi). This result and other examples of action of skeletal muscle could be interpreted in two ways: either the muscle is the target for circulating proctolin (hormonal action), or the muscle is the target of a proctolinergic motoneurone (transmitter action). Distinguishing between these possibilities for the extensor tibialis muscle had to await the development of immunocytochemical methods. Physiological evidence for a transmitter role, however, was provided by the observations of May *et al.* (1979). These authors applied proctolin by microiontophoresis to the locust extensor tibialis muscle (myogenic fibres) and recorded depolarizing iontophoretic potentials. Moreover, they demonstrated hot-spots of proctolin extra-sensitivity localized in the clefts between muscle fibres. Since it is in these clefts that nerve-muscle junctions are thought to be made, this observation suggested that proctolin may be delivered to the muscle not through the hemolymph but by a motoneurone. The existence of proctolinergic innervation of the locust extensor tibialis muscle is now confirmed (Witten *et al.* 1984; Worden *et al.* 1985; O'Shea 1985) by the identification of the proctolinergic nature of the slow extensor tibialis or SETi motoneurone.

Precedence for peptidergic skeletal motoneurones in insects had, however, already been set by studies in the cockroach. In an immunocytochemical survey of the cockroach CNS, Bishop & O'Shea (1982) described the presence of proctolin-immunoreactive axons in the peripheral nerves of thoracic ganglia. This study also revealed a bilaterally symmetrical pair of large immunoreactive neuronal cell bodies on the lateral posterior margins of the dorsal surface of the metathoracic ganglion. These cells were named the giant dorsal bilateral or GDB neurones (Bishop & O'Shea 1982). Subsequently it was realized that these cells were in the same place as the somata of the previously identified slow coxal depressor or Ds motoneurones (Pearson & Iles 1971), which suggested that the GDB and the Ds might be identical.

A combination of intracellular dye injection, immunocytochemistry and electrophysiology confirmed that the immunoreactive GDB neurone was indeed the Ds skeletal motoneurone of Pearson and Iles (O'Shea & Bishop 1982). Moreover the presence of proctolin in this neurone was established by testing for proctolin bioactivity in extracts made from the somata of the Ds motoneurone. Reverse phase HPLC fractionation of the cell extract, combined with bioassay (locust extensor tibialis muscle) provided evidence that the immunoreactivity of the Ds neurone was due to the presence of proctolin (O'Shea & Bishop 1982).

Proctolin was shown subsequently to be released from the Ds motoneurones. This was established by electrically stimulating the neurone, collecting the superfusate from the coxal depressor muscle and applying it to the locust leg muscle bioassay (the myogenic bundle). Release was found to be calcium-dependent (Adams and O'Shea 1983).

The final step in the identification of the proctolinergic nature of the Ds motoneurone was a demonstration of a correspondence between the effects of neuronal stimulation and the response to applied proctolin. When proctolin is applied at nanomolar concentrations the muscle develops a slow maintained dose-dependent contraction. This sustained catch-like proctolin-induced tension could be mimicked by stimulating the Ds motoneurone. In fact the curious catch property of the coxal depressor muscle had already been described (Chesler & Fourtner 1981), though not explained. Now it seemed that the catch property could be explained by the release of proctolin. Together, the evidence on presence, release and action of proctolin established Ds as the first identified peptidergic skeletal motoneurone

Functions and interactions of aminergic (DUM) and peptidergic skeletal motoneurones (SETi and Ds)

The evidence that the third thoracic DUMETi is octopaminergic was reviewed briefly above. Part of the evidence concerned the inhibitory effect of DUMETi on the myogenic rhythm of the ETi muscle. This myogenic rhythm is produced by a very restricted and specialized part of the ETi muscle in the proximal part of the femur. The proximal bundle of myogenic fibres receives branches of the DUMETi axon, but so do other parts of the ETi muscle. Terminals of DUMETi are widely distributed on the ETi muscle so we can conclude that the third thoracic DUMETi neurone has physiological effects unrelated to the myogenic bundle. Such a conclusion can also be extended to the DUMETi neurones of the first and second thoracic ganglia since the ETi muscles of these segments do not possess the myogenic bundle. Also DUM neurones other than those innervating the ETi muscle must have functions unrelated to the control of myogenic oscillations.

A hypothesis put forward in the late 1970s (Evans & O'Shea) was that DUM neurones act as modulators of the actions of the conventional skeletal motoneurones. This arose in part because, when DUMETi was stimulated alone, the only direct effect on the muscle we could observe was the inhibition of the myogenic rhythm. Also, when stimulated alone DUMETi produced no synaptic potential and did not appear to participate directly in the generation of muscle force. So perhaps when stimulated in conjunction with 1 of the 3 other ETi muscle motoneurones (SETi, FETi, CI), DUMETi would have some modulatory effect.Indeed, this was the case. Today DUMETi and its transmitter octopamine are well-established as modulators of neuromuscular transmission in insects.

The primary effect of DUMETi is to modulate SETI-induced contraction (O'Shea & Evans 1979; Evans & Siegler 1982; Evans 1981). When DUMETi is fired in a brief burst, subsequent SETi-induced twitch contractions are modulated in the following ways: the rates of contraction and relaxation are increased and the amplitude of the SETi-induced contraction is amplified. The greatest effect is on the rate of relaxation (which can also be seen for FETi-induced contractions). All of these effects can be produced by application of a brief pulse of octopamine to the nerve-muscle preparation. These modulatory effects are persistent and outlast the duration of the octopamine pulse or the burst of DUMETi activity. Octopamine activated modulatory effects therefore do not require the presence of octopamine.

How can these modulatory effects be explained? Changes in rates of relaxation and contraction are presumably due to the interaction of octopamine with postsynaptic (muscle) octopamine receptors. The increase in the amplitude of the twitch contraction may also be due to postsynaptic receptor activation, but could also be explained presynaptically. Such presynaptic octopamine receptors might, for example, mediate an increase in the amount of glutamate released from SETi terminals. In fact there is some evidence that the increase in amplitude is due to a presynaptic mechanism; for example, a small increase in the amplitude of the SETi post synaptic or junctional potential occurs in response to octopamine. This is not accompanied by a measurable increase in muscle input resistance, and so is presumably due to an increase in the amount of transmitter released (O'Shea & Evans 1979). Moreover there is an octopamine-induced increase in the spontaneous release of transmitter as measured by the frequency of miniature end plate potentials.

So far we have considered the effects of DUMETi and octopamine on the myogenic rhythm and on the SETi-induced twitch contractions. Another important effect of octopamine on SETi evoked contraction has however been described, and this effect provides a link between the octopamine and the proctolinergic systems.

When SETi is stimulated at different frequencies, there is a maintained, frequency-dependent tension. The magnitude of the maintained tension is not, however, simply related to the frequency of SETi stimulation, but also depends on the prior activity of the SETi neurone and the activity of the DUMETi neurone (Evans & Siegler 1982). For example, if SETi is stimulated first at a low frequency (15Hz) then at a higher frequency and then the low frequency again, the maintained tension attained at the low frequency is higher after the high frequency burst than before it. The difference between the before and after maintained force is called the catch tension: it represents a failure of the muscle to relax back to the prior tension at the same low frequency, and appears to depend on events activated during a higher frequency burst. Octopamine selectively reduces the catch tension and causes the muscle to reach the low-frequency dependent maintained tension more rapidly. Octopamine also reduces the level of the frequency dependent maintained tension, while it increases the amplitude of individual SETi twitches. This apparently paradoxical result is not fully understood. It indicates, however, that the absolute amplitude of the SETi twitch does not contribute to the summation mechanism which underlines the frequency-dependent increase in maintained tension. If it did, we would expect that octopamine would increase the maintained tension at a given frequency. One way to explain the observation is to say that the rate of relaxation of each twitch, which increases their duration, is an important parameter in determining the plateau or maintained tension at a given frequency. The slower the relaxation, the more each twitch contributes to the maintained tension. So octopamine, by increasing the rate of relaxation, would produce a lower amplitude maintained tension even though individual twitches are larger in amplitude. This may not, however, be the complete explanation and to gain a full understanding of the action of SETi and its modulation by octopamine perhaps we must invoke a second transmitter in the SETi neurone.

With hindsight, this is easy to propose, because we now have strong evidence that SETi is both glutaminergic and peptidergic (proctolinergic). But perhaps the modulatory effects of octopamine themselves suggest a second transmitter in SETi. Accordingly, the SETi twitch and the catch and part of the frequency dependent tension, could be due to different transmitters; glutamate may be responsible for the twitch tension amplitude and the second transmitter (X) produces the slow and maintained tension, especially when the SETi neurone is activated in a high freuqency burst. Probably, this second transmitter also has an effect on the rate of relaxation of the glutamate-induced twitches. This effect is predicted to be opposite to that of octopamine, i.e. the second transmitter would decrease the rate of relaxation and thereby increase the duration of twitches.

From this point of view we can now explain frequency-dependent maintained tension, catch tension and octopamaine's effect on them in the following way. The SETi motoneurone releases glutamate and transmitter X. Glutamate produces transient twitch contractures and X produces longer lasting effects; a decreased rate of twitch relaxation and a slow sustained muscle contraction. The effects produced by X, but not glutamate, can be abolished by octopamine.

So frequency-dependent maintained force is partly due to the release of transmitter X. Because the muscle-contracting effect of X is long-lasting, it will sum between SETi spikes to produce a frequency-dependent maintained contracture. Transmitter X will also tend to decrease the relaxation rates of twitches, so there will be a corresponding frequency-dependent increase in the twitch duration, which will also contribute to the frequency dependence of the maintained contracture. This is because temporal summation between twitches will be more pronounced for longer duration twitches.

Invoking substance X with the above properties can also explain catch. During the high frequency burst preceding the appearance of catch, more X is released and because it has a long time-constant of action it will delay the return to the steady state conditions at the lower frequency. In this view, catch and frequency-dependent tension are related. Finally, if we say that the X-mediated effects are reversed by the action of octopamine, then we would expect the octopamine-induced reduction in the frequency dependent and the catch tension. Does X exist, what is it, and does it have the properties predicted according to the above hypothesis? Evidence for the peptide proctolin being a second transmitter in SETi has already been discussed above. Thus, for the SETi motoneurone of the locust, the easiest way to understand its physiological action is to assume that 2 transmitters (glutamate and proctolin) are released simultaneously. Glutamate has rapid transient effects 1:1 with motor action potentials, and proctolin has slower and longer-lasting effects which can best be seen when the neurone is activated in a burst.

In many respects the effects of proctoline and octopamine are antagonistic, thus octopamine inhibits the myogenic rhythm and proctolin activates it. Octopamine increases the rate of SETi twitch relaxation but proctolin decreases it. Octopamine causes a relaxation of maintained and catch tensions and proctolin contributes to them (see Adams & O'Shea 1983). In one respect their actions are similar: both proctolin and octopamine increase the amplitude of SETi twitches. Thus proctolin has the properties hypothesized above for substance X. Its direct and modulatory actions as a co-transmitter can perhaps help explain frequency-dependent maintained tension and also catch.

What are the processes that underlie the actions of octopamine and proctolin? A considerable body of evidence suggests that octopamine causes an increase in the muscle intracellular concentration of cyclic AMP (Evans 1985, 1984a, 1984b). It is possible therefore that some, but probably not all, of the physiological actions of octopamine are caused through the activation of an octopamine-dependent adenylate cyclase. Evans provides evidence that cyclic AMP mediates the modulatory action of octopamine but that it is probably not involved in the inhibitory effect of octopamine on the myogenic rhythm.

With respect to the mode of action of proctolin on the ETi muscle very little is known. Evans has suggested that cyclic AMP may be involved in the activation of the myogenic rhythm. Concerning the mechanisms underlying catch and the modulation of SETi twitches, we know very little. It appears, however, that cyclic nucleotides are not involved and extracellular calcium is not necessary (unpublished observations). Proctolin at a concentration (10^{-8}) that produces marked muscle contraction and modulatory effects does not increase levels of either cyclic AMP or cyclic GMP. Indeed it would be difficult to understand the antagonistic physiological effects of octopamine and proctolin if both activated the adenylate cyclase.

The fact that proctolin produces a slow muscle contraction which is independent of extracellular calcium indicates that any increase in intracellular calcium associated with proctolin action is brought about by the releasing of calcium from intracellular stores, such as in the sarcoplasmic reticulum. A mechanism which could mediate the release of intracellular calcium involves the hydrolysis of phosphotidylinositol to yield inositol trisphosphate (IP_3). In many systems, including muscle, IP_3 has been proposed as a second messenger mediating increase in intracellular calcium concentration as a result of transmitter or hormone action (see Berridge 1986). Perhaps there is a proctolin-stimulated increase in phosphotidylinositol hydrolysis resulting in the appearance of IP_3 in the muscle.

Studies on the mode of action of proctolin in the ETi muscle are in their infancy. There is little doubt however that the system provides an excellent model for understanding how peptide transmitters act. There is also little doubt that as we understand more about the modes of action of both octopamine and proctolin, we will discover how these transmitters interact at a subcellular level. Such understanding could provide new insights into how modulatory transmitters function. As we have seen in the ETi system, different modulators probably do not function independently.

Understanding the mechanisms of interactions between second messengers presents us with challenging and important problems. The insect neuromuscular system may help unravel and solve some of them.

Conclusions

Insect neuromuscular systems are pharmacologically complex. This can be turned to an advantage because they are also well characterized and individual motoneurones can be identified uniquely. Octopamine seems to be the primary amine transmitter involved in skeletal muscle innervation and there is little evidence for the presence of other biogenic amines in skeletal motoneurones. On the other hand proctolin may not be the only peptide associated with insect skeletal motoneurones. Although there is no direct evidence for other peptides in skeletal motoneurones, We expect that proctolin is perhaps one of many myoactive peptides involved in muscle innervation. This is suggested by the sensitivity of insect skeletal muscle to other peptides, AKH-like peptides and FMRF–NH_2 (O'Shea *et al.* 1984; Walther *et al.* 1984), for example, and by the fact that the proctolin subpopulation is small (perhaps –10% of the motoneurones). The discovery of other subpopulations of peptidergic skeletal motoneurones will have to await the discovery of new myoactive peptides and the application of the peptide-neurone identification techniques which were used to identify the proctolin subpopulation.

Insect neuromuscular systems have provided us with direct ways to study several important questions in neurobiology in particular with respect to amine and peptide action. With the discovery of dual transmitter motoneurones (glutamate and proctolin) we may also be able to improve our understanding of why neurones operate with more than one transmitter. The possibility that octopamine modulates the action of proctolin may provide an important clue. Perhaps the existence and release of more than one transmitter from the same neurone allows for modulators released by other neurones to act selectively, affecting some but not all of the postsynaptic actions of the multi-transmitter neurone.

References

Adams, M.E. & O'Shea, M. 1983 Peptide cotransmitter at a neuromuscular junction. *Science* **221**:286–289.

Berridge, M. (1986) Cell signalling through phospholipid metabolism. *J. Cell Sci.* in press.

Bishop, C.A. & O'Shea, M. (1982) Neuropeptide Proctolin (H–Arg–Try–Leu–Pro–Thr–OH): Immunocytochemical mapping of neurons in the central nervous system of the cockroach. *J. Comp. Neurol.* **207**:223–238.

Brown, B.E. & Starratt, A.N. (1975) Isolation of proctolin, a myotropic peptide from *Periplaneta americana. J. Insect. Physiol.* **21**:1879–1881.

Chesler, M. & Fourtner, C.R. (1981) Mechanical properties of a slow muscle in the cockroach. *J. Neurobiol.* **12**:391–402.

Christensen, T.A., Sherman, T.G., McCaman, R.E. & Carlson, A.D. (1983) Presence of octopamine in firefly photomotor neurons. *Neuroscience* **9** (1):183–189.

Crossmann, A.R. Kerkut, G.A., Pitman, R.M. & Walker, R.J. (1971) Electrically excitable nerve cell bodies in the central ganglia of two insect species, *Periplaneta americana* and *Schistocerca gregaria*. Investigation of cell geometry and morphology by intracellular dye injection.

Evans, P.D. (1980) Biogenic amines in the insect nervous sytem. In: Advances in Insect Physiology **15**:317–473.

Evans, P.D. (1984a) The role of cyclic nucleotides and calcium in the mediation of the modulatory effects of octopamine on locust skeletal muscle. *J. Physiol.* **348**:325–40.

Evans, P.D. (1985) Octopamine. In: Comprehensive Insect Physiology, Biochemistry and Pharmacology. Eds. G.A. Kerkut & L.I. Gilbert, Pergamon Press, pp. 499–527.

Evans, P.D. & O'Shea, M. (1977) An octopaminergic neurone modulates neuromuscular transmission in the locust. *Nature* **270**:257–259.

Evans, P.D. & O'Shea, M. (1978) The identification of an octopaminergic neurone and the modulation of a myogenic rhythm in the locust. *J. exp. Biol.* **73**:235–260.

Evans, P.D. & Siegler, M.V. (1982) Octopamine mediated relaxation of maintained and catch tension in locust skeletal muscle. *J. Physio.* **324**:93–112.

Hoyle, G., Dagan, D., Moberly, B. & Colquhoun, W. (1974) Dorsal unpaired median insect neurons make neurosecretory endings on skeletal muscle. *J. Exp. Zool.* **187**:159–165.

Hoyle, G. (1974) A function for neurons (DUM) neurosecretory on skeletal muscle of insects. *J. Exp. Zool.* **189**:401–406. *Comp. Biochem. Physiol.* **40A**:579–594.

Hoyle, G. & O'Shea, M. (1974) Intrinsic rhythmic contractions in insect skeletal muscle. *J. Exp. Zool.* **189**:407–412.

Hoyle, G. (1975) Evidence that insect dorsal unpaired median (DUM) neurons are octopaminergic. *J. Exp. Zool.* **1973**:425–431.

Hoyle, G. & Barker, D.L. (1975) Synthesis of octopamine by insect dorsal median unpaired neurons. *J. Exp. Zool.***193**:433–439.

May, T.E., Brown, B.E. & Clements, A.N. (1979) Experimental studies upon a bundle of tonic fibres in the locust extensor tibialis muscle. *J. Insect Physiol.* **25**:169–181.

Molinoff, P.B., Landsberg, L. & Axelrod, J. (1969) An enzymatic assay for octopamine and other -hydroxylated phenylethylamines. *J. Pharmac. exp. Ther.* **170**:253–261.

O'Shea, M. & Evans, P.D. (1977) Synaptic modulation by an identified octopaminergic neuron in the locust. *Soc. Neurosci.* Abstr. 583.

O'Shea, M. & Evans, P.D. (1979) Potentiation of neuromuscular transmission by an octopaminergic neurone in the locust. *J. exp. Biol.* **79**:169–190.

O'Shea, M. & Bishop, C.A. (1982) Neuropeptide proctolin associated with an identified skeletal motoneuron. *J. Neurosci.* **2**:1242–1251.

O'Shea, M., Witten, J. & Schaffer, M. (1984) Isolation and characterization of two myoactive neuropeptides: further evidence of an invertebrate peptide family. *J. Neurosci.* **4**: 521–529.

O'Shea, M. & Schaffer, M. (1985) Neuropeptide function: The Invertebrate Contribution. *Ann. Rev. Neurosci.* **8**171–198.

O'Shea, M. (1985) Neuropeptides in Insects: Possible Leads to New Control Methods. In: Approaches to New Leads for Insecticides. Ed. von Keyserlingk *et al.* Springer Verlag, Heidelberg.

Piek, T. & Mantel, P. (1977) Myogenic contractions in locust muscle induced by proctolin and by wasp, *Philanthus triangulum*, venom. *J. Insect Physiol.* **23**:321–325.

Pearson, K.G. & Iles, J.F. (1971) Innervation of coxal depressor muscle in cockroach *Periplaneta americana*. *J. Exp. Biol.* **54**:215–232.

Plotnikova, S.I. (1969) Effectory neurones with seveal axons in the ventral nerve cord of *Locusta migratoria*. *J. Evol. Biochem. Physiol.* **5**:339–341.

Starratt, A.N. & Brown, B.E. (1975) Structure of the pentapeptide proctolin, a proposed neurotransmitter in insects. *Life Sci.* **17**:1253–1256.

Stuart, A.E., Hudspeth, A.J. & Hall, Z.W. (1974) Vital staining of specific monoamaine-containing cells in the leech nervous system. *Cell Tiss. Res.* **153**:55–61.

Walther, C., Schiebe, M., & Voigt, K.H. (1984) Synaptic and nonsynaptic effects of molluscan cardioexcitatory neuropeptides on locust skeletal muscle. *Neurosci. Lett.* **45**:99–104.

Witten, J.L., Schaffer, M.H., O'Shea, M., Cook, J.C., Hemling, M.E. & Rinehart, K.L. Jr. (1984) Structures of two cockroach neuropeptides assigned by fast atom bombardment mass spectrometry. *Biochem. Biophys. Res. Comm.* **124**:350–358.

Worden, M.K., Witten, J.L. & O'Shea, M. (1985) Proctolin is a co-transmitter for the SETi motoneuron. *Soc. Neurosci.* Abstr. 11(1), 99.9.

Worden, M.K. & O'Shea, M. (1986e) Evidence for stimulation of muscle phosphatidylinositol metabaolism by an identified skeletal amotoneuron. *Soc. Neurosci.* Abstr. 12(2), p. 948.

S.G. WEBSTER & R. KELLER

Physiology and biochemistry of crustacean neurohormonal peptides

1. Introduction

A seemingly bewildering array of factors with putative neurohormonal function have been described in (mainly decapod) Crustacea (see Kleinholz & Keller 1979; Cooke & Sullivan 1982; Keller 1983; Kleinholz 1985). They are implicated in almost every aspect of crustacean physiology, including pigment dispersion and concentration, inhibition of moulting, limb regeneration and gonad development, cardiac control, blood glucose, metabolism and respiratory control, ion and water balance, endogenous rhythmicity and locomotion. Several of these factors are produced by neurosecretory structures in the eyestalk, which can be easily ablated. This accessibility has unfortunately led to a tendency to assign hormonal regulation of physiological mechanisms based solely upon the results of eyestalk removal, often without the rigorous application of deficiency and replacement protocols using physiologically relevant doses of extracts or further purification of the active principle. Thus, apart from the well known neuropeptides, it is not known how many of these described 'factors' genuinely control individual processes and little is known of their precise chemical identity.

Evidence from immunocytochemical studies suggests that many neuropeptides classically known as 'vertebrate' peptides and also neuropeptides that have originally been found in invertebrates (e.g. FMRF amide, proctolin) are ubiquitous in crustaceans (Mancillas *et al.* 1981; Jacobs & Van Herp 1984; Jaros *et al.* 1985; Van Deijnen *et al.* 1985; Stangier *et al.* 1986). However, there is at present little information concerning the role of 'vertebrate-type' peptides in physiological integration.

The purpose of this review is to address problems concerning the nature and action of those crustacean neuropeptides which have been well characterised particularly with regard to recent progress in this field, as well as to draw attention to those factors which in our view lack adequate characterisation.

2. Morphological correlates: neurosecretory structures

Chaigneau (1983) has recently reviewed the morphology of neurohaemal tissues of crustaceans; the sinus gland (SG), pericardial organs (PO), postcommisural organs (PCO), of decapods, and the lateral nerve plexus (LNP) of isopods. With the exception of the SG, where peptidergic input is mainly, but not exclusively, associated with adjacent tissues of the medulla terminalis X-organ (Jaros 1978; Andrew & Saleuddin 1978; Andrew *et al.* 1978), other neurohaemal structures such as the PO receive peptidergic input from other parts of the CNS such as the segmental nerves and thoracic ganglion (Maynard 1961; Cooke & Sullivan 1982). Whilst several of the neuropeptides from the SG and PO have been characterised, little is known about the neurosecretory products of the PCO and LNP except that the PCOs contain chromatophorotropins (Carlisle & Knowles 1959). Thus, this review will be restricted to the neurosecretory products of the SG and PO in decapod crustaceans (see also Chapter 4).

3. Chromatophorotropins

Chromatophorotropins are neurohormones that induce pigment migration in chromatophores of the epidermis and of some internal organs (which adapt the colour of the organism to the environment), as well as certain ommatidial cells in which pigment migration adapts the compound eye to light and dark conditions (see Rao 1985). Single cell chromatophores (the most common) are classified as melanophores, leucophores, erythrophores and xanthophores according to the colour of their pigment granules. In addition to such monochromatic chromatophores, in some crustaceans several different chromatophore types may form polychromatic clusters or chromatosomes. In the compound eye, different cells carry screening pigments. It has long been recognised that the control of pigmentary effectors is bihormonal, involving both pigment dispersing and concentrating hormones (see Rao 1985). One question has been whether the control of the different effectors in a single species is brought about by different effector-specific hormones, or whether a single hormone can act on different types of effectors. Two alternative hypotheses have been proposed: Kleinholz (1976) outlines "The unitary hypothesis", in which only two antagonistic hormones regulate different effectors, and "The multiple hormone hypothesis" which involves several different effector-specific hormones. Because only poorly characterised active fractions were used in earlier bioassays, no clear cut evidence favouring either of these hypotheses has been obtained. The availability of pure or chemically-synthesised chromatophorotropins has made unequivocal studies possible. Pure DRPH (distal retinal pigment hormone, light adapting) not only effects pigment migration in the distal retinal pigment cells, but also leads to melanophore dispersion of the integumentary chromatophores (Kleinholz 1976; Kleinholz & Keller 1979). Thus, DRPH and melanophore or black pigment dispersing hormone (MDH,BPDH) appear to be identical in effect and indeed, in structure, as described later. Additionally, Kleinholz (1975) has shown that in *Uca pugilator* DRPH effects dispersion of the leucophores as well as the erythrophores. Since the results lend support to the "unitary hypothesis", the acronym PDH (pigment dispersing hormone) has recently come into use (Rao *et al.* 1985).

For pigment concentrating hormones, the balance of available evidence also points (though less clearly so) towards the "unitary hypothesis". Although at least two type-specific concentrating hormones, the white pigment concentrating (WPCH) and red pigment concentrating (RPCH) hormones have been suggested (Kleinholz 1970; Skorkowski 1971, 1972), it has been shown that synthetic RPCH concentrates the erythrophores, leucophores and melanophores of *Crangon crangon* (Skorkowski & Biegniewska 1981). However, in *Cambarellus shufeldtii* and *Uca* only the erythrophores were affected (Fingerman 1973). Further, it is not known whether RPCH acts antagonistically to PDH by moving pigment in the distal retinal cells to the dark adapted position.

In *Uca* PDH and RPCH are antagonistic; concentration of pigment by RPCH is markedly reduced by simultaneous injection of an assumed physiological dose of PDH (Liu & Keller, unpublished observations). Clearly, such experiments using pure hormones support the "unitary hypothesis" but, as yet, the existence of several effector type-specific effector hormones cannot be ruled out. The question remains as to how independent regulation of the different pigmentary effectors can occur without type-specific hormones. It seems possible that regulation may be effected by different sensitivities of different chromatophore types to differing concentrations of dispersing and concentrating hormones. Clearly, further studies in this area could be rewarding.

Pigment dispersing hormones (PDHs)

The first isolation of a PDH (as DRPH) was reported by Fernlund (1976) from *Pandalus borealis* eyestalks. The structures of PDHs from *Uca* and *Cancer magister* have recently been reported (Rao *et al.* 1985; Kleinholz *et al.* 1986) and are shown below.

Pandalus
Asn–Ser–Gly–Met–Ile–Asn–Ser–Ile–Leu–Gly–Ile–Pro–Arg–Val–Met–Thr–Glu–Ala–NH_2

Uca, Cancer
Asn–Ser–Glu–Leu–Ile–Asn–Ser–Ile–Leu–Gly–Leu–Pro–Lys–Val–Met–Asn–Asp–Ala–NH_2

Species or systematic group-related molecular variation is reflected by different residues at positions 3,4,11,13,16,17, (underlined) and by differences in interspecific biological activity. For example, *Pandalus* PDH is 21 times less potent in *Uca* than is *Uca* PDH (Rao *et al.* 1985). Although both hormones show the familiar

Table 1. *Biological activity of native and non-amidated Pandalus–PDH and of N– and C–terminal deletion analogs**

Compound	Relative activity
1–18–NH_2	100
1–18–OH	0.3
2–18–NH_2	15
3–18–NH_2	0.9
4–18–NH_2	0.27
5–18–NH_2	0.003
6–18–NH_2	0.03
1–17–NH_2	0.9
1–16–NH_2	0.8
1–15–NH_2	0.1
1–14–NH_2	0.09
1–13–NH_2	0.009
1–12–NH_2	0.003
1–10–NH_2	0.001
1– 9–NH_2	0.001
1– 8–NH_2	0.0
1– 6–NH_2	0.0
1– 4–NH_2	0.0

*Bioassays were performed on melanophores of *Uca*. (Data from Riehm & Rao 1982; and Riehm *et al.* 1985)

theme of invariant and variable partial sequences, it remains to be seen whether this is validated in other species.

Chromatographic or electrophoretic separation of eyestalk extracts yields 3–7 zones of PDH bioactivity in one species (Kleinholz 1972 amongst others). The structures shown above are those of the major forms in *Pandalus, Uca* and *Cancer*. Little is known concerning the identity of the other forms, but preliminary evidence from amino acid analysis (Keller & Kegel 1984) suggests that pigment dispersing substances with rather different structures may exist.

Using structural analogs of *Pandalus* PDH (Table 1), Riehm and Rao (1982) and Riehm *et al.* (1985) have shown the importance of the C–terminal amide in maintaining biological activity of PDH. Although the N–terminal residue is not a particularly strict requirement for biological activity, progressive truncation from the

Table 2. *Structure function relationships of RPCH, AKH I and synthetic analogs, tested on erythrophores of* Leander adspersus

	Compound	Relative activity
RPCH	pGlu–Leu–Asn–Phe–Ser–Pro–Gly–Trp–NH_2	100
	pGlu–Leu–Asn–Tyr–Ser–Pro–Gly–Trp–NH_2	400
	pGlu–Leu–Asn–Phe–Ser–Pro–Gly–NH_2	0.001
	pGlu–Leu–Asn–Phe–Ser–Pro–Gly	0.001
	pGlu–Leu–Asn–Phe–Ser–Pro–Gly–Trp	0.04
AKH I	pGlu–Leu–Asn–Phe–Ser–Pro–Asn–Trp–Gly–Thr–NH_2	2
	pGlu–Leu–Asn–Phe–Ser–Pro–Gly–D–Trp–NH_2	0.1
	pGlu–Pro–Gy–Trp–NH_2	1
	pGlu–Trp–NH_2	0.01
	Gly–Trp–NH_2	0.01
	Ac–Gly–NH_2	0.00001
	Ac–Gly–Trp–NH_2	0.001

From: Christensen, Carlsen & Josefsson 1978, 1979.

N–terminal end leads to a loss of activity and the area between residues 6–9 of the central, invariant hexapeptide core sequence, seems to be of particular importance (Riehm *et al.* 1985).

Pigment concentrating hormones

The red pigment or erythrophore concentrating hormone (RPCH) from the eyestalks of *Pandalus borealis* is the only hormone of this group whose primary structure has been determined (Fernlund & Josefsson 1972) and, indeed, was the first invertebrate neurohormone to be fully characterised. Later research revealed that RPCH belongs to the adipokinetic hormone group (AKHs) whose members are hyperlipaemic, hypertrehalosaemic, myoactive and cardioactive in insects (see Chapters 5 and 7). Indeed, AKH family members from insects cause red pigment concentration in crustaceans. The overlapping biological activity of AKHs and RPCH is a direct consequence of their related structures, as illustrated on Tables 2 and 3.

Studies using synthetic analogs (Christensen *et al.* 1978, 1979) have shown that the C–terminal amide and Trp are important for biological activity. Interestingly, for RPCH activity, substitution of Phe^4 for Tyr^4 increases the potency. For AKHs both the N– and C–termini are also important for activity. It is not known at present, whether different species or taxonomic group-related RPCH molecules occur in crustaceans. Considering the variety of AKH-like peptides in insects (see Chapters 5

Table 3. *RPCH and the other presently known members of the AKH-family of peptides*

RPCH[1]	pGlu–Leu–Asn–Phe–Ser–Pro–Gly–Trp–NH_2
AKH II *(Schistocerca*[2]*)*	pGlu–Leu–Asn–Phe–Ser–Thr–Gly–Trp–NH_2
AKH II *(Locusta)*[2]	pGlu–Leu–Asn–Phe–Ser–Ala–Gly–Trp–NH_2
M I *(Periplaneta)*[3]	pGlu–Val–Asn–Phe–Ser–Pro–Asn–Trp–NH_2
M II *(Periplaneta)*[3]	pGlu–Leu–Thr–Phe–Thr–Pro–Asn–Trp–NH_2
Manduca AKH[4]	pGlu–Leu–Thr–Phe–Thr–Ser–Ser–Trp–Gly–NH_2
AKH I[5]	pGlu–Leu–Asn–Phe–Thr–Pro–Asn–Trp–Gly–Thr–NH_2

References: [1]Fernlund & Josefsson 1982; [2]Siegert *et al.* 1985; [3]Scarborough *et al.* 1984; [4]Ziegler *et al.* 1985; [5]Stone *et al.* 1976.

and 7), this is a strong possibility and preliminary evidence (Mangerich *et al.* 1986) suggests that *Carcinus maenas* RPCH may differ from that of *Pandalus*.

4. The crustacean hyperglycemic hormone

Abramowitz *et al.* (1944) first demonstrated that a 'diabetogenic factor' (now called the crustacean hyperglycemic hormone: CHH) from the eystalks of *Callinectes sapidus* caused rapid rises in blood glucose levels. It has generally been observed that eyestalk ablation results in hypoglycaemia which can be reversed by injection of eyestalk or sinus gland extracts; this is a very rapid response, occurring within minutes of injection, reaching a peak within 2h, and thereafter declining rapidly (Keller & Andrew 1973). The physiological significance of hyperglycaemia is not yet clear. However, stress (Dean & Vernberg 1965; Salminen & Linquist 1975) induces hyperglycaemia, nocturnal activity is coincident with hyperglycaemia (Gorgels–Kallen & Voorter 1985) and stressors such as rapid salinity and temperature changes result in increased titres of circulating CHH and subsequently hyperglycemia in *Carcinus* (Orth 1985). It is possible, therefore, that this response plays a central role in physiological adaptation during periods of high energy demand. Purification of CHH has been described by Kleinholz and Keller (1973), Keller and Wunderer (1978), Keller (1981), and Keller and Kegel (1984). The CHH neuropeptides are all similar in terms of molecular weight (6–7,000), contain many acidic residues (Asx,Glx), possess low isoelectric points, are moderately hydrophobic, and are N–terminally blocked (Keller & Wunderer 1978; Keller 1981; Newcomb 1983; Keller & Kegel 1984; Martin *et al.* 1984a). Amino acid

Table 4. *Amino acid compositions of CHHs of* Carcinus, Orconectes, Porcellio *and of MIH from* Carcinus.

Amino acid	Hyperglycemic hormone (CHH)			Moult-inhibiting hormone (MIH)
	Carcinus[1]	*Orconectes*[1]	*Porcellio*[2]	*Carcinus*[3]
Asx	9	8	7	9
Thr	2	2	2	2
Ser	4	3	2	2
Glx	5	6	5	7
Pro	1	1	1	1
Gly	1	2	2	4
Ala	4	3	4	2
1/2Cys	4	4	4	4
Val	4	5	3	4
Met	3	1	2	1
Ile	1	3	1	3
Leu	5	5	4	5
Tyr	4	5	2	1
Phe	2	3	3	3
Trp	1	nd	nd	3
His	1	O	1–2	2
Lys	2	3	3	2
Arg	4	4	4	6
Total	57	58	50–52	61

Data from [1]Keller 1981; [2]Martin *et al.* 1984a; [3]Webster & Keller 1986; n.d. = not determined.

compositions of CHHs from several crustaceans are shown on Table 4. Some investigators have reported hyperglycaemic factors with somewhat different molecular weights (Skorskowski *et al.* 1977; Trausch & Bachau 1982; Van Wormhoudt *et al.* 1984), but these substances have not been investigated in detail.

One of the most striking features of CHH concerns its marked specificity. In general, CHHs from Brachyura are ineffective in Astacura and *vice versa* (Keller 1969; Leuven *et al.* 1982; Keller *et al.* 1985). This phenomenon is even more noticeable with *Porcellio dilatatus* CHH, which is hardly effective in decapods (and *vice versa*). This is reflected by the somewhat different amino acid composition of CHH from this animal (Table 4) compared with decapod CHH (Martin *et al.* 1984a).

Comparative electrophoretic studies on CHHs from 27 species of decapods (Keller 1975; Keller *et al.* 1985) highlight the 'group specificity' of CHH, which has also been demonstrated in HPLC studies (Keller & Kegel 1984). Clearly, CHH has undergone a considerable degree of molecular evolution in the Crustacea.

The abundance of CHH in the SG of decapod crustaceans – up to 10% of the total protein in *Carcinus* SG (Keller & Wunderer 1978) – facilitated the production of polyclonal antisera directed against CHH. This, together with the preparation of [^{125}I]–CHH, has allowed immunocytochemical (Jaros & Keller 1979a; Gorgels–Kallen & Van Herp 1981; Gorgels–Kallen *et al.* 1982; Keller *et al.* 1985) and radioimmunochemical (Jaros & Keller 1979b, 1981) studies. Immunocytochemical investigations have re-affirmed the sinus gland as the major site of storage of CHH (see also section 9). Radioimmunoassay has shown that the SG of medium sized *Carcinus* contain about 1.2 µg of CHH, in contrast to the medulla terminalis (3–4ng), and that the only other tissues containing low but significant quantities of CHH are the POs (4ng) (Keller *et al.* 1985). Levels of circulating CHH in the haemolymph are variable, but generally low (30 fmol.ml^{-1}), rising to peak levels of 1.4 pmol.ml^{-1} under stressful conditions, and dropping to undetectable levels following eyestalk removal (Orth 1985). As would be expected, structural differences between different CHHs are confirmed by a marked decrease in cross-reactivity with heterologous antigens in the CHH RIA, for example *Orconectes limosus* CHH has only a 2% cross-reaction with *Carcinus* CHH antibodies (Keller *et al.* 1985); for *Porcellio*, cross-reactivity is only 0.1% (Martin *et al.* 1984a).

Recent reviews by Sedlmeier (1987) and Keller & Sedlmeier (1987) have discussed in detail the mode of action of CHH upon the target organs (hepatopancreas and abdominal muscle). Briefly, it appears that CHH binds to specific membrane receptors, leading to production of cAMP and cGMP, which in turn bind to protein kinases, activating phosphorylase (which hydrolyses glycogen) and inhibiting glycogen synthetase. Recently, a further species specific effect of CHH has been described (Sedlmeier 1987), whereby amylase release from the hepatopancreas of *Orconectes* is stimulated by physiological levels of hormone. This action may again point to a function related to glucose metabolism.

5. The moult-inhibiting hormone

Crustaceans express growth overtly by periodically shedding the exoskeleton (ecdysis). This process is hormonally controlled by ecdysteroids, the moulting hormones of all arthropods (Hoffmann & Porchet 1984), and the moult-inhibiting hormone (MIH). It is now generally acepted that moulting processes are initiated and sustained by increases in ecdysteroid synthesis by the Y-organs, resulting in increased haemolymph ecdysteroid titres. This process is thought to be suppressed during intermoult by MIH produced by the eystalk neurosecretory complex (see Skinner 1985a for review). Thus, it has been known for many years that eyestalk removal results in accelerated moulting, presumably as a direct result of decreased MIH levels. Moreover, this is associated with increases in ecdysteroid synthesis and haemolymph ecdysteroid titre (Chang *et al.* 1976; Keller & Schmid 1979; Jegla *et al.* 1983) which can be prevented by injection of sinus gland extract (Keller & O'Connor 1982; Bruce & Chang 1984; Snyder & Chang 1986).

Previous attempts at characterising MIH, mostly based upon classical bioassay, deficiency and replacement experiments, have been reviewed by Skinner (1985a,b) and Webster (1986). Most of these have shown (not always convincingly) that MIH is probably a peptide of relative M.W. > 5,000.

A major deterrent to the isolation of MIH has been the lack of a physiologically relevant, rapid and sensitive bioassay, as well as the absence of a suitable working hypothesis on which to base such experiments (Skinner 1985a,b; Webster 1986; Webster & Keller 1986). Because one of the actions of MIH appears to be by the repression of ecdysteroid synthesis, suitable bioassays would involve measurement of ecdysteroid titres *in vivo* following eyestalk removal and/or injection of sinus gland extracts (Keller & O'Connor 1982; Bruce & Chang 1984) or measurement of the inhibition of ecdysteroid synthesis by Y-organs *in vitro* in the presence of sinus gland extracts (Soumoff & O'Connor 1982; Mattson & Spaziani 1985a,b; Webster 1986; Webster & Keller 1986). This latter assay method has allowed rapid progress towards the identification of MIH.

(a) *The putative moult-inhibiting hormone of* Carcinus maenas

MIH is mostly found in the sinus glands of *Carcinus* although small quantities are found in the medulla terminalis, and is a heat stable, trypsin sensitive peptide, M.W. > 6,000, which appears not to be species specific within the Brachyura (Webster 1986). Further studies (Webster & Keller 1986) have shown that, in addition to this novel peptide, CHH is also active in repressing ecdysteroid synthesis *in vitro*. The recently identified putative MIH is, however, many times more potent than CHH in repressing ecdysteroid synthesis (minimum effective concentration < 2.5×10^{-10} mol/l). It may be related to CHH since there are some important similarities in amino acid composition (see Table 4), although MIH is slightly larger (61 residues, M.W. 7,200, as opposed to CHH, 57 residues, M.W. 6,700). MIH shows no immunochemical cross-reaction to CHH, and is not hyperglycaemic, suggesting major structural differences which may preclude the possibility that the activity of CHH in respressing ecdysteroid synthesis is due to receptor cross-reactivity. The physiological significance of CHH mediated reduction of ecdysteroid synthesis thus remains unclear, but may be related to stress-mediated inhibition of moulting in crustaceans, a commonly observed phenomenon. Indeed, Mattson and Spaziani (1985c) have produced evidence for a role of eyestalk neurohormones in stress-mediated inhibition of ecdysteroid synthesis in *Cancer anternnarius*.

(b) *The putative moult-inhibiting hormone of* Cancer antennarius

In *Cancer*, a factor in the eyestalks, which may be related to the vertebrate vasopressins, inhibits ecdysteroid synthesis by Y-organs *in vitro* (Mattson & Spaziani 1985b). Authentic vasopressins (particularly lysine–vasopressin) are active in repressing ecdysteroid synthesis and preincubation of eyestalk extract with vasopressin antisera abolishes inhibitory activity, suggesting a specific response. However, concentrations of up to 10^{-6} mol/l Lys– and Arg–vasopressins are totally ineffective in reducing ecdysteroid production of Y-organs from *Carcinus* (unpublished observations).

6. Peptides of the pericardial organs – cardioactive peptides

The pericardial organs (POs) consist of paired nerve plexuses located in lateral positions in the pericardial cavity and are formed by nerve fibres originating from perikarya in the thoracic ganglion. These fibres project through openings of the branchiocardiac veins into the pericardium. Their nature as neurohaemal organs releasing cardioactive substances was first recognised by Alexandrowicz (1953a,b) and Alexandrowicz and Carlisle (1953). Subsequently, it was shown that the POs store and release several amines such as serotonin, dopamine and octopamine which are cardioactive and may affect other, distal targets (see Cooke & Sullivan 1982).

The presence of cardioactive peptides in the PO which can be released by electrical stimulation of the PO was suggested by Maynard and Welsh (1959), Cooke (1964) and Berlind and Cooke (1970), and biochemical studies (Belamarich 1963; Belamarich & Terwilliger 1966) suggested the presence of at least two small peptides. Recent studies, which will be discussed in detail in the following paragraphs have identified two cardioactive peptides, namely proctolin and crustacean cardioactive peptide. However, it is not yet clear whether these two peptides are identical to those proposed in earlier studies, or whether other cardioactive peptides exist.

(a) *Proctolin*

The structure of proctolin, originally isolated from cockroach hindguts (Brown & Starratt 1975; Starratt & Brown 1975) is Arg–Tyr–Leu–Pro–Thr–OH. Proctolin was discovered as a myotropic agent in insects, and later studies confirmed its role as a transmitter and/or neuromodulator, but it is not yet clear whether it functions as a blood-born neurohormone in insects (Miller 1979) (see also Chapter 8). In crustaceans, it is active on several neuromuscular preparations and on the heart (Kravitz *et al.* 1980; Miller & Sullivan 1981; Sullivan *et al.* 1981; Sullivan & Miller 1984). Sullivan (1979), using TLC and electrophoretic techniques, first suggested that proctolin was present in the POs of *Cardisoma carnifex*. This has been supported by radioimmunochemical, immunocytochemical and bioassay techniques for *Homarus americanus* (Schwarz *et al.* 1984) and, finally, by isolation and microsequencing from *Carcinus* POs (Stangier *et al.* 1986). Thus proctolin is now firmly established as one of the cardioactive peptides in the POs of crustaceans. The studies cited above point to a role as a neurohormone, and this is clearly supported by immunocytochemical results (Schwarz *et al.* 1984; Stangier *et al.* 1986). Other studies have shown that proctolin (or a substance closely related to it) in the nervous system of *Limulus polyphemus* induces myogenic activity in *Limulus* heart muscle (Watson & Augustine 1982; Watson *et al.* 1983; Watson & Hoshi 1985). Thus, it appears that proctolin may have a more general distribution among invertebrates than was previously expected.

(b) *Crustacean cardioactive peptide*

A recent HPLC study on cardioactive substances in the POs of *Carcinus* in our laboratory demonstrated a second zone of activity quite separate from proctolin. The primary structure of this material (known as crustacean cardioactive peptide: CCAP) has been elucidated by manual and automated gas-phase sequencing, and confirmed by synthesis (Stangier *et al.* 1987). The structure of this peptide, (shown below) does not resemble any of the known invertebrate or vertebrate neuropeptides.

CCAP: Pro–Phe–Cys–Asn–Ala–Phe–Thr–Gly–Cys–NH_2

At present, little is known of its physiological significance. At a dose of 1×10^{-9} mol/l it produces a marked accelerating action (inotropic and chronotropic) on the semi-isolated heart, and therefore meets the requirement for hormonal status in that it is active at low concentrations. The yield of extractable CCAP from both POs of a crab was 30–40 pmol (proctolin: 5.4 pmol), and the bioassay results indicate that the release of less than 1% of this amount would produce a haemolymph level of CCAP in the pericardial cavity sufficient to cause cardioacceleration. This is compatible with a neurohormonal role for this peptide. A non-amidated synthetic analog of CCAP was approximately 50 times less effective than the native, C-terminally amidated form (Stangier *et al.* 1987). Clearly, further work is required to establish a physiological role for this peptide.

7. Neurodepressing hormone

Eyestalk removal generally results in an increase in activity and loss of endogenous rhythmicity in crustaceans (see Aréchiga & Naylor 1976). This has led to the concept of a neurodepressing hormone (NDH) which lowers spontaneous neural activity in crustaceans. Purification of this factor from the eyestalks of *Procambarus bouveri* has been reported by Huberman *et al.* 1979 (see Aréchiga *et al.* 1985); NDH appears to be a heat-stable pronase-sensitive peptide, M.W. 1,000–1,300, which displays little species specificity. It shows no net charge when subjected to high voltage paper electrophoresis, and has recently been separated from other sinus gland neuropeptides of similar properties (e.g. RPCH) by HPLC (Aréchiga *et al.* 1985).

Non-specific effects in the bioassay used due to the inhibitory action of GABA, glycine and taurine, present in eyestalk extracts can be blocked with picrotoxin, suggesting a specific response to NDH (Aréchiga *et al.* 1985). Its actions seem to be unusually widespread, in that sensory receptors, inter- and motorneurones are all profoundly inhibited by this hormone, while the abdominal stretch receptors seem to be especially sensitive. NDH appears to act by stimulation of electrogenic Na^+ pumps (Aréchiga & Huberman 1980) and by activating Na^+–K^+–ATPase in crayfish abdominal ganglia (Aréchiga & Cerbón 1981). However, some doubt has been cast upon the existence of this factor as a neuropeptide (Cooke & Sullivan 1982) because it is only effective in the bioassay used (the abdominal stretch receptor of *Procambarus*) at relatively high concentrations (4–400 nmol/l). Furthermore, electrical stimulation of isolated sinus glands apparently removes *all* measurable NDH activity (Aréchiga *et al.* 1985), a feature which seems highly unlikely when compared with other, well-characterised crustacean neuropeptides, where similar operations (electrical stimulation or exposure to high K^+ solutions) releases only 10–20% of the extractable hormone (see Cooke & Sullivan 1982). Clearly, further critical work is required on this hormone, to investigate further these surprising results.

8. Other neuropeptides

Whilst the hormonal bases of colour change, blood glucose and cardiac regulation, and moult-inhibition are relatively well known, there are several other hormonal 'factors' which are at present poorly defined (see also Kleinholz & Keller 1979; Cooke & Sullivan 1982; Kleinholz 1985).

(a) *The gonad (or vitellogenesis) inhibiting hormone*

Evidence based primarily upon eyestalk removal experiments, suggests that an eyestalk neurohormone (GIH or VIH) inhibits gonad maturation and secondary vitellogenesis in decapod crustaceans (see Adiyodi & Adiyodi 1970; Adiyodi 1985). The precise nature of this factor, although no longer in question, has remained obscure. Several studies (Bomirski *et al.* 1981; Quackenbush & Herrnkind 1983; Van Deijnen 1986) have demonstrated that this substance is a peptide, M.W. 2–7,000. Jugan and Soyez (1985) have recently demonstrated that in *Macrobrachium rosenbergii*, GIH acts by inhibiting endocytotic uptake of lipovitellin by the oocytes. By measurement of the reduction of lipovitellin binding to oocytes of *Macrobrachium*, in conjunction with an *in vivo* assay which measures inhibition of oocyte growth in *Atyaephyra desmaresti*, a putative GIH from HPLC purified extracts of *Homarus gammarus* sinus glands of M.W. 7,000 has recently been isolated (Soyez & Van Deijnen, personal communcation). Since such assays seem to be sensitive and rapid, we expect that important new evidence on the nature of this hormone is imminent.

(b) *Osmoregulatory neurohormones*

Crustaceans exhibit a variety of osmoregulatory patterns, ranging from osmoconformation to hyper- and hypo-osmoregulation. These processes are apparently under neurohormonal control. Recent detailed reviews (Kamemoto & Oyama 1984; Mantel 1985) have highlighted the difficulties in studying such processes: depending on the animal or experimental protocol used, practically the entire neurosecretory system can be implicated in ion and water balance. The precise nature of these factors remain poorly defined (see Kleinholz & Keller 1979; Mantel & Farmer 1983 for specific examples). The diversity of responses observed has illuminated the need for rapid, specific and sensitive assays and better purification techniques. Promising assay systems have been reported by Charmantier *et al.* (1984a,b) using *Homarus americanus* larvae, and by Berlind & Kamemoto (1977) and Kamemoto & Oyama (1984) using an isolated gill perfusion system.

(c) *Limb growth inhibiting factor*

A heat-stable, pronase-sensitive eyestalk peptide of relative M.W. 1000 apparently inhibits proecdysial limb growth in *Gecarcinus lateralis* (Hopkins *et al.* 1979). Further characterisation of this factor has not been reported.

(d) *Other factors*

Some, rather insubstantial evidence has implicated the neurosecretory system in the control of lipid and protein metabolism, respiration and hypoglycaemia (see Cooke & Sullivan 1982). All these factors require critical re-examination.

9. Immunocytochemical studies

As far as the immunocytochemical localization of neuropeptides in the crustacean nervous system is concerned, we may distinguish between 'native' and 'naturalised' peptides (Greenberg & Price 1983). In the present case 'native' peptides are the typical crustacean peptides described above. They may, of course, occur in other invertebrates, or perhaps even in vertebrates. 'Naturalised' peptides are substances that have traditionally been considered to be 'vertebrate' peptides, such as enkephalins, substance P, somatostatin, ACTH, vasopressin etc. Many of these are now considered to be widespread in the animal kingdom, although their immunocytochemical or radioimmunochemical demonstration have rarely been corroborated by rigorous structural analysis. Thus, unspecific cross-reactivity cannot be ruled out. The presence of a considerable number of such 'vertebrate-type' peptides in the crustacean nervous system has been demonstrated (Van Deijnen *et al.* (1985), amongst others), but their function presently remains unknown (see also Chapters 2, 3, 4, 6 and 12).

Immunocytochemical studies of 'native' peptides have been hampered by lack of antisera, and it is only recently that antisera against CHH, RPCH, and PDH have been produced. An antiserum directed against CHH has been raised in rabbits using native unconjugated hormone (Jaros & Keller 1979b; Keller *et al.* 1985) whereas a synthetic antigen consisting of the N–terminal tetrapeptide sequence common to AKH I, II and RPCH, coupled to thyroglobulin (Schooneveld & Veenstra 1985) has been used to raise an antiserum suitable for detection of putative RPCH-containing structures. Recently, an antiserum directed against PDH has been raised using synthetic *Uca* PDH (Rao *et al.* 1985) coupled to thyroglobulin (Dircksen *et al.* 1987). In addition an antiserum against proctolin (Eckert *et al.* 1981), whose status as a 'native' peptide might be questioned because it was primarily isolated from insects, has been employed.

Fig. 1a summarises the results of the studies using CHH and RPCH antisera. In each case, a complete neurosecretory pathway has been elucidated. CHH immunoreactivity was found in a group of large (typically 40–50 µm) perikarya in the medulla terminalis X-organ (XO) in several decapod species (see Keller *et al.* 1985), and could also be traced through a major tract of secretion carrying axons that connect the X-organ perikarya with the neurohaemal organ, the sinus gland (SG). In the SG, a large number of CHH positive axon endings were found which were also identified by electron microscopical immunocytochemistry (see Keller *et al.* 1985). In the isopod *Porcellio dilatatus* only one pair of CHH cells in the anterior part of each of the frontal lobes of the protocerebrum were found (Martin *et al.* 1984b, see also Chapter 4). CHH is mainly restricted to the eyestalk neurosecretory system as very little radioimmunoreactivity or bioassayable hormone has been found in other parts of the nervous system. This agrees with immunocytochemical results. Only the POs contain low but significant quantities of hormone (Keller *et al.* 1985).

Immunocytochemical studies on structures containing putative RPCH in *Orconectes* and *Carcinus* have revealed a neurosecretory pathway consisting of a group of small perikarya close to the CHH cells of the XO with axonal processes which join the principal XO–SG tract and terminate in the SG (Fig. 1a,b). In contrast to CHH, RPCH immunoreactive elements (mostly cells, but also fibres) are found in other locations in the eyestalk (Fig. 1a) as well as in the brain and thoracic ganglion. However, no immunocytochemical or biological activity was found in the POs (Mangerich *et al.* 1986). RPCH activity in other parts of the nervous system has previously been demonstrated by bioassay (Fingerman & Couch 1967).

A recent study in our laboratory employing the new PDH antiserum has revealed an architecture of PDH perikarya and fibres quite different from that of the CHH and RPCH elements. Briefly, cells or groups of cells are scattered more widely in the optic ganglia and positive elements are also found in other parts of the nervous system (Mangerich *et al.* 1987; Dircksen *et al.* 1987). This is in line with earlier findings that bioassayable PDH is widespread in the nervous system.

Studies on the occurrence of proctolin in the POs have been accompanied by

Fig. 1a. Diagram of an eyestalk of *Carcinus maenas* demonstrating a: CHH containing perikarya (white cells in the XO) and RPCH immunoreactivity (group of black perikarya distal to the XO. The axons of both cell types form a tract that ends in axon terminals in the SG (note central haemolymph lacuna in the SG). The axons give rise to collaterals in the neuropile of the MT. Note single black RPCH cells in peripheral locations adjacent to the MT, MI and ME, and a small group with collateral forming axons in a proximal location in the MT. b: PAP-stained CHH cells in the XO (arrowheads) and silver enhanced immunogold stained RPCH cells (darkly stained cells to the left of the XO) on the same histological section. Note other immunonegative cells of intermediate size and unknown function in the XO (arrows) (× 280). c: Part of the ventral trunk of a pericardial organ of *Carcinus maenas* with proctolin immunoreactivity visualised by immunofluorescence in a whole mount preparation showing beaded cortical fibres and terminals (× 250).

LG lamina ganglionaris; ME, MI, MT, medulla externa, interna and terminalis; XO medulla terminalis X-organ; ON optic nerve; SG sinus gland. a: Mangerich *et al.* 1986. b: preparation and photographs by S. Mangerich. c: Stangier *et al.* 1986.

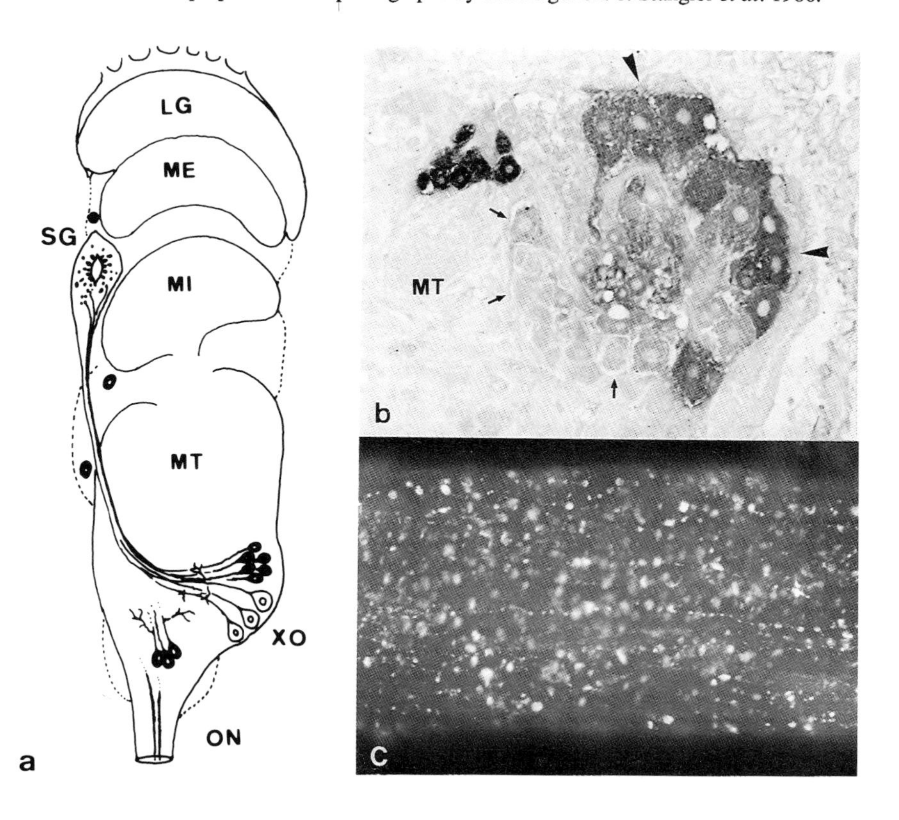

immunocytochemical demonstrations (Schwarz *et al.* 1984; Stangier *et al.* 1986). Fig. 1c shows a whole mount preparation of part of a longitudinal trunk of the POs of *Carcinus*, demonstrating the abundance of proctolin-containing fibres, and notably, terminals in close proximity to the epineurium.

10. Future perspectives

Compared with our knowledge of vertebrate neuroendocrine systems as well as those of insects and molluscs, it is apparent that many aspects of crustacean endocrinology are still poorly understood. Some of the difficulties encountered in identifying and purifying defined neuropeptides are due to the minute amount of starting material available, and to a paucity of sufficiently sensitive and rapid bioassays. For this reason little is known concerning the precise mode of action, target organs/receptors, biosynthesis and release of neuropeptides in crustaceans. However, to counter this somewhat pessimistic view, it should be mentioned that microanalytical techniques such as HPLC, manual and automated gas-phase peptide sequencing, and fast atom bombardment (FAB) mass spectrometry are increasingly

Fig. 2. HPLC elution profile of a 2 mol/1 acetic acid extract of 160 *Carcinus* sinus glands chromatographed on a Waters μ Bondapak Phenyl column (30 × 0.39 cm). Solvents were, A: 0.11% trifluoaroacetic acid, B: 60% acetonitrile in 0.1% trifluoroacetic acid. Gradient conditions 30–80% B in A over 1h. Flow rate 0.9 ml min^{-1} at 20°C. RPCH = red pigment concentrating hormone, PDH = pigment dispersing hormone(s), MIH = moult-inhibiting hormone, (arrowed), CHH = crustacean hyperglycemic hormone. AUFS = absorbtion units, full scale (210nm).

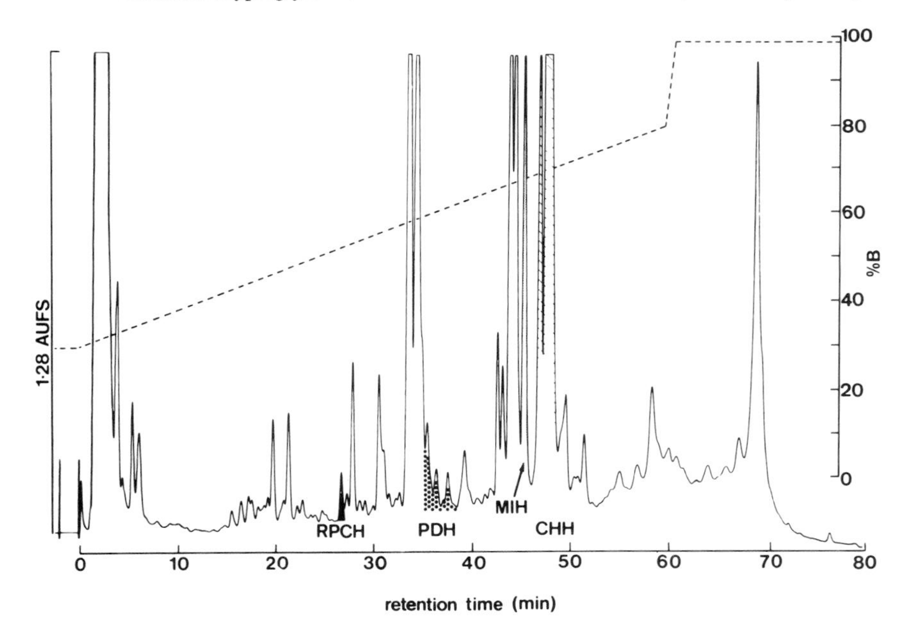

being used to isolate and sequence minute amounts of peptide. In particular, HPLC has allowed rapid purification of crustacean sinus gland neuropeptides (see Keller & Kegel 1984; Newcomb *et al.* 1985). An example of such an analysis obtained in our laboratory in which RPCH, PDH, MIH and CHH peptides have been profiled is shown on Fig. 2. Normally, material pure enough for amino acid analysis, or even sequencing, can be obtained after only two HPLC steps. Once a peptide has been purified, and preferably sequenced and synthesised, new areas of research such as immunocytochemical localisation, measurement of circulating hormone titres by radioimmunoassay and studies on hormone receptors may be initiated and progress can be made towards solving the problems outlined above. Another area in which we expect to see progress is in the identification of functions for peptides immunocytochemically similar to vertebrate peptides. The physiological significance of such compounds has yet to be demonstrated. Finally, the importance of molecular biological techniques in neuroendocrinology must be emphasised. As yet these powerful tools have only just begun to be used in crustaceans; their importance in shaping much of the field of current vertebrate neuroendocrinology cannot be underestimated.

Acknowledgements

Our work reported in this article was supported by grants from the Deutsche Forschungsgemeinschaft (Ke 206/7–1;6–5) to R.K. and by a Royal Society European Fellowship to S.G.W.

References

Abramowitz, A.A., Hisaw, F.L. & Papandrea, D.N. (1944) The occurrence of a diabetogenic factor in the eyestalks of crustaceans. *Biological Bulletin,* **86**, 1–5.

Adiyodi, R.G. (1985) Reproduction and its control. In: *The Biology of Crustacea.* **9**, ed. D.E. Bliss, pp. 147–215. New York: Academic Press.

Adiyodi, K.G. & Adiyodi, R.G. (1970) Endocrine control of reproduction in decapod Crustacea. *Biological Reviews of the Cambridge Philosophical Society,* **45**, 121–65.

Alexandrowicz, J.S. (1953a) Nervous organs in the pericardial cavity of the decapod Crustacea. *J. Marine Biol. Assoc. UK.*, **31**, 563–80.

Alexandrowicz, J.S. (1953b) Notes on the nervous system of the Stomatopoda. *Pubblicazioni della Stazione Zoologica di Napoli*, **24**, 29–45.

Alexandrowicz, J.S. & Carlisle, D.B. (1953) Some experiments on the function of the pericardial organs in Crustacea. *J. Marine Biol. Assoc. UK.*,**32**, 175–92.

Andrew, R.D. & Saleuddin, A.S.M. (1978) Structure and innervation of a crustacean neurosecretory cell. *Can. J. Zool.*, **56**, 235–46.

Andrew, R.D., Orchard, I. & Saleuddin, A.S.M. (1978) Structural re-evaluation of the neurosecretory system in the crayfish eyestalk. *Cell & Tiss. Res.*, **190**, 235–46.

Aréchiga, H. & Cerbón, J. (1981) The influence of temperature and deuterium oxide on the spontaneous activity of crayfish motoneurons. *Comp. Biochem. & Physiol.*, **69A**, 631–6.

Aréchiga, H., Flores–López, J. & Garcia, U. (1985) Control of biosynthesis and release of the crustacean neurodepressing hormone. In: *Comparative Endocrinology Symposium,* ed. B. Lofts & D. Chan (in press).

Aréchiga, H., García, U. & Rodríguez–Sosa, L. (1985) Neurosecretory role of crustacean eyestalk in the control of neuronal activity. In: *Model Neural Networks and Behavior.* Ed. Allen I. Selverston, pp. 361–79. New York, Boston: Plenum Publishing Company.

Aréchiga, H. & Huberman, A. (1980) Peptide modulation of neuronal ativity in crustaceans. In: *The Role of Peptides in Neuronal Function.* ed. J.F. Barker & T.G. Smith, pp. 317–49. New York: Marcel Dekker Inc.

Aréchiga, H. & Naylor, E. (1976) Endogenous factors in the control of rhythmicity in decapod crustaceans. In: *Biological Rhythms in the Marine Environment.* ed. P.J. De Coursey, pp. 1–16. University of South Carolina Press.

Belamarich, F. (1963) Biologically active peptides from the pericardial organs of the crab *Cancer borealis. Biol. Bull.*, **124**, 9–16.

Belamarich, F.A. & Terwilliger, R.C. (1966) Isolation and identification of cardioexcitor hormone from the pericardial organs of *Cancer borealis. Am. Zool.,***6**, 101–6.

Berlind, A. & Cooke, I.M. (1970) Release of a neurosecretory hormone as peptide by electrical stimulation of crab pericardial organs. *J. Exp. Biol.*, **53**, 679–86.

Berlind, A. & Kamemoto, F.I. (1977) Rapid water permeability changes in eyestalkless euryhaline crabs and in isolated perfused gills. *Comp. Biochem. & Physiol.*, **58a**, 383–5.

Bomirski, A., Arendarczyk, M., Kawinska, E. & Kleinholz, L.H. (1981) Partial characterization of crustacean gonad–inhibiting hormone. *Int. J. Invert. Reprod.*, **3**, 213–9.

Brown, B.E. & Starratt, A.N. (1975) Isolation of proctolin, a myotropic peptide from *Periplaneta americana. J. Insect Physiol.*, **21**, 1879–81.

Bruce, M.J. & Chang, E.S. (1984) Demonstration of a molt-inhibiting hormone from the sinus gland of the lobster *Homarus americanus. Comp. Biochem. Physiol.*, **79A**, 421–4.

Carlisle, D.B. & Knowles, F. (1959) *Endocrine Control in Crustaceans.* Cambridge: Cambridge University Press.

Chaigneau, I. (1983) Neurohemal Organs in Crustacea. In: *Neurohemal Organs in Arthropods.* ed. A.P. Gupta, pp. 53–89. Illinois: Charles C. Thomas.

Chang, E.S., Sage, B.A. & O'Connor, J.D. (1976) The qualitative and quantitative determination of ecdysones in tissues of the crab *Pachygrapsus crassipes,* following molt induction. *Gen. & Comp. Endocrinol.*, **30**, 21–33.

Charmantier, G.M., Charmantier–Daures, M. & Aiken, D.E. (1984a) Neuroendocrine control of hydromineral regulation in the American lobster, *Homarus americanus.* H. Milne–Edwards 1837 (Crustacea, Decapoda). 1. Juveniles. *Gen. & Comp. Endocrinol.*, **54**, 8–19.

Charmantier, G.M., Charmantier–Daures, M. & Aiken, D.E. (1984b). Neuroendocrine control of hydromineral regulation in the American lobster, *Homarus americanus.* H. Milne–Edwards 1837 (Crustacea, Decapoda). 2. Larval and postlarval stages *Gen. & Comp. Endocrinol.*, **54**, 20–34.

Christensen, M., Carlsen, J. & Josefsson, L. (1978) Structure-function studies on red-pigment concentrating hormone. The significance of the terminal residues. *Hoppe–Seyler's Zeitschrift für Physiologische Chemie,* **359**, 813–9.

Christensen, M., Carlsen, J. & Josefsson, L. (1979) Structure–function studies on red-pigment concentrating hormone, II. The significance of the C–terminal tryptophan amide. *Hoppe–Seyler's Zeitschrift für Physiologische Chemie*, **360**, 1051–61.

Cooke, I.M. (1964) Electrical activity and release of neurosecretory material in crab pericardial organs. *Comp. Biochem. & Physiol.*, **13**, 353–66.

Cooke, I.M. & Sullivan, R.E. (1982) Hormones and Neurosecretion. In: *The Biology of Crustacea* **3**, ed. D.E. Bliss, pp. 205–90. New York: Academic Press.

Dean, J.M. & Vernberg, F.J. (1965) Effects of temperature acclimation on some aspects of carbohydrate metabolism in decapod crustacea. *Biol. Bull.*, **129**, 87–94.

Dircksen, H., Zahnow, C.A., Gaus, G., Keller, R., Rao, K.R. & Riehm, J.P. (1987) The ultrastructure of nerve endings containing pigment dispersing hormone (PDH) in crustacean sinus glands. Identification by an antiserum against a synthetic PDH. *Cell & Tiss. Res.*, **250**, 377–387.

Eckert, M., Agricola, H. & Penzlin, H. (1981) Immunocytochemical identification of proctolin-line immunoreactivity in the terminal ganglion and hindgut of the cockroach *Periplaneta americana*. *Cell & Tiss. Res.*, **217**, 633–45.

Fernlund, P. (1976) Structure of a light-adapting hormone from the shrimp *Pandalus borealis*. *Biochem. et Biophys. Acta.*, **439**, 17–25.

Fernlund, P. & Josefsson, L. (1972) Crustacean color-change hormone: Amino acid sequence and chemical synthesis. *Science*,**177**, 173–4.

Fingerman, M. (1973) Behavior of the chromatophores of the fiddler crab *Uca pugilator* and the dwarf crayfish *Cambarellus schufeldtii* in response to synthetic *Pandalus* red-pigment concentrating hormone. *Gen. & Comp. Endocrinol.*, **20**, 289–92.

Fingerman, M. & Couch, E.F. (1967) The red pigment dispersing hormone of the abdominal nerve cord and its contribution to the physiology of the prawn, *Palaemonetes vulgaris*. *Rev. of Can. Biol.*, **26**, 109–17.

Gorgels–Kallen, J.L., Van Herp, F. & Leuven, R.S.E.W. (1982) A comparative immunocytochemical investigation of the crustacean hyperglycemic hormone (CHH) in the eyestalk of some decapod Crustacea. *J. Morph.*, **174**, 161–8.

Gorgels–Kallen, J.L. & Voorter, C.E.M. (1985) The secretory dynamics of the CHH-producing cell group in the eyestalk of the crayfish, *Astacus leptodactylus* in the course of the day/night cycle. *Cell & Tiss. Res.*, **241**, 361–6.

Greenberg, M.J. & Price, D.A. (1983) Invertebrate neuropeptides: Native and naturalised. *Ann. Rev. of Physiol.*, **45**, 271–8.

Hoffmann, J. & Porchet, M. (1984) *Biosynthesis, Metabolism and Mode of Action of Invertebrate Hormones*. Berlin: Springer–Verlag.

Hopkins, P.M., Bliss, D.E., Sheehan, S.W. & Boyer, J.R. (1979) Limb growth-controlling factors in the crab *Gecarcinus lateralis* with special reference to the limb growth-inhibiting factor. *Gen. & Comp. Endocrinol.*, **39**, 192–207.

Huberman, A., Aréchiga, H., Cimet, A., De La Rosa, J. & Arámburo, C. (1979) Isolation and purification of a neurodepressing hormone from the eyestalk of *Procambarus bouvieri* (Ortmann). *Europ. J. Biochem.*, **99**, 192–207.

Jacobs, A.A.C. & Van Herp, F. (1984) Immunocytochemical localization of a substance in the eyestalk of the prawn *Palaemon serratus* reactive with an anti-FMRF-amide rabbit serum. *Cell & Tiss. Res.*, **235**, 601–5.

Jaros, P.P. (1978) Tracing of neurosecretory neurons in crayfish optic ganglia by cobalt iontophoresis. *Cell & Tiss. Res.*, **194**, 297–302.

Jaros, P.P., Dircksen, H. & Keller, R. (1985) Occurrence of immunoreactive enkephalins in a neurohemal organ and other nervous structures in the eyestalk of the shore crab *Carcinus maenas* L. *Cell & Tiss. Res.*, **241**, 111–7.

Jaros, P.P. & Keller, R. (1979a) Immunocytochemical identification of hyperglycemic hormone-producing cells in the eyestalks of *Carcinus maenas*. *Cell & Tiss. Res.*, **204**, 379–85.

Jaros, P.P. & Keller, R. (1979b) Radioimmunoassay of an invertebrate peptide hormone – the crustacean hyperglycemic hormone. *Experientia*, **35**, 1252–3.

Jaros, P.P. & Keller, R. (1981) Improvement and first application of a RIA for an invertebrate neurosecretory peptide hormone, the crustacean hyperglycemic hormone. In: *Neurosecretion*. eds. D.S. Farner & K. Lederis, pp. 517–8, New York, Boston: Plenum Publishing Company.

Jegla, T.C., Ruland, K., Kegel, G. & Keller, R. (1983) The role of the Y-organ and cephalic gland in ecdysteroid production and control of molting in the crayfish *Orconectes limosus*. *J. Comp. Physiol. B*, **152**, 91–5.

Jugan, P. & Soyez, D. (1985) Démonstration in vitro de l'inhibition de l'endocytose ovocytaire par un extrait de glandes du sinus chez la crevette *Macrobrachium rosenbergii*. *Comptes Rendus de l'Académie des Sciences. Paris*, **300**, 705–9.

Kamemoto, F.I. & Oyama, S.N. (1984) Neuroendocrine influence on effector tissues of hydromineral balance in crustaceans. In: *Ninth International Symposium on Comparative Endocrinology Proceedings*. ed. B. Lofts, Hong Kong University Press.

Keller, R. (1969) Untersuchungen zur Artspezifität eines Crustaceen hormons. *Z. vergl. Physiol.*, **63**, 137–45.

Keller, R. (1981) Purification and amino acid composition of the hyperglycemic neurohormone from the sinus gland of *Orconectes limosus* and comparison with the hormone from *Carcinus maenas*. *J. Comp. Physiol.*, **41**, 445–50.

Keller, R. (1983) Biochemistry and specificity of the neurohemal haormones in Crustacea. In: *Neurohemal Organs of Arthropods*. ed. A.P. Gupta, pp. 118–48. Illinois: Charles C. Thomas.

Keller, R. & Andrew, E.M. (1973) The site of action of the crustacean hyperglycemic hormone. *Gen. & Comp. Endocrinol.*, **20**, 572–8.

Keller, R. & O'Connor, J.D. (1982) Neuroendocrine regulation of ecdysteroid production in the crab *Pachygrapsus crassipes*. *Gen. & Comp. Endocrinol.*, **46**, 384.

Keller, R. & Kegel, G. (1984) Studies on crustacean eyestalk neuropeptides by the use of high performance liquid chromatography. In: *Biosynthesis, Metabolism and Mode of Action of Invertebrate Hormones*. ed. J. Hoffmann & J. Porchet, pp. 145–54. Heidelberg: Springer–Verlag.

Keller, R., Jaros, P.P. & Kegel, G. (1985) Crustacean hyperglycemic neuropeptides. *Am. Zool.*, **25**, 207–21.

Keller, R. & Schmid, E. (1979) In vitro secretion of ecdysteroids by Y-organs and lack of secretion by mandibular organs of the crayfish following molt induction. *J. Comp. Physiol. B*, **130**, 347–53.

Keller, R. & Sedlmeier, D. (1987) A metabolic hormone in crustaceans: The hyperglycemic neuropeptide. In: *Invertebrate Endocrinology*. **1**, ed. G.H. Downer & H. Laufer. New York: Alan R. Liss. (in press).

Keller, R. & Wunderer, G. (1978) Purification and amino acid composition of the neurosecretory hyperglycemic hormone from the sinus gland of the shore crab *Carcinus maenas*. *Gen. & Comp. Endocrinol.*,**34**, 328–35.

Kleinholz, L.H. (1966) Hormonal regulation of retinal pigment migration in crustaceans. In: *Proceedings of the International Symposium on the Functional Organization of the Compound Eye. Stockholm, 1965*. Oxford, New York: Pergamon Press.

Kleinholz, L.H. (1970) A progress report on the separation and purification of crustacean neurosecretory pigmentary effector hormones. *Gen. & Comp. Endocrinol.*, **14**, 578–88.

Kleinholz, L.H. (1972) Comparative studies of crustacean melanophore stimulating hormones. *Gen. & Comp. Endocrinol.*, **19**, 473–83.

Kleinholz, L.H. (1975) Purified hormones from the crustacean eyestalk and their physiological specificity. *Nature*, **258**, 256–7.

Kleinholz, L.H. (1976) Crustacean neurosecretory hormones and physiological specificity. *Am. Zool.*, **16**, 151–66.

Kleinholz, L.H. (1985) Biochemistry of crustacean hormones. In: *The Biology of Crustacea*. **9**, ed. D.E. Bliss, pp. 463–522. New York: Academic Press.

Kleinholz, L.H. & Keller, R. (1973) Comparative studies in crustacean hyperglycemic hormones. I. The initial survey. *Gen. & Comp. Endocrinol.*, **21**, 554–64.

Kleinholz, L.H. & Keller, R. (1979) Endocrine regulation in Crustacea. In: *Hormones and Evolution*. **1**, ed. E.T.W. Barrington, pp. 159–213. New York: Academic Press.

Kleinholz, L.H., Rao, K.R., Riehm, J.P., Tarr, G.E., Johnson, L. & Norton, S. (1986) Isolation and sequence analysis of a pigment dispersing hormone from eyestalks of the crab *Cancer magister*. *Biol. Bull.*, **170**, 135–43.

Kravitz, E.A., Glussman, S., Harris–Warrick, R.M., Livingstone, M.S., Schwarz, T. & Goy, M.F. (1980) Amines and a peptide as neurohormones in lobsters: Actions on neuromuscular preparations and preliminary behavioural studies. *J. Exp. Biol.*, **89**, 159–75.

Leuven, R.S.E.W., Keller, R., Van Herp, F. & Jaros, P.P. (1982) Species or group specificity in biological and immunological studies of crustacean hyperglycemic hormone. *Gen. & Comp. Endocrinol.*, **46**, 288–96.

Mancillas, J.R., McGinty, J.F., Selverston, A.I., Karten, H. & Bloom, F.E. (1981) Immunocytochemical localization of enkephalin and substance P in retina and eyestalk neurons of lobster. *Nature*, **293**, 576–8.

Mangerich, S., Dircksen, H. & Keller, R. (1986) Immunocytochemical identification of structures containing putative red pigment-concentrating hormone in two species of decapod crustacean. *Cell & Tiss. Res.*, **245**, 377–86.

Mangerich, S., Keller, R., Dircksen, H., Rao, K.R. & Riehm, J.P. (1987) Localization of pigment-dispersing hormone (PDH) and coexistence with FMRFamide immunoreactivity in the eystalks of two decapod crustaceans. *Cell & Tiss. Res.*, **250**, 365–375.

Mantel, L.H. (1985) Neurohormonal integration of osmotic and ionic regulation. *Am. Zool.*,**25**, 253–63.

Mantel, L.H. & Farmer, L.L. (1983) Osmotic and ionic regulation. In: *The Biology of Crustacea*. **5**, ed. D.E. Bliss, pp. 54–161. New York: Academic Press.

Martin, G., Keller, R., Besse, G. & Jaros, P.P. (1984a) The hyperglycemic neuropeptide of the terrestrial isopod, *Porcellion dilatatus*. I: Isolation and characterisation. *Gen. & Comp. Endocrinol.*,**55**, 208–16.

Martin, G., Jaros, P.P., Besse, G. & Keller, R. (1984b) The hyperglycemic neuropeptide of the terrestrial isopod, *Porcellio dilatatus*. II.: Immunocytochemical demonstration in neurosecretory structures of the nervous system. *Gen. & Comp. Endocrinol.*,**55**, 217–26.

Mattson, M.P. & Spaziani, E. (1985a) Characterization of molt-inhibiting hormone (MIH) action on crustacean Y-organ segments and dispersed cells in culture and a bioassay for MIH activity. *J. Exp. Zool.*, **236**, 93–101.

Mattson, M.P. & Spaziani, E. (1985b) Functional relations of crab molt-inhibiting hormone and neurohypophysial peptides. *Peptides,* **6**, 635–40.

Mattson, M.P. & Spaziani, E. (1985c) Stress reduces hemolymph ecdysteroid levels in the crab: Mediation by the eyestalks. *J. Exp. Zool.,* **234**, 319–23.

Maynard, D.M. (1961) Thoracic neurosecretory structures in Brachyura. II. Secretory neurons. *Gen. & Comp. Endocrinol.,***1**, 237–63.

Maynard, D.M. & Welsh, J.H. (1959) Neurohormones of the pericardial organs of brachyuran Crustacea. *J. Physiol.,* **149**, 215–27.

Miller, T. (1979) Nervous versus neurohormonal control of insect heartbeat. *Am. Zool.,* **19**, 77–86.

Miller, M.W. & Sullivan, R.E. (1981) Some effects of proctolin on the cardiac ganglion of the marine lobster, *Homarus americanus.* (Milne–Edwards). *J. Neurobiol.,* **12**, 629–39.

Newcomb, R.W. (1983) Peptides in the sinus gland of *Cardisoma carnifex:* Isolation and amino acid analysis. *J. Comp. Physiol. B,* **153**, 207–21.

Newcomb, R.W., Stuenkel, E.L. & Cooke, I.M. (1985) Characterization, biosynthesis and release of neuropeptides from the X-organ-sinus gland system of the crab, *Cardisoma carnifex. Am. Zool.,* **25**, 157–71.

Orth, H.P. (1985) Untersuchungen zur Physiologie des Hyperglykämischen Hormons der Crustaceen (CHH) bei *Carcinus maenas.* Diplomarbeit, Universität Bonn.

Quackenbush, L.S. & Herrnkind, W.F. (1983) Partial characterization of eyestalk hormones controlling molt and gonadal development in the spiny lobster *Panulirus argus. J. Crust. Biol.,* **3**, 34–44.

Rao, K.R. (1985) Pigmentary effectors. In: *The Biology of Crustacea.* **9**, ed. D.E. Bliss, pp. 395–462. New York: Academic Press.

Rao, K.R., Riehm, J.P., Zahnow, C.A., Kleinholz, L.H., Tarr, G.E., Johnson, L., Norton, S., Landau, M., Semmes, O.J., Sattelberg, R.M., Jorenby, W.H. & Hintz, M.F. (1985) Characterization of a pigment-dispersing hormone in eyestalks of the fiddler crab, *Uca pugilator. Proc. Natl. Acad. Sci. USA,* **82**, 5319–23.

Riehm, J.P. & Rao, K.R. (1982) Structure-activity relationships of a pigment dispersing crustacean neurohormone. *Peptides,* **3**, 643–7.

Riehm, J.P., Rao, K.R., Semmes, O.J., Jorenby, W.H., Hintz, M.F. & Zahnow, C.A. (1985) C-terminal deletion analogs of a crustacean pigment-dispersing hormone. *Peptides,* **6**, 1051–6.

Salminen, I. & Lindquist, O.V. (1975) Effect of temperature change on the blood glucose level in the crayfish *Astacus astacus.* In: *Freshwater Crayfish Papers from the Second International Symposium on Freshwater Crayfish.* ed. J.W. Avault, pp. 203–9. Baton Rouge: Louisiana State University.

Scarborough, R.M., Jamieson, G.C., Kalish, F., Kramer, S.J., McEnroe, G.A., Miller, C.A. & Schooley, D.A. (1984) Isolation and primary structure of two peptides with cardioacceleratory and hyperglycemic activity from the corpora cardiaca of *Periplaneta americana. Proc. Natl. Acad. Sci. USA,* **81**, 5575–80.

Schooneveld, H. & Veenstra, J.A. (1985) Insect neuroendocrine cells and neurons containing various adipokinetic hormone (AKH)-immunoreactive substances. In: *Neurosecretion and the Biology of Neuropeptides,* ed. H. Koyabashi *et al.*, pp. 425–34, Springer: Berlin, Japan Scientific Society Press, Tokyo.

Schwarz, T.L., Lee, G.M.H., Siwiki, K., Standaert, D.G. & Kravitz, E.A. (1984) Proctolin in the lobster: The distribution, release and chemical characterisation of a likely neurohormone. *J. Neuroscience,* **4**, 1300–11.

Sedlmeier, D. (1987) The crustacean hyperglycemic hormone (CHH): A regulatory secretagogue of crayfish midgut gland amylase secretion. *Regulatory Peptides*. (in press).

Siegert, K., Morgan, P. & Mordue, W. (1985) Primary structures of locust adipokinetic hormones II. *Hoppe–Seyler's Zeischrift für Biologische Chemie*, **366**, 723–7.

Skinner, D.M. (1985a) Molting and Regeneration. In: *The Biology of Crustacea*. **9**, ed. D.E. Bliss, pp. 43–146. New York: Academic Press.

Skinner, D.M. (1985b) Interacting factors in the control of the crustacean molt cycle. *Am. Zool.*, **25**, 275–84.

Skorkowski, E.F. (1971) Isolation of three chromatophorotropic hormones from the eyestalk of the shrimp *Crangon crangon*. *Marine Biol.*, **8**, 220–23.

Skorkowski, E.F. (1972) Separation of three chromatophorotropic hormones from the eyestalk of the crab *Rhithropanopeus harrisi* (Gould). *Gen. & Comp. Endocrinol.*, **18**, 329–34.

Skorkowski, E.F. & Biegniewska, A. (1981) Neurohormones and control of physiological processes in Crustacea. *Adv. in Physiol. Sci.*, **23**, 419–32.

Skorkowski, E.F., Rykiert, M. & Lipinska, B. (1977) Hyperglycemic hormone from the eyestalk of the shrimp *Crangon crangon*. *Gen. Comp. Endocrinol.*, **33**, 460–7.

Snyder, M.J. & Chang, E.S. (1986) Effects of sinus gland extracts of larval moulting and ecdysteroid titers of the American lobster, *Homarus americanus*. *Biol. Bull.*, **170**, 244–54.

Soumoff, C. & O'Connor, J.D. (1982) Repression of Yorgan secretory activity by molt-inhibiting hormone in the crab *Pachygrapsus crassipes*. *Gen. & Comp. Endocrinol.*, **48**, 432–9.

Stangier, J., Dircksen, H. & Keller, R. (1986) Identification and immunocytochemical localization of proctolin in pericardial organs of the shore crab, *Carcinus maenas*. *Peptides*, **7**, 67–72.

Stangier, J., Hilbich, C., Beyreuther, K. & Keller, R. (1987) A novel cardioactive peptide (CCAP) from pericardial organs of the shore crab *Carcinus maenas*. *Proc. Natl. Acad. Sci. USA.*, **84**, 575–9.

Starratt, A.N. & Brown, B.E. (1975) Structure of the pentapeptide, proctolin, a proposed neurotransmitter in insects. *Life Sciences*, **17**, 1253–56.

Stone, J.V., Mordue, W., Battley, K.E., Morris, H.R. (1976) Structure of the locust adipokinetic hormone, a hormone that regulates lipid utilisation during flight. *Nature*, **263**, 207–11.

Sullivan, R.E. (1984) A proctolin-like peptide in crab pericardial organs. *J. Exp. Zool.*, **210**, 543–53.

Sullivan, R.E. & Miller, M.W. (1984) Dual effects of proctolin on the rhythmic burst activity of the cardiac ganglion. *J. Neurobiol.*, **15**, 173–96.

Sullivan, R.E., Tazaki, K. & Miller, M.W. (1981) Effects of proctolin on the lobster cardiac ganglion. *Soc. Neurosci. Abst.*, **7**, 253.

Trausch, G. & Bachau, A. (1982) Biological activity of eyestalk extracts from the lobster, *Homarus americanus*. *Gen. & Comp. Endocrinol.*, **46**, 385.

Van Deijnen, J.E. (1986) Structural and biochemical investigations into the neuro-endocrine system of the optic ganglia of decapod Crustacea. Proefschrift, Katholieke Universiteit Nijmegen.

Van Deijnen, J.E., Vek, F. & Van Herp, F. (1985) An immunocytochemical study of the optic ganglia of the crayfish *Astacus leptodactylus* (Nordmann 1842) with antisera against biologically active peptides of vertebrates and invertebrates. *Cell & Tiss. Res.*, **240**, 175–83.

Van Wormhoudt, A., Van Herp, F., Bellon–Humbert, C. & Keller, R. (1984) Changes and characteristics of the crustacean hyperglycemic hormone (CHH material) in *Palaemon serratus* (Crustacea, Decapoda, Natantia) during the different steps of the purification. *Comp. Biochem. & Physiol.*, **79B**, 353–60.

Watson, W.H. & Augustine, G.J. (1982) Peptide and amine modulation of the *Limulus* heart: A simple neural network and its target tissue. *Peptides*, **3**, 485–92.

Watson, W.H., Augustine, G.T., Benson, J.A. & Sullivan, R.E. (1983) Proctolin and an endogenous proctolin-like peptide enhance the contractility of the *Limulus* heart. *J. Exp. Biol.*, **103**, 55–74.

Watson, W.H. & Hoshi, T. (1985) Proctolin induces rhythmic contractions and spikes in *Limulus* heart muscle. *Am. J. Physiol.*, **249**, 490–5.

Webster, S.G. (1986) Neurohormonal control of ecdysteroid biosynthesis by *Carcinus maenas* Y-organs *in vitro*, and preliminary characterisation of the putative moult-inhibiting hormone (MIH). *Gen. & Comp. Endocrinol.*, **61**, 237–47.

Webster, S.G. & Keller, R. (1986) Purification, characterisation and amino acid composition of the putative moultzinhibiting hormone (MIH) of *Carcinus maenas* (Crustacea, Decapoda). *J. Comp. Physiol. B*, **156**, 617–624.

Ziegler, R., Eckart, K., Schwarz, H. & Keller, R. (1986) Amino acid sequence of *Manduca sexta* adipokinetic hormone elucidated by combined fast atom bombardment (FAB)/Tandem Mass Spectrometry. *Biochem. & Biophys. Res. Comm.*, 337–42.

PART III

Neurohormones in Coelenterates, Annelids and Protochordates

C.J.P. GRIMMELIKHUIJZEN, D. GRAFF &
A.N.SPENCER

Structure, location and possible actions of Arg–Phe–amide peptides in coelenterates

Introduction

Coelenterates have the simplest nervous system in the animal kingdom, and it was probably within this group of animals that nervous systems first evolved. Extant coelenterates are diverse and comprise two phyla. The classes Hydrozoa (for example hydroids and their medusae), Cubozoa (box jellyfishes), Scyphozoa (true jellyfishes), and Anthozoa (for example sea anemones and corals) constitute the phylum Cnidaria. A companion phylum is that of the Ctenophora (comb jellies), or Acnidaria. The general plan of the coelenterate nervous system has often been described as a nerve net. This is an oversimplification because many species show condensation of neurones to form linear or circular tracts. Linear tracts occur in the stem and tentacles of physonectid siphonophores (Mackie 1973; Grimmelikhuijzen *et al.* 1986), between the ocelli and marginal nerve rings of anthomedusae (Singla & Weber 1982; Grimmelikhuijzen & Spencer 1984), and at the bases of mesenteries in sea anemonies (Batham *et al.* 1960). Circular tracts, or nerve rings, have been found in the bell margin of hydrozoan medusae (Hertwig & Hertwig 1878; Jha & Mackie 1967; Spencer 1979; Grimmelikhuijzen *et al.* 1986) and at the base of the hypostome in some hydrozoan and cubozoan polyps (Werner *et al.* 1976; Grimmelikhuijzen 1985). Presumably, these tracts have evolved to form pathways for rapid conduction of information. The marginal nerve rings of hydromedusae, in addition, form a circular CNS, which is capable of both integrating a variety of sensory inputs, and of transmitting the input rapidly throughout the margin (Spencer & Arkett 1984).

Ultrastructural studies have shown that many synapses in the coelenterate nervous system are chemical. Dense-cored vesicles (70–150 nm), often associated with pre- and postsynaptic membrane specializations, have been seen at synapses of all coelenterate classes (Horridge & Mackay 1962; Jha & Mackie 1967; Westfall 1973a,b; Hernandez–Nicaise 1973; Spencer 1979). Physiological studies demonstrating synaptic blockage by depletion of Ca^{2+} or by addition of excess Mg^{2+}, also suggested that classical, exocytotic release of transmitter substances occurs (McFarlane 1973; Spencer 1982; Satterlie 1979; Anderson & Schwab 1982; Spencer

& Arkett 1984). Finally, simultaneous intracellular recordings at both pre- and postsynaptic neurones in the hydromedusa *Polyorchis* (Spencer 1982; Spencer & Arkett 1984) and the scyphomedusa *Cyanea* (Anderson 1985), displayed the expected E.P.S.P.s (excitatory post synaptic potentials) with constant latency from the presynaptic spike.

In addition to chemical synapses, electrical synapses have also been found. The presence of these synapses has been indicated by a variety of experimental approaches including the demonstration of electrical coupling between neurones, the spread of dye, and the presence of structures which had the morphological characters of gap junctions (Anderson & Mackie 1977; Spencer 1978, 1981; Spencer & Satterlie 1980; Westfall *et al.* 1980; Satterlie & Spencer 1983; Spencer & Arkett 1984; Satterlie 1985). In some cases, such as in the nerve net and 'giant axon' of the stem of physonectid siphonophores, the neurons have apparently fused with each other and true syncytia might exist (Mackie 1973; Grimmelikhuijzen *et al.* 1986).

Despite overwhelming evidence that coelenterate neurones form chemical synapses, no transmitter substance has been unequivocally identified in these animals (see review by Martin & Spencer 1983). Recently, however, using immunocytochemistry and radioimmunoassays, we demonstrated in the nervous systems of coelenterates the presence of several immunoreactive substances related to vertebrate and invertebrate neuropeptides. Among them were peptides resembling bombesin/gastrin-releasing-peptide, oxytocin/vasopressin and the molluscan neuropeptide Phe–Met–Arg–Phe–amide (Grimmelikhuijzen *et al.* 1981, 1982a,b). In this review we would like to restrict ourselves to a discussion of peptides related to Phe–Met–Arg–Phe–amide (FMRFamide; Price & Greenberg 1977). The reader is referred to Grimmelikhuijzen (1984) for a review of the other neuropeptides in coelenterates.

Immunocytochemistry

Numerous polyclonal antisera against FMRFamide have been prepared, which gave strong staining* of neurones in a wide variety of coelenterates. These antisera did not show any affinity to FMRFamide-related peptides such as gastrin/cholecystokinin (carboxyterminus: WMDFamide), or met–enkephalin–Arg^{6}–Phe^{7} (carboxyterminus: FMRF), nor to many unrelated peptides such as bombesin, gastrin-releasing-peptide, oxytocin and vasopressin (Grimmelikhuijzen 1983, 1984). FMRFamide antisera did, however, generally cross-react with peptides bearing the Arg–Phe–amide or Arg–Tyr–amide moiety (Grimmelikhuijzen 1983,1984; Triepel &

* In this context, 'staining' means binding of a primary antiserum together with a secondary antiserum which contains a chromophore. This is part of a standard immunocytochemical technique (see Chapters 2, 3 and 6).

Fig. 1. Whole-mount staining of *Hydra* with RFamide antiserum 146II (cf. Grimmelikhuijzen 1985). (a) The hypostome of *Hydra attenuata*, showing an agglomeration of sensory neurones around the mouth opening (×200). (b) Low magnification micrograph of *Hydra oligactis*, showing a nerve ring at the base of the hypostome (×85). (c) Higher magnification of the hypostomal nerve ring of *Hydra oligactis*. Also a cluster of sensory neurones (somewhat out of focus) can be seen around the mouth opening (×200).

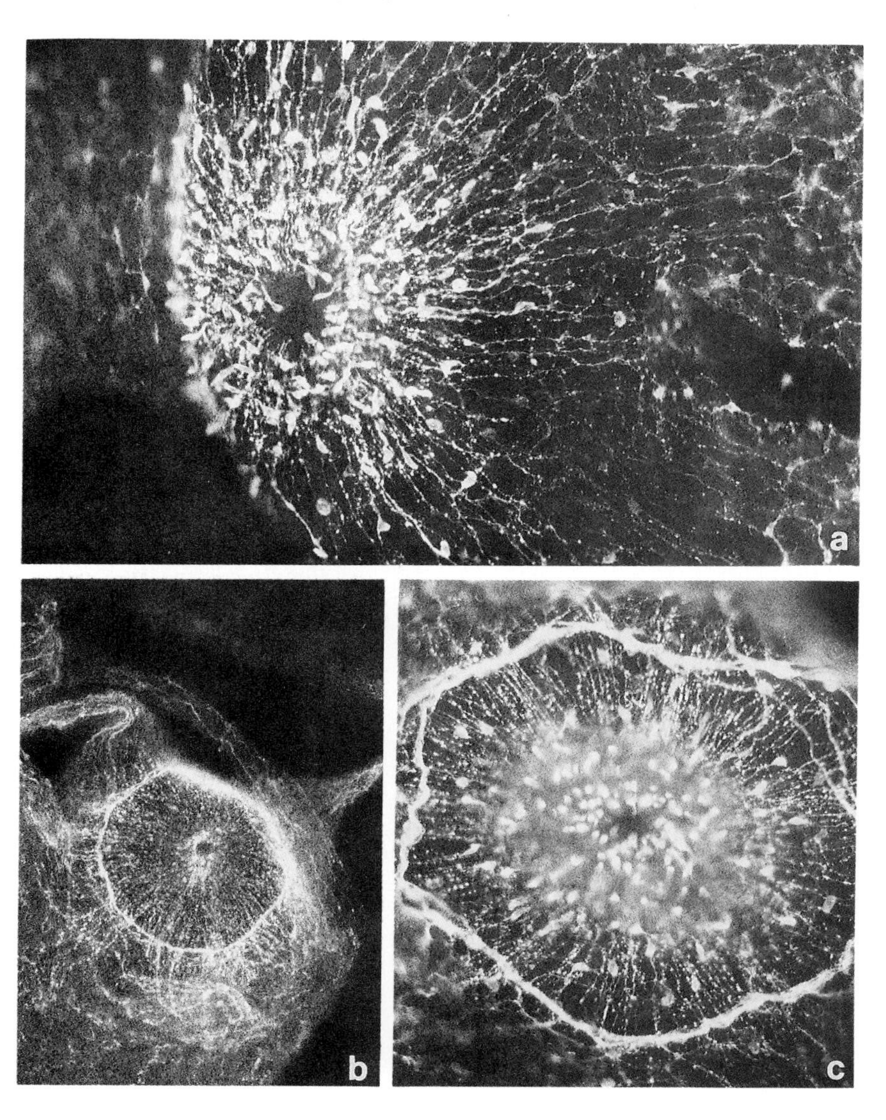

Fig. 2. Whole mount of polyps of *Hydractinia echinata* stained with RFamide antiserum 146II. (a) The upper half of a gastrozooid. The hypostome has been broken somewhat by pressure of the coverslip. Note the dense plexus of neuronal processes throughout the hydranth and sensory neurones around the mouth opening (×140). (b) Gonophores are covered by highly immnoreactive neurones (×150). Adapted from Grimmelikhuijzen 1985.

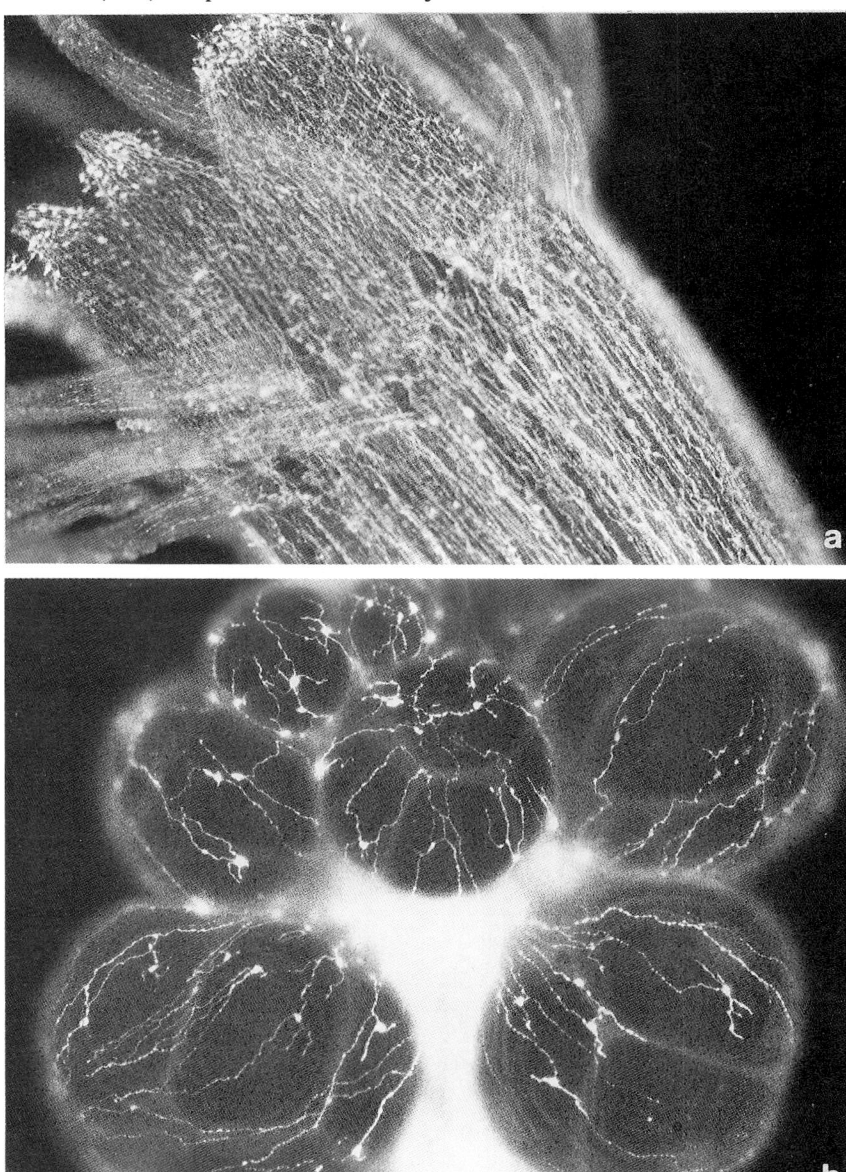

Grimmelikhuijzen 1984a). Peptides containing the Arg–Tyr–amide moiety belong to the family of pancreatic polypeptide (Lin & Chance 1974; Kimmel *et al.* 1975; Tatemoto 1982a,b). Although all rabbit antisera to FMRFamide cross reacted to peptides of this family, some of the antisera raised in guinea pigs showed no affinity to bovine and avian pancreatic polypeptide (Grimmelikhuijzen & Graff 1985; Grimmelikhuijzen 1986) and to neuropeptide Y (Triepel & Grimmelikhuijzen 1984), which is another member of the pancreatic polypeptide family. Since these guinea-pig antisera were still found to react with all neuronal structures identified by the other FMRFamide antisera in coelenterates, the indications are that the coelenterate FMRFamide-like peptides are not pancreatic polypeptide-like.

If one assumes that the FMRFamide-like peptides of coelenterates have only a part of the sequence Phe–Met–Arg–Phe–amide in common with FMRFamide itself, then antisera against fragments might react with coelenterate neurones much better than antisera against the complete peptide. Indeed, we found that antisera against the carboxyterminal fragment RFamide stained more strongly than FMRFamide antisera. These findings indicate that the coelenterate FMRFamide-like peptides have the sequence RFamide in common with FMRFamide, but that the amino terminus is different.

RFamide peptides in the nervous system of hydrozoans

Whole-mount incubations of fixed *Hydra* with an FRamide antiserum, revealed neuronal centralizations in the nerve net of the hypostome (Grimmelikhuijzen 1985). The centralization in the hypostome of *Hydra attenuata* consists of a strong aggregation of sensory neurons and neuropile around the mouth opening (Fig. 1a). In *Hydra oligactis* there is an obvious nerve ring at the base of the hypostome (Fig. 1b,c). With conventional histological and ultrastructural techniques, these two centralizations have not been demonstrated in their full extent before (cf. Davis *et al.* 1968; Kinnamon & Westfall 1982; Matsuno & Kageyama 1984).

Staining of other hydroid polyps, such as those of *Hydractinia schinata*, does not show such a clear neuronal condensation in the hypostome. Instead, there is a dense plexus of neurites throughout the ectoderm of the hydranth (Fig. 2a). This neuronal plexus, again, has not been found before with conventional techniques (cf. Stokes 1974). Similarly, neurones covering the gonophores of *Hydractinia* (Fig. 2b) have remained undiscovered (cf. Hertwig & Hündgen 1984). These neurones are probably involved in the light-induced or circadian release of oocytes and sperm (Grimmelikhuijzen 1985).

Hydromedusae show strongly immunoreactive neurites in the marginal nerve rings (Fig. 3a). Most processes are located within the outer nerve ring (exumbrellar side of the velum), although some processes are also associated with the swimming motor neurones of the inner nerve ring (subumbrellar side; Grimmelikhuijzen & Spencer

Fig. 3. Whole-mount staining of hydrozoan medusae and scyphozoan ephyra larvae with RFamide antiserum 146II. Mn, manubrium; NR, nerve ring; rN, radial nerve; Te, tentacle. (a) A 2-day-old medusa of *Podocoryne carnea* (Anthomedusa). The nerve rings, radial nerves and the nerve nets of tentacles and manubrium all show immunoreactivity (×120). (b) A 2-day-old medusa of *Eirene sp.* (Leptomedusa). Note the immunoreactive subumbrellar nerve net (×100). (c) The subumbrellar nerve net of an ephyra larva of *Pelagia sp.* (×200). (d) The neuronal network in the base of a lappet of *Pelagia* (×200).

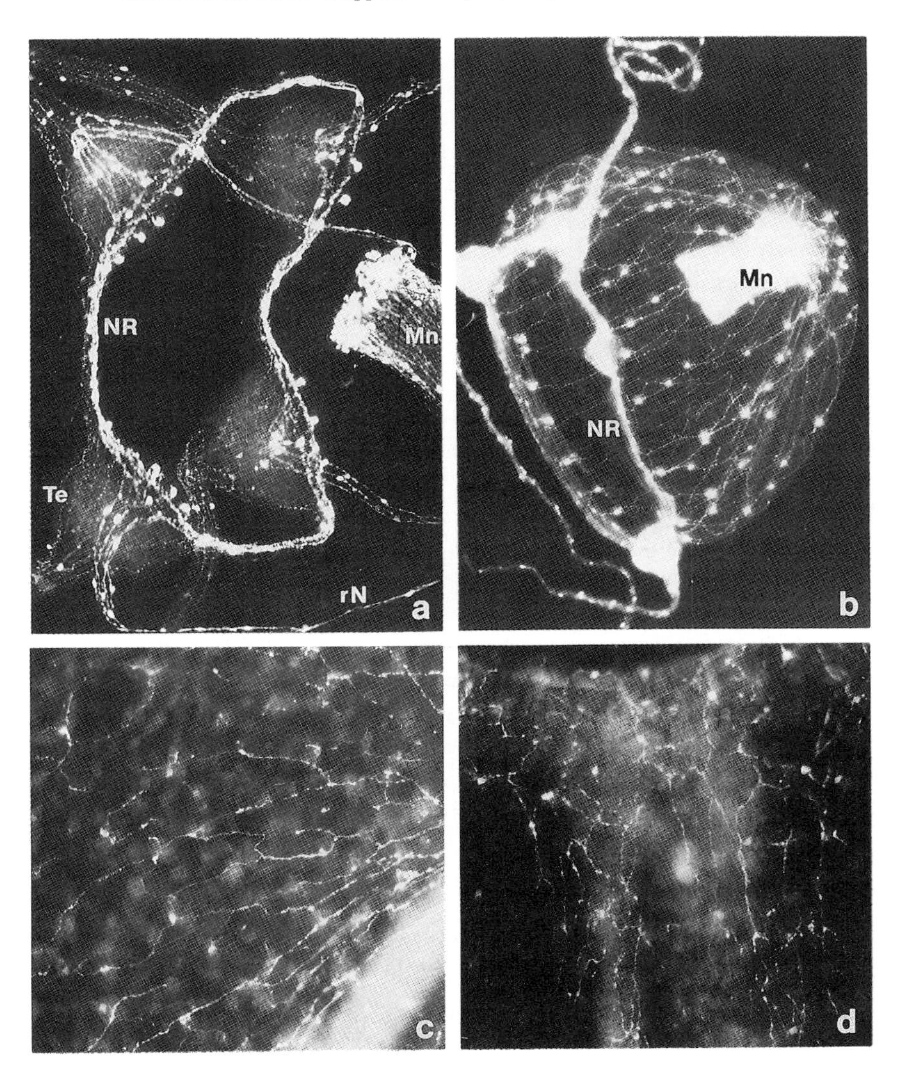

1984). The ocelli of anthomedusae include two peripheral groups of immunoreactive neurones, which are connected to the outer nerve ring by the ocellar nerves. (Grimmelikhuijzen & Spencer 1984). In all hydrozoan medusae, immunoreactive nerve nets are found in the ectoderm of tentacles and manubrium (Fig. 3). In anthomedusae, the four subumbrellar radial nerves often contain immunoreactive fibres, which connect the inner nerve ring to the manubrial nerve net (Fig. 3a). In leptomedusae, the subumbrellar nerve net, which replaces the radial nerves (Hertwig & Hertwig 1878; Spencer & Schwab 1982), becomes very obvious when stained with an RFamide antiserum (Fig. 3b).

In physonectid siphonophores, immunoreactive nerve nets occur in the stem, gastrozooids, dactylozooids, tentacles and pneumatophore. A marginal nerve ring is found in the nectophores, a double nerve tract in the tentacles, and several neuronal bands in the stem and between the various individuals (Grimmelikhuijzen *et al.* 1986). Several of these neuronal structures remained undiscovered until RFamide antisera were used.

RFamide peptides in the nervous systems of scyphozoans, anthozoans and ctenophores

Immunocytochemistry has shown the presence of RFamide peptides in the nervous system of every coelenterate species that we investigated. In scyphozoans, whole mounts of ephyra larvae of *Cassiopeia* and *Pelagia* show RFamide-positive nerve nets in the ectoderm of the subumbrella, manubrium, tentacles and lappets (Fig. 3c,d).and oral disc, and a dense network of neuronal processes is associated with all ecto- and endodermal muscles (Grimmelikhuijzen 1983; Fig. 4b,c).

Anthozoans are generally not transparent enough to be inspected as whole mounts, although compressed preparations of small specimens can be used (Fig. 4a). In sectioned material immunoreactive nerve nets can be seen in the ectoderm of tentacles

The ctenophore *Beroe* possesses RFamide-positive sensory neurons in the ectoderm of the stomodaeum, and a diffuse nerve net in the ectoderm of the body surface (Fig. 4d,e).

Structure of a coelenterate RFamide peptide

We have developed a radioimmunoassay for the dipeptide RFamide (Fig. 5,6). This assay recognizes free RFamide and aminoterminal elongated peptides containing the RFamide moiety (Fig. 6). Peptides terminating with RYamide can also be measured, but with a much lower efficiency (100× less than peptides containing RFamide). Only peptides containing arginine, followed by an amidated aromatic amino acid are effectively recognized in the assay. FMKFamide, a FMRFamide analogue, in which the arginine is replaced by the other positively charged amino acid, lysine, reacts poorly (2000× less efficient than RFamide peptides). The

Fig. 4. Staining of anthozoans and ctenophores with RFamide antiserum 146II. Me, mesentery; cM, circular muscle. (a) A compressed tentacle (whole mount) of the sea anemone *Actinia equina*, showing a typical nerve net (×200). (b) Section of the sea anemone *Calliactis parasitica*. The mesenteric muscles and the circular muscle of the body column show dense innervation (×140). (c) Section of a retractor muscle of *Calliactis*. Many immunoreactive neurites are associated with muscle fibres (×230). (d) A sensory neuron in the stomodaeum of the ctenophore *Beroe* (section; ×350). (e) A diffuse nerve net with round perikarya (arrows) in the ectoderm of the body surface of *Beroe* (section; ×350).

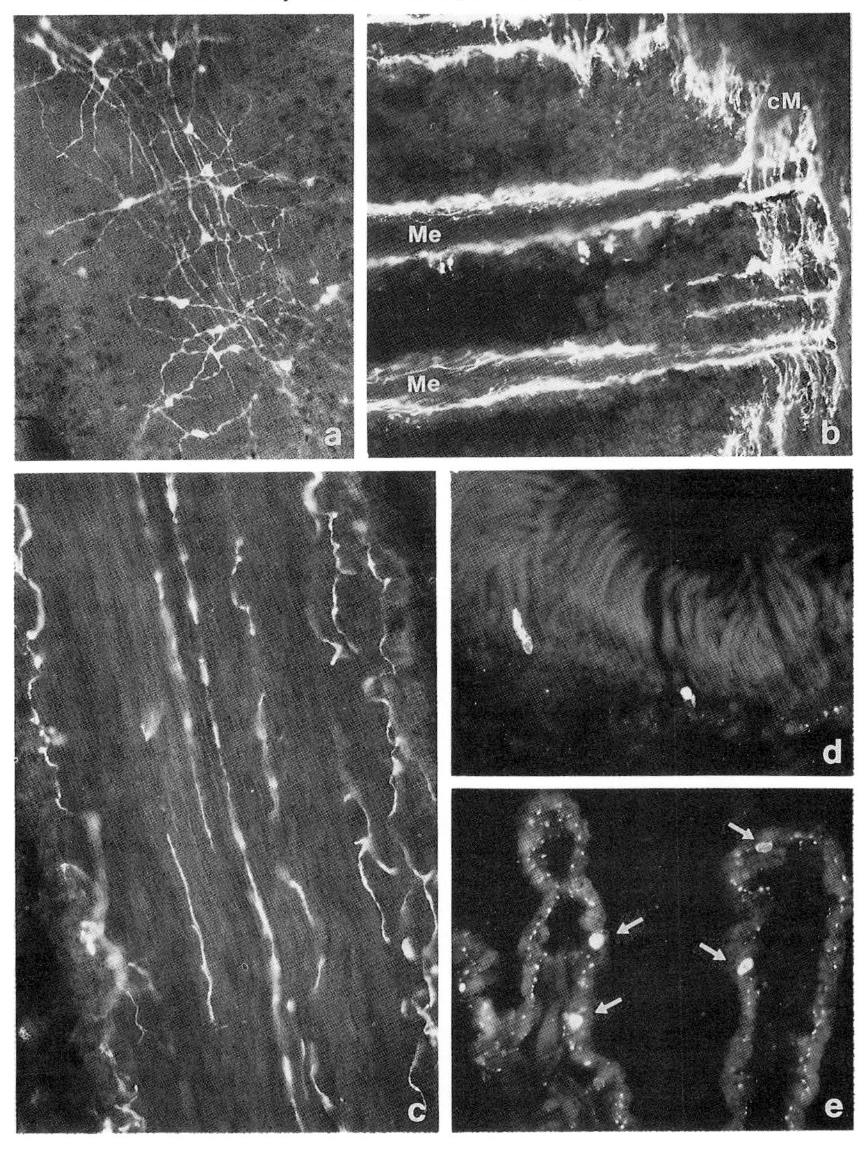

carboxyterminus of gastrin/cholecystokinin, WMDFamide, which contains a negatively charged amino acid, aspartate, in conjunction with Famide, cannot be measured in the radioimmunoassay (Fig. 6). The same holds for vasopressin (CYFQNCPRGamide), which contains an arginine followed by an amidated non-aromatic amino acid, and FMRF, in which the terminal amino acid is non-amidated.

Using our radioimmunoassay as a monitoring system, we have purified the RFamide peptide from an acetic-acid extract of the sea anemone *Anthopleura elegantissima* (Fig. 5). Initial purification, desalting and concentration of the immunoreactive material involved passage through Sep–pak reversed-phase cartridges (Waters). A successful subsequent purification method was cation-exchange chromatography on CM–Sephadex C–25, using a salt and a pH gradient (Fig. 7). As could be expected from a positively-charged peptide (arginine), the immunoreactive material was strongly bound to the resin and could only be eluted at 0.7 M ammonium acetate (pH 5.3), when most of the contaminating substances had been removed. After further cation-exchange chromatography at pH 7 and pH 8.5 (Figs. 8,9) the purification of the peptide was nearly completed. Subsequent reversed-phase HPLC yielded one homogenous peak of pure immunoreactive material (Fig. 10).

Fig. 5. RFamide radioimmunoassay of an acetic-acid extract of the sea anemone *Anthopleura elegantissima* (0.4 gram wet weight extracted per ml 0.5 M acetic acid).[J–125]–YFMRFamide was used as a tracer, added to RFamide antiserum 145 IV (diluted 1:10 000 in phosphate buffer), together with 0.01–10 μl equivalents of the sea-anemone extract. After 2 days at 4 °C, bound (B) and free (F) tracer were separated. From the radioimmunoassay it can be calculated that *Anthopleura* contains 10 nmol RFamide equivalents per gram wet weight. After Grimmelikhuijzen & Graff 1985, 1986.

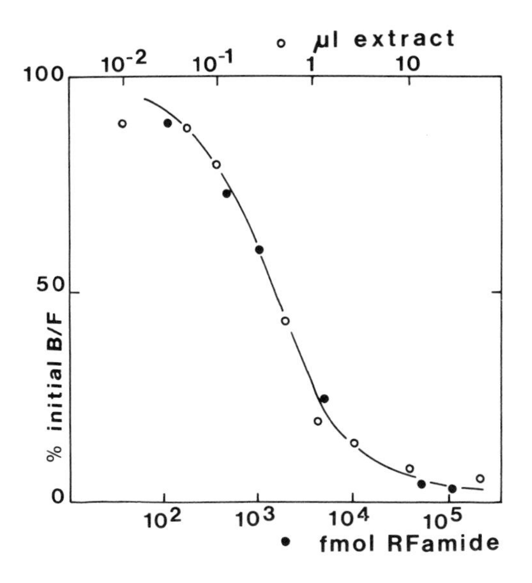

Fig. 6. Specificity of the RFamide radioimmunoassay of Fig. 2. (Crosses) RFamide; (open square) FMRFamide; (filled squares) LPLRFamide; (filled trangles) bovine pancreatic polypeptide–(31–36)–hexapeptide (LTRPRYamide); (filled circles) FMKFamide; (open circles) FMRF, cholecystokinin–(10–33)–tetrapeptide (WMDFamide), or [Arg^{8}]–vasopressin (CYFQNCPRGamide). Note that all peptides containing the RFamide moiety are equally well recognized. After Grimmelikhuijzen & Graff 1985.

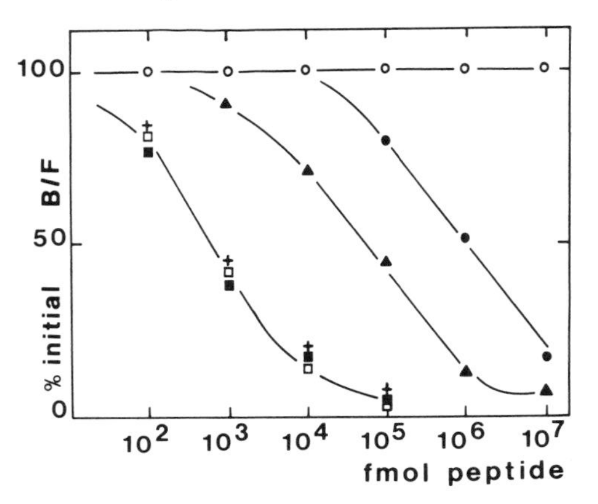

Fig. 7. Cation-exchange chromatography on CM–Sephadex C–25 of an acetic-acid extract of 400 gram *Anthopleura* (after desalting on Sep–pak and elution with 60% methanol). The column (2.6 x 39 cm) was equilibrated with 0.3 M ammonium acetate, pH 5. Subsequently, 70 ml of the extract was applied, and a linear gradient was started from 0.3 M ammonium acetate, pH 5 to 1.0 M ammonium acetate, pH 7 (24 ml/h, 20 h). After this gradient, the column was washed with 250 ml 2 M ammonium acetate, pH 7. The immunoreactive material was eluted around 0.7 M ammonium acetate. After Grimmelikhuijzen & Graff 1986.

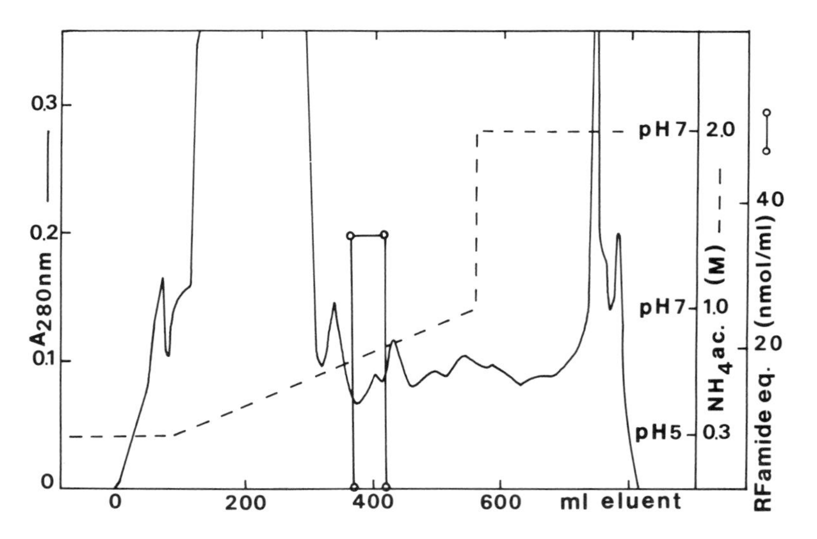

A part of the HPLC purified material was hydrolyzed in 6 N HCl and subsequently derivatised using dansyl chloride. Separation of the dansylated products on small polyamide sheets yielded the following amino acids: Glu, Gly, Arg, and Phe. An endgroup determination with dansylchloride did not yield any reactive endgroup. Because glutamate was present, the endgroup could be the unreactive pyroglutamate (pGlu), a cyclisation product of glutamine which lacks a free, primary amine. This was confirmed by treatment with the enzyme pyroglutamate aminopeptidase and subsequent purification by HPLC. The new peak of immunoreactive material contained Gly, Arg and Phe, but not Glu. Thus only one glutamate was present in the peptide, being in the form of an aminoterminal pGlu. Using an amino-acid analyzer, the stoichiometry of the amino acids in the intact peptide was determined as 1:1:1:1. As the peptide only contains one Glu, the other amino acids must also be present as single copies. From our knowledge of the specificity of the radioimmunoassay (see above), by which only Gly–Arg–Phe–amide can be recognized and no other amidated combination of the three carboxyterminal amino acids, the structure of the intact peptide had to be pGly–Gly–Arg–Phe–amide. This sequence was subsequently confirmed by both the dansyl chloride/isothiocyanate procedure (Gray & Smith 1970) and the more recently developed DABITC method (Chang *et al.* 1978).

Fig. 8. (left) Cation-exchange chromatography of the combined immunoreactive fractions of Fig. 7 (after desalting) on CM–Sephadex C–25 (column dimensions 2.6 x 42 cm). The column was equilibrated with 0.5 M ammonium acetate, pH 7. After application of the sample, a linear gradient of 0.5 M ammonium acetate, pH 7 to 0.8 M ammonium acetate, pH 7 was started (24 ml/h, 20 h). The immunoreactive material was eluted at 0.6–0.7 M ammonium acetate.

Fig. 9. (right) Cation-exchange chromatography of the combined immunoreactive fractions of Fig. 8 (after desalting) on a CM–Sephadex C–25 column (2.6 x 39 cm). After application of the sample, a linear gradient of 0.5 M ammonium acetate, pH 8.5 to 0.8 M ammonium acetate, pH 8.5 was started (24 ml/h, 20 h). The immunoreactive material was eluted at 0.6–0.7 M ammonium acetate.

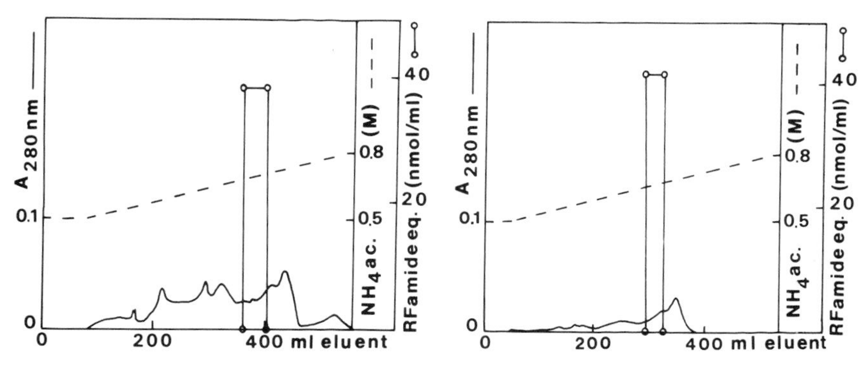

Action of RFamide peptides

The presence of a transmitter-like substance in a neurone does not necessarily mean that it is being used as a transmitter or modulator. It is therefore essential that immunocytochemical studies localizing peptides in specific neurones are followed by a rigorous physiological examination of the effects of the peptides on potential postsynaptic cells. A suitable animal for such studies is one in which it is possible to make intracellular or patch-clamp recordings from cells which might be targets for the transmitter substance. The hydromedusa *Polyorchis penicillatus* meets these requirements. There are substantial background data available on the morphology and cellular physiology of the *Polyorchis* nervous system (Anderson & Mackie 1977; Spencer 1978, 1979, 1981, 1982; Singla 1978; Spencer & Arkett 1984). Three morphologically and physiologically identifiable nerve nets are present in the nerve rings and their peripheral extensions. These are the swimming motor

Fig. 10. Reversed-phase HPLC of 30 nmol RFamide equivalents of the immunoreactive material purified in Fig. 9 (after desalting). An analytical C–18 column (Spherisorb ODS–2, 5 μm particle size, 80 Å pore diameter, dimensons 4 x 250 mm) was equilibrated with 10% acetonitrile in 0.1% TFA. Subsequently a gradient was started from 10% to 60% acetonitrile in 0.1% TFA (1 ml/min, 20 min). The RFamide peptide was eluted at 8 min (arrow). This material was found to be very pure, both by subsequent HPLC and amino-acid analyses. After Grimmelkhuijzen & Graff 1986.

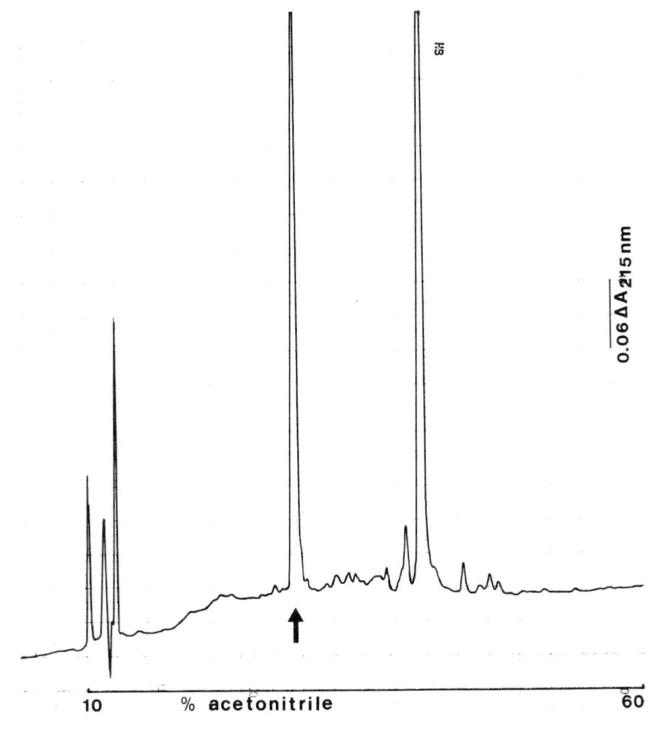

neurones (SMNs), which innervate the circular muscle of the subumbrella, the B system, which is presynaptic to the SMNs, and the O system, which is a primary photoreceptor system. RFamide immunoreactive neurites are found in the region of the SMNs (Grimmelikhuijzen & Spencer 1984), although SMNs themselves do not stain with the antiserum against RFamide. Therefore it is possible that the swimming motor neurone network is postsynaptic to neurones releasing an RFamide peptide. Preliminary double labelling experiments (Lucifer yellow/RFamide antiserum) suggest that neither the B nor the O system contains immunoreactivity. This would indicate that the RFamide-positive neurones belong to a separate system.

Without any neuronal involvement, the whole ectodermal epithelium of the jellyfish is capable of propagating action potentials (King & Spencer 1981) that can be generated by epithelial stimulation. Nevertheless, muscular portions of this epithelium are innervated by the nervous system. For example, the B system innervates the tentacle longitudinal muscle.

An ideal experimental situation would be to isolate each type of potential postsynaptic cells, both epithelial and neuronal, and then examine whole-cell currents or postsynaptic potentials in response to putative transmitters applied locally through a micropipette. Such experiments are necessary if one wishes to ensure that the effect of the transmitters are direct and not via di- or polysynaptic pathways. Here we report on some preliminary experiments using intact preparations (Spencer & Arkett 1984) of the marginal region. Because neither authentic *Polyorchis*, nor *Anthopleura* RFamide peptides were available during the time of our experiments, related peptides, such as FMRFamide and FLRFamide have been used.

It is possible to identify activity in each of the SMN, B, and O systems from extracellular suction-electrode recordings of the nerve rings. Action potentials in the epithelium overlying the nerve rings and action potentials in the swimming muscles are also recorded by these electrodes, and can be identified. Thus, if peptides are applied beneath the epithelium of the margin, or into the ring canal, their effects on each system can be detected in intact preparations. Both FMRFamide and FLRFamide (doses 5×10^{-10} mol) have excitatory effects on the B system, with FMRFamide also causing increased firing of the epithelial system (Fig. 11). The time course of their actions on the B system is quite different, with FMRFamide causing a short burst of spikes, while the influence of FLRFamide lasts several minutes. Intracellular recordings indicate that FLRFamide (10^{-8} mol) may also have direct excitatory effects on the swimming motor neurones (Fig. 12), though it is difficult to separate this direct depolarization from the depolarization that results from the summing, B-induced E.P.S.P.s. Incubation of FMRFamide with an antiserum (117) raised against it, removes its excitatory effects (not shown). These preliminary experiments, although not carried out with the authentic RFamide peptide of *Polyorchis*, give us some reason to suppose that RFamide peptides can have transmitter-like roles in coelenterates.

Fig. 11. Effect of FLRFamide and FMRFamide on the activity of identifiable neurones and epithelia in the margin of the hydromedusa *Polyorchis penicillatus*. Recordings were made by suction electrodes attached near the outer nerve ring in one dissected marginal quadrant. Amplification was by AC-coupled differential amplifiers. Synthetic peptides were dissolved in Millipore filtered sea water and applied by pressure injection (at 70kPa for 500 ms) through micropipettes under the epithelium covering the nerve ring. B, B system spikes; C, epithelial action potentials; S, action potentials of the swimming muscle. (a) FLRFamide (5×10^{-10} mol) causes a maintained increase in the firing rate of the B system. There is no increase in the rate of discharge of the overlying epithelium. (b) Upper trace: FMRFamide (5×10^{-10} mol) causes a burst of B system spikes followed by a burst of epithelial action potentials. Lower trace: Injection of a sea water control shows no change in the firing frequency of all systems.

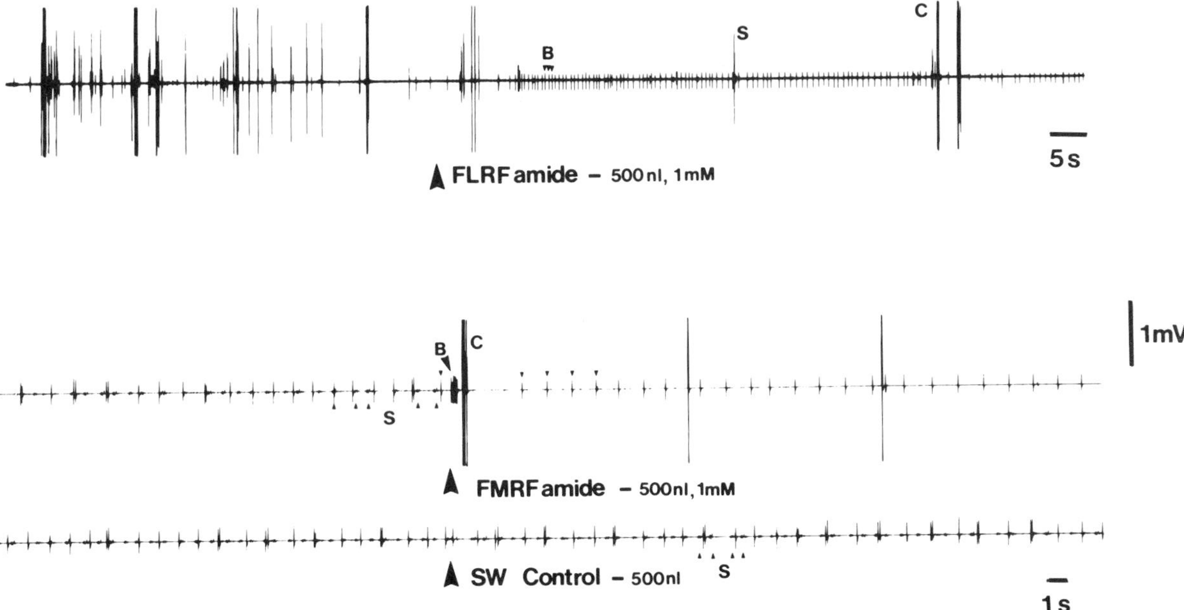

Fig. 12. Effect of FLRFamide on the swimming motor neurones (SMNs) of *Polyorchis penicillatus*. B, B system spikes; S, action potentials of the swimming muscle. Simultaneous intracellular recording (see Spencer & Arkett 1984) from the SMNs (upper trace) and suction electrode recording from the inner nerve ring (lower trace). FLRFamide prepared as in the previous figure was injected into the ring canal close to the recording electrode for the period indicated by the arrows (10^{-8} mol). This peptide causes a maintained depolarization and repetitive discharge of the SMNs which begins after a delay of about 10s, presumably due to the time taken for the peptide to diffuse into the nerve rings. A marked increase in the frequency of B system spikes accompanies the depolarization.

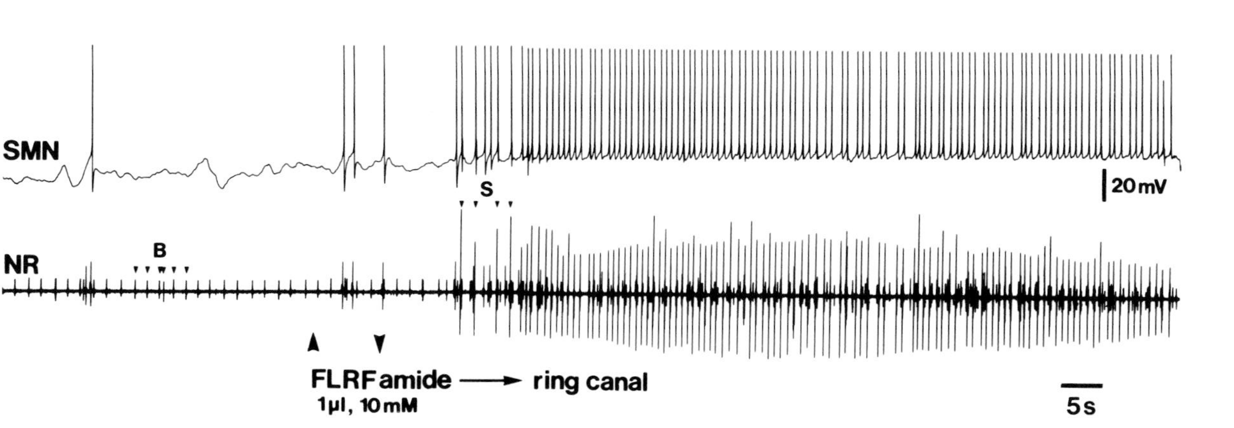

Conclusions

RFamide peptides are very common in the nervous systems of coelenterates, which suggests that they perform a crucial role. From the morphology and distribution of immunoreactive neurones, we can predict that RFamide peptides must be involved in feeding (Fig. 1), in light perception (Grimmelikhuijzen & Spencer 1984), in muscle contraction (Fig. 4b,c) and in reproduction (Fig. 2b). Whether RFamide peptides are really acting as neurotransmitters in these systems, has yet to be established. Our preliminary, electrophysiological experiments with RFamide analogues, however, suggest that RFamide peptides could have such function.

The RFamide peptide from the sea anemone *Anthopleura* has now been sequenced (pGlu–Gly–Arg–Phe–amide), while purification and sequencing of homologous peptides from *Polyorchis* and other coelenterates is underway. The availability of these authentic RFamide peptides will allow more conclusive experiments about the action of these substances in coelenterates. Such work will certainly contribute to our knowledge of primitive nervous systems. Besides, this work might also reveal principles of neuronal communication, which have been conserved throughout evolution, and which, because of the complexity of higher nervous systems, have not been recognised so far.

Acknowledgements

We thank Mrs S.L. Himanen for typing the manuscript, and the Deutsche Forschungsgemeinschaft (Gr 762/4, 762/6) and the National Science and Engineering Research Council of Canada (AO419) for financial support. C.J.P.G. is the recipient of a Heisenberg Fellowship (Gr 762/1). Some of the studies described here were carried out at Bamfield Marine Station, Canada.

Abbreviations

Abbreviations of amino acids: C, cysteine; D, aspartic acid; F, phenylalanine; G, glycine; K, lysine; L, leucine; M, methionine; N, aspargine; P, proline; Q, glutamine, R, arginine, W, tryptophan, Y, tyrosine.

References

Anderson, P.A.V. 1985 Physiology of a bidirectional, excitatory, chemical synapse. *J. Physiol.*, **53**, 821–835.

Anderson, P.A.V. & Mackie, G.O. (1977) Electrically coupled, photosensitive neurons control swimming in a jellyfish. *Science*, **197**, 186–188.

Anderson, P.A.V. & Schwab, W.E. (1982) Recent advances and model systems in coelenterate neurobiology. *Prog. in Neurobiol.*, **19**, 213–236.

Batham, E.J., Pantin, C.F.A. & Robson, E.A. (1960) The nerve net of the sea anemone *Metridium senile*: the mesenteries and the column. *Quart. J. Micro. Sci.*, **101**, 487–510.

Chang, J.Y., Brauer, D. & Wittmann–Liebold, B. (1978) Microsequence analysis of peptides and proteins using 4–NN–dimethylaminoazobenzene 4'–

isothiocyanate/phenylisothiocyanate double coupling method. *FEBS Lett.*, **93**, 205–214.

Davis, L.E., Burnett, A.L. & Haynes, J.F. (1968) Histological and ultrastructural study of the muscular and nervous system in *Hydra*. II. Nervous system. *J. Exp. Zool.*, **167**, 295–332.

Gray, W.R. & Smith, J.F. (1970) Rapid sequence analysis of small peptides. *Analyt. Biochem.*, **33**, 36–42.

Grimmelikhuijzen, C.J.P. (1983) FMRFamide immunoreactivity is generally occurring in the nervous systems of coelenterates. *Histochem.*, **78**, 361–381.

Grimmelikhuijzen, C.J.P. (1984) Peptides in the nervous system of coelenterates. In: *Evolution and Tumor Pathology of the Neuroendocrine System*, eds. S. Falkmer, R. Haanson & F. Sundler, pp. 39–58. Amsterdam: Elsevier.

Grimmelikhuijzen, C.J.P. (1985) Antisera to the sequence Arg–Phe–amide visualize neuronal centralization in hydroid polyps. *Cell & Tiss. Res.*, **241**, 171–182.

Grimmelikhuijzen, C.J.P. (1986) FMRFamide-like peptides in the primitive nervous systems of coelenterates and complex nervous systems of higher animals. In: *Handbook of Comparative Opioid and related Neuropeptide Mechanisms*, ed. G. Stephano, pp. 103–115. Boca Raton: CRC Press.

Grimmelikhuijzen, C.J.P., Dierickx, K. & Boer, G.J. (1982a) Oxytocin/vasopressin-like immunoreactivity is present in the nervous system of hydra. *Neurosci.*,**7**, 3191–3199.

Grimmelikhuijzen, C.J.P., Dockray, G.J. & Schot, L.P.C. (1982b) FMRFamide-like immunoreactivity in the nervous system of hydra. *Histochem.*, **73**, 499–508.

Grimmelikhuijzen, C.J.P., Dockray, G.J. & Yanaihara, N. (1981) Bombesin-like immunoreactivity in the nervous systems of hydra. *Histochem.*, **73**, 171–180.

Grimmelikhuijzen, C.J.P. & Graff, D. (1985) Arg–Phe–amide-like peptides in the primitive nervous systems of coelenterates. *Peptides*, **6**, Supplement 3, 477–483.

Grimmelikhuijzen, C.J.P. & Graff, D. (1986) Isolation of < Glu–Gly–Arg–Phe–amide (Antho–RFamide), a neuropeptide from sea anemones. *Proc. Natl. Acad. Sci. USA.*, **83**, 9817–9821.

Grimmelikhuijzen, C.J.P. & Spencer, A.N. (1984) FMRFamide immunoreactivity in the nervous system of the medusa *Polyorchis penicillatus*. *J. Comp. Neurol.*, **230**, 361–371.

Grimmelikhuijzen, C.J.P., Spencer, A.N. & Carré, D. (1986) Organization of the nervous system of physonectid siphonophores. *Cell & Tiss. Res.*, **246**, 463–479.

Hernandez–Nicaise, M.L. (1973) The nervous system of Ctenophora. III. Ultrastructure of synapses. *J. Neurocytol.*, **2**, 249–263.

Hertwig, O. & Hertwig, R. (1878) *Das Nervensystem und die Sinnesorgane der Medusen*. Leipzig: Vogel.

Hertwig, I. & Hündgen, M. (1984) The development of gonophores and germ cells in *Hydractinia echinata*. Fleming 1828 (Hydrozoa, Athecata). *Zoologisches Jahrbuch der Anatomie*, **112**, 113–136.

Horridge, G.A. & Mackay, B. (1962) Naked axons and symmetrical synapses in an elementary nervous system. *Nature*, **193**, 899–900.

Jha, R.K. & Mackie, G.O. (1967) The recognition, distribution and ultrastructure of hydrozoan nerve elements. *J. Morph.*, **123**, 43–62.

Kimmel, J.R., Hayden, L.J. & Pollock, H.G. (1975) Isolation and characterization of a new polypeptide hormone. *J. Biol. Chem.*, **250**, 9369–9376.

King, M.G. & Spencer, A.N. (1981) The involvement of nerves in the epithelial control of crumpling behaviour in a hydrozoan jellyfish. *J. Exp. Biol.*, **94**, 203–218.

Kinnamon, J.C. & Westfall, J.A. (1981) A three-dimensional serial reconstruction of neuronal distributions in the hypostome of a *Hydra*. *J. Morph.*, **168**, 321–329.

Lin, T.M. & Chance, R.E. (1974) Bovine pancreatic polypeptide (BPP) and avian pancreatic polypeptide (APP): candidate hormones of the gut. *Gastroenterol.*, **67**, 737–738.

McFarlane, I.D. (1973) Spontaneous contractions and nerve-net activity in the sea anemone *Calaliactis parasitica*. *Mar. Behaviour & Physiol.*, **2**, 97–113.

Mackie, G.O. (1973) Report on giant nerve fibres in *Nanomia*. *Publ. of Seto Marine Labs.*, **20**, 745–756.

Martin, S.M. & Spencer, A.N. (1983) Neurotransmitters in coelenterates. *Comp. Biochem. & Physiol. C.*, **74**, 1–14.

Matsuno, T. & Kageyama, T. (1984) The nervous system in the hypostome of *Pelmatohydra robusta*: the presence of a circumhypostomal nerve ring in the epidermis. *J. Morph.*, **182**, 153–168.

Price, D.A. & Greenberg, M. (1977) Structure of a molluscan cardioexcitatory neuropeptide. *Science*, **197**, 670–671.

Satterlie, R.A. (1979) Central control of swimming in the cubomedusan jellyfish *Charibdea rastonii*. *J. Comp. Physiol.*, **133**, 357–367.

Satterlie, R.A. (1985) Central generation of swimming activity in the hydrozoan jellyfish *Aequorea aequorea*. *J. Neurobiol.*, **16**, 41–55.

Satterlie, R.A. & Spencer, A.N. (1983) Neuronal control of locomotion in hydrozoan medusae. *J. Comp. Physiol.*, **150**, 195–206.

Singla, C.L. (1978) Fine structure of the neuromuscular system of *Polyorchis penicillatus* (Hydromedusae, Cnidaria). *Cell & Tiss. Res.*, **193**, 163–174.

Singla, L. & Weber, C. (1982) Fine structure studies of the ocelli of *Polyorchis penicillatus* (Hydrozoa, Anthomedusae) and their connections with the nerve ring. *Zoomorph.*, **99**, 117–129.

Spencer, A.N. (1978) Neurobiology of *Polyorchis*. I. Function of effector systems. *J. Neurobiol.*, **9**, 143–157.

Spencer, A.N. (1979) Neurobiology of *Polyorchis*. II. Structure of effector systems. *J. Neurobiol.*, **10**, 95–117.

Spencer, A.N. (1981) The parameters and properties of a group of electrically coupled neurones in the central nervous system of a hydrozoan jellyfish. *J. Exp. Biol.*, **93**, 33–50.

Spencer, A.N. (1982) The physiology of a coelenterate neuromuscular synapse. *J. Comp. Physiol.*, **148**, 353–363.

Spencer, A.N. & Arkett, S.A. (1984) Radial symmetry and the organization of central neurones in a hydrozoan jellyfish. *J. Exp. Biol.*, **110**, 69–90.

Spencer, A.N. & Satterlie, R.A. (1980) Electrical and dye coupling in an identified group of neurons in a coelenterate. *J. Neurobiol.*, **11**, 13–19.

Spencer, A.N. & Schwab, W.E. (1982) Hydrozoa. In: *Electrical Conduction and Behaviour in "Simple" Invertebrates*, ed. G.A.B. Shelton, pp. 73–148. Oxford: Oxford University Press.

Stokes, D.R. (1974) Morphological substrates of conduction in the colonial hydroid *Hydractinia echinata*. I. An ectodermal nerve net. *J. Exp. Zool.*, **190**, 19–46.

Tatemoto, K. (1982a) Isolation and characterization of peptide YY (PYY), a candidate gut hormone that inhibits pancreatic exocrine secretion. *Proc. Natl. Acad. Sci. USA*, **79**, 2514–2518.

Tatemoto, K. (1982b) Neuropeptide Y: complete amino acid sequence of the brain peptide. *Proc. Natl. Acad. Sci. USA.*, **79**, 5485–5489.
Triepel, J. & Grimmelikhuijzen, C.J.P. (1984a) A critical examination of the occurrence of FMRFamide immunoreactivity in the brain of guinea pig and rat. *Histochem.*, **80**, 63–71.
Triepel, J. & Grimmelikhuijzen, C.J.P. (1984b) Mapping of neurons in the central nervous system of the guinea pig by use of antisera specific to the molluscan neuropeptide FMRFamide. *Cell & Tiss. Res.*, **237**, 575–586.
Werner, B., Chapman, D.M. & Cutress, C.E. (1976) Muscular and nervous systems of the cubopolyp (Cnidaria). *Experientia*, **32**, 1047–1049.
Westfall, J.A. (1973a) Ultrastructural evidence of neuromuscular systems in coelenterates. *Am. Zool.*, **13**, 237–246.
Westfall, J.A. (1973b) Ultrastructural evidence for a granule containing sensory-motor-interneuron in *Hydra littoralis*. *J. Ultrastructural Res.*, **42**, 268–282.
Westfall, J.A., Kinnamon, J.C. & Sims, D.E. (1980) Neuro–epitheliomuscular cell and neuro-neuronal gap junctions in *Hydra*. *J. Neurocytol.*, **9**, 725–732.

M. PORCHET & N. DHAINAUT–COURTOIS

Neuropeptides and monoamines in annelids

Amongst invertebrates, few bioactive peptides have been sequenced and, of these, most are from molluscs or arthropods and there are, as yet, no sequenced neuropeptides from the annelids. In spite of this paucity of information, we expect that a range of model systems will be developed within this phylum. The failure to sequence any annelid peptide is therefore unfortunate in view of the experimental opportunities provided by the nervous system (NS) of various species (leeches and polychaetes). From an evolutionary point of view, annelids are basic to the phylogeny of numerous animal groups. Knowledge of the structure of their neuropeptides is therefore of interest for comparison with those of invertebrates either more primitive (*Hydra,* Schaller *et al.* 1984, see also Chapter 10 this volume) or more advanced (molluscs and arthropods, see Chapters 5–9 and 13–15 this volume).

Studies on neuropeptides are approached in two ways: the first involves a neurophysiological and neurobiological analysis directed towards an understanding of neurone physiology and more particularly the role of peptides in neurotransmission and neuromodulation (O'Shea & Schaffer 1985; see also Chapters 8 and 14 this volume). The pioneering work of Stent has established the leech as an important annelid model for the study of the cellular basis of behaviour (Stent & Kristan 1981) and as a model for neuronal development (Stent *et al.* 1982). The second approach is neuroendocrinological and is based upon a study of the production of peptides by neurosecretory centres in the central nervous system (CNS). Annelids have no discrete endocrine glands except for the infracerebral region (ICR), a neurohaemal area (fig. 5) found in several polychaetes (Dhainaut–Courtois 1970; Golding & Whittle 1977; Olive 1980; Durchon 1984).

We can distinguish two strategies in the characterization of peptides. A number of investigators (see below and Table I) have successfully used immunocytochemistry to demonstrate that annelid neurons contain peptides structurally related to a variety of mammalian neuropeptides. The list of apparent homologues is long. However, there is no example of such a peptide being isolated from an annelid. A second strategy is the characterization of native annelid peptides by conventional biochemical techniques (Porchet *et al.* 1985).

Table I. *Annelid neuropeptides*

Neuropeptides Neurohormones	Isolated from: Animal (organ)	Isolated from: Technique	Identified action	Chemical characteristic	References
Polychaetes					
Nereidin	*Perinereis cultifera* (brain)	HPLC mass spectrometry	Inhibition of spermatogenesis and epitoky	rich in glycine and proline	(38–7)
Dynorphin-like substance	*Nereis diversicolor* (brain, NHA)	RIA	?	peptidic nature	(a)
Spawning hormone (SH)	*Nephtys hombergii* (brain)	Methanolic extraction	Spawning	peptidic nature	(34)
Meiotic reinitiation substance	*Arenicola marina*	crude extract	meiotic reinitiation	?	(33)
Oligochaetes					
Brain hormones	*Eisenia foetida* (brain)	HPLC	clitellogenesis	peptidic nature	(30)
Met–, Leu–enkephalin-like substances	*Lumbricus terrestris* (nerve, gut)	RIA	intestinal contractile properties (neuro-modulation of dopaminergic contraction)	peptidic nature	(43)
Hirudinae					
FRBH Hydric balance regulation hormone	*Theromyzon tessulatum*	HPLC	hydric regulation	?	(27–29)

(a) Dhainaut–Courtois, Croix, Masson & Bulet, to be published.

Biochemical studies

Table I details the various attempts to isolate brain neurohormones in annelids. A number of groups have been involved in this work which has been largely directed towards the identification of peptides involved in the reproductive process, spawning and osmoregulation. The endocrine system of the Nephtyidae (Polychaeta) secretes a gonadotropic hormone (Bentley & Olive 1982) and a neuropeptide which controls the spawning mechanism (Bentley *et al.* 1984). In Lumbricidae (Oligochaeta), a peptidic brain hormone, probably elaborated by neurosecretory cells of the supraoesophageal ganglion, stimulates clitellogenesis and the testicular secretion of androgen, masculinizes the testis and maintains clitellum turgescence (Marcel & Cardon 1982, 1983; Lattaud & Marcel 1983). We can also draw attention to the work of Rzasa *et al.* (1984) which identifies opioids in *Lumbricus* by RIA, although it should be pointed out that RIA alone is not sufficient for identification and rather more vigorous methods are needed as has been shown in molluscs (Leung & Stephano 1984; see also Chapters 2 and 14 this volume).

The present authors are particularly concerned with the isolation of a brain neuropeptide from *Nereis*, 'nereidin', which inhibits the maturation of spermatocytes (Porchet 1984). A bioassay using the culture of male parapodia *in vitro* is used to monitor nereidin activity in extracts. In the presence of nereidin, germinal cell evolution is stopped; without nereidin, spermatozoa are produced over a period of 5 to 7 days. *Nereis* brains were extracted in methanol and this was followed by several chromatographic procedures. The biologically active fraction was further purified by HPLC, and an amino acid analysis performed by an Edman degradation sequence analysis coupled to mass spectrometry. The major peptide found (613 molecular weight) is rich in glycine and proline and the following sequence was proposed: NH_2 – Pro – Gly – Pro – Pro – Gly. Different peptides have been synthesized and only the sequence: NH_2 – Gly – Pro – Pro – Pro shows weak biological activity *in vitro*. In addition, during the purification, an unidentified product appears which has a molecular weight of 117, is acidic, and possesses an amine group. Its biological function is unknown.

Recently, an alternative extraction technique has been used (Bulet & Porchet 1986). Using an acidic medium, only weak biological activity is recovered (Table II). As the isoelectric point of nereidin seems to be near 6.3 to 6.4, a neutral extraction medium has been assayed. In such conditions an equivalent of 4 brains/ml gives an inhibition of spermatogenesis (Table II).

An alternative method for nereidin isolation involves the incubation of brains *in vitro*. Using a slightly modified cell culture medium based on amino acids, sugars, nucleotides, antibiotics and a TC 199 mixture, incubated for 5 days at 14°C, an equivalent of less than 0.16 brain/ml is necessary for biological activity. Such a complex medium is unfortunately not suitable for peptide purification. To overcome this problem, a simplified medium has been developed. Secretion of nereidin is

Table II. *Cerebral hormone recovery (brain equivalent/ml) in the new techniques of extraction of brain neurohormone (nereidin) in* Nereis

	Preparation medium		Medium pH		Brain equivalent (minimum dose for spermatogenesis inhibition)
Acid Medium	AcAc 2M	/EtOH	3	3.2	absence of inhibition
	AcAc 0.1M/EtOH		3.5	3.6	3 to 6
	TDW neutral medium		6.8		4
"in vitro" preparation	3 days				0.8
	5 days		7.4		<0.16
	8 days				0.5

similar in the two media and this simplified medium forms the basis of a new biochemical analysis of brain neuropeptide (Bulet & Porchet 1986).

Immunochemical studies

In annelids, methods classically used for studying neurosecretions and neurotransmitters [for reviews, see Golding & Whittle 1977; Dhainaut–Courtois *et al.* 1979a, b, 1985a, 1986 (Polychaeta); Herlant–Meewis 1983 (Oligochaeta); Malecha 1983, Kai–Kai 1984 (Hirudinea)] have not permitted the detection of the exact site of synthesis of neuroendocrine factors whose involvement in several biological phenomena has been experimentally demonstrated (Table I).

However, significant progress in the knowledge of the NS has recently been made with the development of immunochemical techniques. In this way, as in other groups, antibodies raised against invertebrate or vertebrate peptides have been used to characterize neurosecretory cell types. As in other phyla (see also Chapters 3 and 6 this volume), it is important to recognize that for full identity to be achieved complete chemical characterization of the native peptide product is essential. Nevertheless, when considered as markers, the numerous positive results obtained with antisera raised against vertebrate peptides may give very valuable information about neuronal circuitry.

The aim of this chapter is to report and discuss the principal data which were obtained with immunocytochemical and RIA methods.

Localization

Invertebrate peptides. Until now, only two antibodies have been used. Strong immunoreactivity was obtained with antibody (gift of Professor H. Boer, Amsterdam, Netherlands) raised against FMRF–amide, the molluscan cardioactive tetrapeptide (Schot & Boer 1982) in the whole CNS of *Nereis* (Figs. 1–3). In the same way, FMRF–amide and proctolin-like immunoreactivities were found respectively in several species of leech (Flanagan *et al.* 1985; Flanagan & Zipser 1985; Li & Calabrese 1985; O'Shea & Schaffer 1985).

Figs. 1–4. Paraffin cross-section (Fig. 1) and parasagittal sections (Figs. 3, 4) through the brain of *Nereis* immunostained for FMRF–amide (Figs. 1, 3) and insulin (Fig. 4). Fig. 2: paraffin horizontal section through the VNC of *Nereis* immunostained for FMRF–amide.

Note the unipolar and bipolar neurons both immunostained for FMRF–amide (Figs. 1, 3) and the numerous fibres immunostained for FMRF–amide in the neuropile (NP) of the brain and the VNC (Figs. 1, 2) and for insulin in the peduncles of corpora pedunculata (CP) (small arrows). B, brain; Ep, epidermis; LGF, lateral giant fibre; S, septum. The Arabic numerals show the different nuclei. Figs. 1 & 2, × 230; Figs. 3 & 4, × 375. (N. Dhainaut–Courtois, M. Masson & H. Azeddoug, unpublished results.)

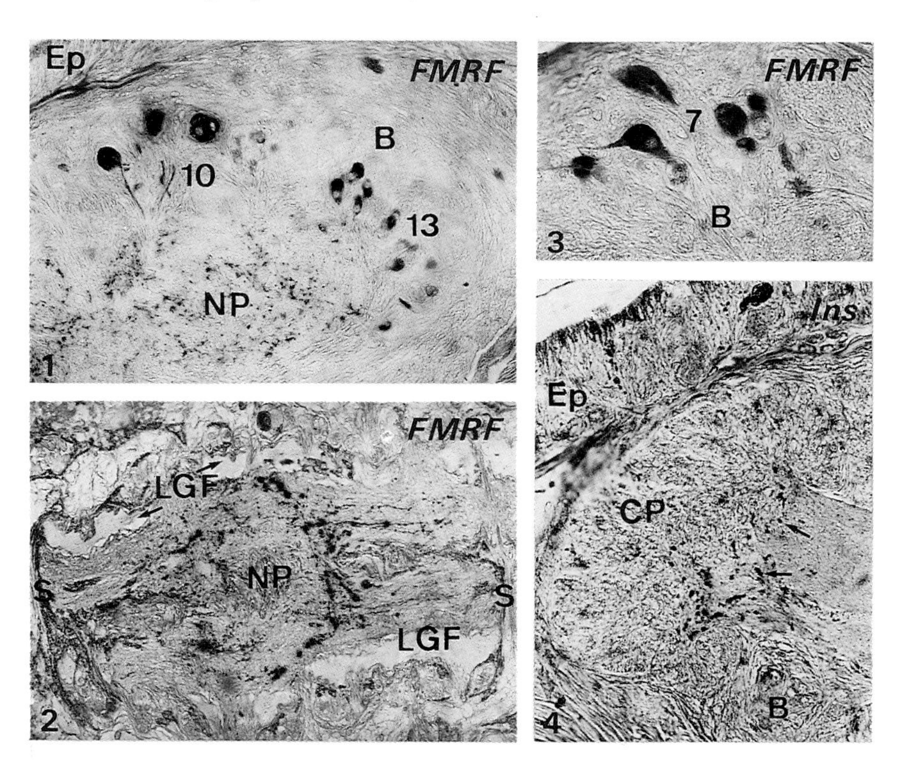

Table III. *Distribution in annelids of substances immunologically related to vertebrate peptides and neurotransmitters (with special reference to the nervous system).*

Substances	Distribution						Genus–Authors
	CNS				PNS	Other tissues	
	Brain	NHA	SOG	VNC			
Peptide families							
Opiates							
Leu–enkephalin	+	*	+	+		oocytes	P Nereis – 11 – a
	+						O Lumbricus – 1
	+					gut	O.Lumbricus – 43
	+						O.Eisenia – 26
				+			H.Haemopis – 50
				+			H.Haemopis – 17
				+			H.Haemopis – 16
				+			H.Hirudo – 16
				+			H.Macrobdella – 16
	+		+	+		proboscis	H.Theromyzon – b
Met–enkephalin	+	*		+			P.Nereis – 13
	+					gut	O.Lumbricus – 43
	+						O.Eisenia – 26
				+			H.Hirudo – 35
				+		gut	H.Hirudo – 23
	+		+	+		proboscis	H.Theromyzon – b
α–endorphin	+				+		P.Nereis – 11
			+				O.Dendrobaena – 39
β–endorphin	+			+			P.Nereis – 11 – 13
	+						O.Lumbricus – 1
α–neoendorphin	+		+				P.Nereis – 13
	+		+	+		proboscis	H.Theromyzon – b
Dynorphin 1–17	+	*		+		spermatocytes	P.Nereis – 13 – a
	+						O.Eisenia – 26
	+		+				H.Theromyzon – b
Hypothalamic neurohormones							
Somatostatin	+		+	+		oocytes	P.Nereis – 13
				+		gut; heart	H.Hirudo – 23
GRF	+		+	+			P.Nereis – 11 – 13
	+						O.Eisenia – 26
	+		+			proboscis	H.Theromyzon – b
CRF	+		+	+			P.Nereis – 11
	+						O.Dendrobaena – 40
	+						O.Eisenia – 26
	+		+			proboscis	H.Theromyzon – b

Table III (contd)

Substances	Distribution						Genus–Authors
	CNS				PNS	Other tissues	
	Brain	NHA	SOG	VNC			
LHRH	–?						P.Nereis – 11
	–?						O.Eisenia – 26
	+		+			proboscis	H.Theromyzon – b
Vasopressin	+		+	+			P.Nereis – a
	+		+	+			H.Theromyzon–28–29
Oxytocin	+		+	+			H.Theromyzon–28–29
Antehypophysial hormones							
β–MSH	+		+		+		P.Nereis – 11
γ–MSH	+		+			proboscis	H.Theromyzon – b
ACTH	–?						P.Nereis – 11
	+						O.Lumbricus – 2
Gastro–intestinal hormones							
CCK	+		+	+	+	gut	P.Nereis – 15 – 12
						gut	O.Lumbricus – 42
	+		+	+			O.Eisenia – 12 – 26
				+			H.Hirudo – 35
	+		+	+		proboscis	H.Theromyzon – 12
Substance P	+		+	+			P.Nereis – 11
	+						O.Lumbricus – 2
	+						O.Eisenia – 26
				+			H.Hirudo – 35
				+			H.Hirudo – 16
				+		gut; heart	H.Hirudo – 23
	+		+			proboscis	H.Theromyzon – b
PP	+		+	+			O.Lumbricus – 48
VIP	+		+	+			P.Nereis – 11
			+	+			O.Lumbricus – 48
				+			H.Hirudo – 35
				+		gut; heart	H.Hirudo – 23
	+		+			proboscis	H.Theromyzon – b
Motilin	+						P.Nereis – a
	+		+				H.Theromyzon – b
Insulin	+		+			proboscis	H.Theromyzon – b
GRP	+		+			proboscis	H.Theromyzon – b
Miscellaneous							
Angiotensin	+						P.Nereis – a
	+		+			proboscis	H.Theromyzon – b

Table III (contd)

Substances	Distribution						Genus–Authors
	CNS				PNS	Other tissues	
	Brain	NHA	SOG	VNC			
Bombesin				+			H.Hirudo – 35
Neurotransmitter families							
5–HT	+		+	+	+		P.Nereis – 13
	+		+	+			O.Eisenia – 26
				+			H.Hirudo – 35
				+			H.Hirudo – 16
				+	+		H.Hirudo – 22
			+	+			H.Theromyzon – b
Dopamine–β–hydroxylase				+			H.Hirudo – 35

a – Dhainaut–Courtois, Masson, Azeddoug (unpublished).
b – Verger–Bocquet, Malecha, Wattez (unpublished).

CNS, central nervous system; H, Hirudinea; NHA, neurohemal area; NS, nervous system;), Oligochaeta; P, Polychaeta, PNS, peripheral nervous system; SOG, suboesophageal ganglion; VNC, ventral nerve cord.

+, present; *, found in a neurohaemal area; –?, not found on paraffin sections.

Vertebrate peptides and neurotransmitters. The principal results obtained by immunocytochemical procedures are summarized in Table III. As can be seen, the list of the represented families is already very long. Nevertheless, it is certain that this is still incomplete and new immunochemical reactions will be observed as new antibodies become available, or, more simply, assays previously done on paraffin sections will be performed on frozen tissue sections. For instance in the case of *Nereis,* it has been found that antisera to met–enkephalin, vasopressin and oxytocin, give positive results only on frozen tissue sections (Dhainaut–Courtois *et al.* 1985a; N. Dhainaut–Courtois, M. Masson & H. Azeddoug, unpublished observations). A comparison of the different results (Table III) and particularly of those which were obtained with the same antisera (gifts of G. Tramu, Lille and M–P. Dubois, Nouzilly, France) in *Nereis, Eisenia* and *Theromyzon* allows a number of conclusions to be made concerning the abundance and distribution of these molecules.

In all species, strong immunoreactivity was observed with several antibodies (e.g. antibodies raised to CCK, substance P, CRF, GRF, α–neoendorphin, dynorphin). In contrast, negative reactions were obtained for anti-LHRH in *Nereis* and *Eisenia* and for anti-ACTH in *Nereis* while positive results were observed respectively in *Theromyzon* and *Lumbricus*. However, these data should be treated with caution since the antibodies were applied only to paraffin sections.

In most cases, the search for immunoreactivity was carried out in the CNS only. However, for several antibodies (e.g. antisera raised to CCK, insulin, somatostatin, leu–enkephalin, dynorphin) positive reactions were also observed in the peripheral nervous system (PNS) and even in other tissues such as the gut and sexual products (Table III). Leu–enkephalin- and dynorphin-like immunoreactivities were observed in *Nereis* oocytes and spermatocytes, respectively (N. Dhainaut–Courtois, M. Masson & H. Azeddoug, unpublished data). Enkephalin-like immunoreactivity was also found in ganglia that innervate the sex organs of the leech (Zipser 1980). Moreover, the presence of met– and leu–enkephalin-like immunoreactivity in nerve, gut, seminal vesicle and body wall tissues of *Lumbricus* has been demonstrated by means of RIA (Rzasa *et al.* 1984).

A study of the whole NS in *Nereis,* revealed different distributions according to the antibodies used. Here, CCK-like immunoreactivity is abundant in the whole CNS, while vasopressin-, CRF- and GRF-like immunoreactivities are more concentrated in the ventral nerve cord (VNC) and opioid-like peptides, particularly a dynorphin-like substance, are essentially located in the brain (Figs. 5,6).

In annelids, most immunoreactivity is observed in perikarya as well as in fibres located in the neuropile and sometimes in perioesophageal connectives and nerves. In addition, in *Nereis,* numerous positive reactions were observed in the peduncles of the corpora pedunculata (CP) (Fig. 4), and a number of axons immunoreactive with leu–, met–enkephalin and dynorphin antisera were observed in the neurohaemal area (ICR) (Figs. 5,7,8).

Immunoreactivities for one given antibody are often obtained in several cell types (Fig. 1,3) (see Dhainaut–Courtois *et al.* 1985b). On the other hand, dynorphin-like immunoreactivity is observed quite exclusively in C II cells in the caudal part of the *Nereis* brain (Figs. 5,6). In addition, in *Nereis,* several molecules (such as molecules immunochemically related to CCK and CRF/GRF or 5–HT) may coexist in a single neuron (Dhainaut–Courtois *et al.* 1985a,b, 1986). Unfortunately, at the present time, little is known concerning the chemical nature of the molecules detected by immunocytochemistry (Table I). Recent investigations into the presence of opioids in *Nereis* have, however, provided some useful information (N. Dhainaut–Courtois, D. Croix, M. Masson, P. Bulet, unpublished data). RIA on prostomia from *Nereis* or on the artificial sea water in which prostomia were incubated for 4 days, confirmed the presence of dynorphin-like substance in both young and old worms. The binding curve dilutes in parallel with the standard dynorphin curve and there is no cross-reactivity with leu– or met–enkephalin. The preliminary results are shown in Table IV.

These results must be supported by other similar studies for further conclusions concerning the evolution of the rate of synthesis and release according to the age of the worms. Nevertheless, they indicate, at least, that a dynorphin-like molecule is present in the brain of both male and female *Nereis* and its release continues in old animals.

Roles. In most cases, the roles of the detected peptide-like factors are unknown. Nevertheless, in leeches, there is a good deal of circumstantial evidence available which allows some attempt at role identification. For example, one of the larger proctolinergic neurons corresponds in cell body position to an identified inhibitory motoneuron. In addition, neurons immunoreactive to FMRF–amide include identified efferent neurons, the HE and HA cells which are known to be involved in the control of cardiac muscle (O'Shea & Schaffer 1985). When applied to the heart, FMRF–amide produced an increasing frequency of spontaneous myogenic contractions. A similar effect is produced by tonic activation of the HA cells. It should be pointed out, however, that the presence of authentic FMRF–amide or proctolin in the leech or the existence of altered leech forms of these peptides have not yet been established. The experimental results of Rosca (1972) which demonstrate the role of oxytocin on water

Fig. 5. Parasagittal view of the *Nereis* brain showing the localization of several ganglionic nuclei and the infracerebral region (ICR). The large arrow shows the antero-posterior direction. Immunoreactivities were observed in the neuropile (NP), in the peduncles of the CP, and in most of the nuclei shown here (stars). However, perikarya in CP and nucleus 16 were never immunostained. The principal results related to the caudal part of the brain (which contains the classical neurosecretory cells) and the ICR are reported. Ep, epidermis; ICC, infracerebral cells; Ng, neuroglia; OC, optic commissure; V, blood vessel. The Arabic numerals show the different nuclei. (Dhainaut–Courtois *et al.* 1986.)

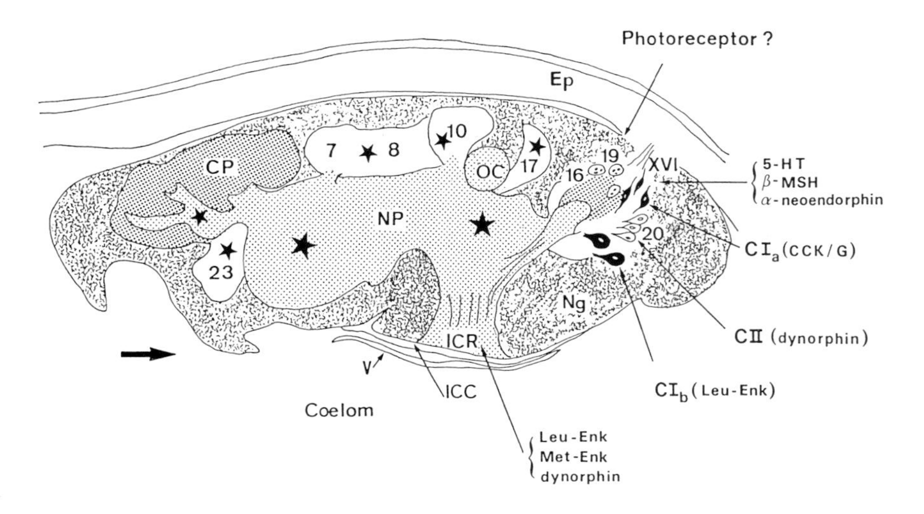

Table IV. *Dynorphin concentrations in* Nereis

Prostomia

Females : oocyte diameter lower than 80 μm: 41.25 ng ± 2.40/prostomium

: oocyte diameter higher than 180 μm: 51.00 ng ± 3.03/prostomium

Males : (with spermatocytes only) : 37.75 to 48.25 ng/prostomium.

Incubation medium (for 4 days)

Females : oocyte diameter lower than 80 μm : 8.19 ng/prostomium
: oocyte diameter higher than 180 μm : 22.72 ng/prostomium

flux in *Hirudo* are supported by the immunocytochemical detection of a molecule related to this hormone in a leech (Malecha 1983; Malecha *et al.* 1986). However, possible relationships between the observed oxytocin- and vasopressin-like immunoreactivity and hydromineral regulation remain to be explained.

In *Nereis,* the presence of numerous immunoreactive fibres in the neuropile suggests a role as neurotransmitters, as is postulated for other invertebrates. Furthermore, the abundant immunoreactivity seen in the peduncles of the corpora pedunculata allow us to speculate that the latter play a role in integration. Since CCK-like immunoreactivity exists not only in the CNS and in the gut, as in vertebrates, but also in the PNS, near the epidermis, it might act there with other peptides such as β–MSH- and insulin-like substances. Here, we might also recall the induction of supernumerary structures in the presence of insulin obtained in *Nereis pelagica* by Boilly–Marer (1983).

The findings of immunoreactive axons with met–, leu–enkephalin and dynorphin antisera in the infracerebral region are also of interest. In view of the abundance of dynorphin detected by immunohistochemical and RIA techniques, experimental studies using an organ culture method (see above) were undertaken to compare its action with that of nereidin. Parapodia of young males were incubated in artificial sea water containing dynorphin (10^{-5}, 10^{-6}, 10^{-7} M) or in artificial sea water where prostomia had previously been maintained for 4 days and in which an antibody dynorphin was then added (dilution: 1/100 – 1/400). No effect on spermatogenesis was observed. Thus, it seems likely that dynorphin-like peptides do not act directly as nereidin. Their presence in the neurohaemal area suggests that they only control the indicate perhaps a potential role for opiates in reproductive events, as recently shown in a mollusc (Marchand & Dubois 1986).

Neurotransmitters

Antibodies raised against 5–HT give positive reactions in the 3 classes of annelida (Table III). These results confirm previous data which were obtained with histofluorescence and autoradiographic methods (for reviews, see Dhainaut–Courtois *et al.* 1979b; Dhainaut–Courtois *et al.* 1985a). In addition immunoreactivities for dopamine–β–hydroxylase were obtained in neurons located in the VNC of *Hirudo* which were previously thought to be octopaminergic (Osborne *et al.* 1982). Further,the presence of octopamine has been shown by means of radio–chemical–enzymatic assay in *Lumbricus* and several leeches (Tanaka & Webb 1983).

Figs. 6–7. Paraffin parasagittal sections through the brain of *Nereis* immunostained for dynorphin. In nucleus 20, the C II cells are immunoreactive while the small fuchsinophilic C Ia cells and the large cells in nucleus 22 show negative results (Fig. 6). A number of axonal endings (small arrows) are visible in infracerebral region (ICR) (Fig. 7), B, brain; Ep, epidermis; M, muscles; V, blood vessel. Fig. 6, × 535; Fig. 7, × 340.

Fig. 8. Thin section through an axonal ending of a C II cell (C II ae) treated by the immunogold method for dynorphin. Note the 5nm gold particles specifically labeling the secretory granules. ca, Collagen capsule surrounding the brain; Ng, neuroglia. × 40,700. (N. Dhainaut–Courtois, J.C. Beauvillain, G. Tramu & H. Azeddoug, unpublished data.)

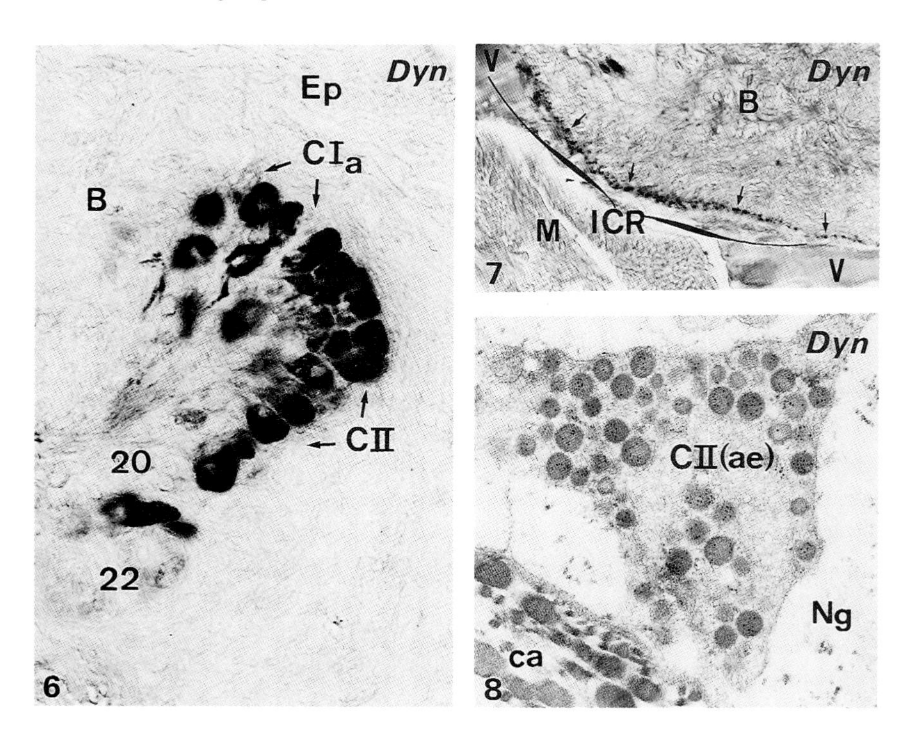

Conclusion

We now have at our disposal a considerable body of information on the distribution and immunocytochemical localization of peptide-like molecules in annelids. This should act as an incentive for rapid progress in biochemical and functional studies on the annelid NS directed towards a better understanding of the neuroendocrine physiology of this group.

Acknowledgements

We thank F. Bonet (C.N.R.S.) and A. Auger for their technical assistance. We also would like to thank N. Barrett for reviewing the English text.

References

1. Alumets, J., Håkanson, R., Sundler, F. & Thorell, J. (1979) Neuronal localisation of immunoreactive enkephalin and β–endorphin in the earthworm. *Nature,* **279**, 805–806.
2. Aros, B., Wenger, T. & Vigh–Teichmann, I. (1980) Immunohistochemical localization of substance P and ACTH-like activity in the central nervous system of the earthworm *Lumbricus terrestris* L. *Acta Histochem.,* **66**, 262–268.
3. Bentley, M.G. & Olive, P.J.W. (1982) An *in vitro* asssay for gonadotrophic hormone in the polychaete *Nephtys hombergii* Sav. (Nephtyidae). *Gen. Comp. Endocrinol.,* **47**, 467–474.
4. Bentley, M.G., Olive, P.J.W., Garwood, P.R. & Wright, N. (1984) The spawning and spawning mechanism of *Nephtys caeca* and *N. hombergii* (Annelida: Polychaeta). *Sarsia,* **69**, 63–68.
5. Boilly–Marer, Y. (1983) Culture de segments et induction de morphogenèses surnuméraires en présence d'insuline chez *Nereis pelagica* Linné (Néréidien à épitoquie, Annélide Polychète). *C. R. Acad. Sc. Paris,* **297**, 213–218.
6. Bulet, P. & Porchet, M. (1986) A new purification procedure for the neuropeptide controlling gametogenesis in Nereids (Annelida Polychaeta). *4th Intern. Symp. ISIR,* Lille, 1–5/09/86.
7. Cardon, C., Durchon, M. & Porchet, M. (1981) Purification par chromatographie liquide de haute pression HPLC de l'hormone cérébrale de *Nereis diversicolor* et de *Perinereis cultrifera. Reprod. Nutr. Dévelop.,* **21**, 383–390.
8. Dhainaut–Courtois, N. (1970) Contribution à l'étude morphologique des processus secrétoires dans le système nerveux central et au niveau de la glande infracérébrale des *Nereidae* (Annélides Polychètes). Thèse Doctorat Etat Sc. Nat. Lille, 1–191.
9. Dhainaut–Courtois, N., Engelhardt, R.–P. & Dhainaut, A. (1979a) Etude cytophysiologique des systèmes monoaminergiques et cholinergique des *Nereis* (Annélides Polychètes). I. Système nerveux périphérique et jonctions neuromusculaires. *Arch. Biol. (Bruxelles),* **90**, 225–244.
10. Dhainaut–Courtois, N., Engelhardt, R.–P. & Dhainaut, A. (1979b) Etude cytophysiologique des systèmes monoaminergiques et cholinergique des *Nereis* (Annélides Polychètes). II. Système nerveux central. *Arch. Biol. (Bruxelles),* **90**, 273–288.
11. Dhainaut–Courtois, N., Dubois, M.–P., Tramu, G. & Masson, M. (1985a) Occurrence and coexistence in *Nereis diversicolor* O.F. Müller (Annelida Polychaeta) of substances immunologically related to vertebrate neuropeptides. *Cell Tissue Res.,* **242**, 97–108.

12. Dhainaut–Courtois, N., Tramu, G., Marcel, R., Malecha, J., Verger–Bocquet, M., Andriès, J.–C., Masson, M., Selloum, L., Belemtougri, G. & Beauvillain, J.C. (1985b) Cholecystokinin in the Nervous System of Invertebrates and Protochordates. Immunohistochemical localization of a cholecystokinin–8-like substance in Annelids and Insects. 1st Int. Conf. Neuronal Cholecystokinin 3–6/06/1984, Brussels, Belgium, *An. N.Y. Acad. Sc.*, **448**, 167–187.
13. Dhainaut–Courtois, N., Tramu, G., Beauvillain, J.–C. & Masson, M. (1986) A qualitative approach of the *Nereis* neuropeptides by use of antibodies to several vertebrate peptides. *Neurochem. Int.*, **8**, 327–338.
14. Durchon, M. (1984) Peptidic hormones in Annelids. In *Biosynthesis, Metabolism and Mode of action in Invertebrate Hormones*, pp. 10–18, Springer–Verlag, Berlin.
15. Engelhardt, R.–P., Dhainaut–Courtois, N. & Tramu, G. (1982) Immunohistochemical demonstration of a CCK-like peptide in the nervous system of a marine annelid worm, *Nereis diversicolor* O.F. Müller. *Cell Tissue Res.*, **227**, 401–411.
16. Flanagan, T. & Zipser, B. (1985) Varicosity-associated antigens define neuropile subfields in the leech central nervous system. *J. Neurocytol.*, **14**, 653–672.
17. Flanagan, T., Schley, C. & Zipser, B. (1985) Antibody staining reveals novel aspects of segmentation within the leech central nervous system. *Brain Res.*, **345**, 147–152.
18. Golding, D.W. & Whittle, A.C. (1977) Neurosecretion and related phenomena in Annelids. *Int. Rev. Cytol.*, Suppl. 5, 189–302. Acad. Press Inc. New York.
19. Herlant–Meewis, H. (1983) Etude ultrastructurale des ganglions cérébroïdes d'*Eisenia foetida*. Les cellules névrogliques, les neurones banaux et les neurones aminergiques. *Arch. Biol.*, **94**, 347–378.
20. Kai–Kai, M.A. (1984) Ultrastructural localization of [^{3}H] dopamine in neurons of leech (*Haemopis sanguisuga*) abdominal ganglia. *Comp. Biochem. Physiol.*, **78C**, 363–367.
21. Lattaud, C. & Marcel, R. (1983) Stimulation *in vitro* de la sécrétion de l'androgène testiculaire par une fraction purifiée de cerveaux chez *Eisenia foetida* (Annélide Oligochète). *J. Can. Zool.*, **61**, 2399–2404.
22. Leake, L.O., Griffith, S.G. & Burnstock, G. (1985) 5–Hydroxytryptamine-like immunoreactivity in the peripheral and central nervous system of the leech *Hirudo medicinalis. Cell Tissue Res.*, **239**, 123–130.
23. Leake, L.O., Growe, R. & Burnstock, G. (1986) Localization of substance P–, somatostatin–, vasoactive intestinal polypeptide– and met–enkephalin immunoreactive nerves in the peripheral and central nervous systems of the leech *Hirudo medicinalis. Cell Tissue Res.*, **243**, 345–351.
24. Leung, M.K. & Stephano, G.B. (1984) Isolation and identification of enkephalins in pedal ganglia of *Mytilus edulis* (Mollusca). *Proc. Natl. Acad. Sci. USA*, **81**, 955–958.
25. Li, C. & Calabrese, R.L. (1985) Evidence for proctolin-like substances in the central nervous system of the leech *Hirudo medicinalis. J. comp. Neurol.*, **232**, 414–424.
26. Lkhider, M., Marcel, R. & Tramu, G. (1987) Etablissement d'une carte des neurones du cerveau d'*Eisenia fetida* Sav. (Annélide Oligochète) contenant des substances immunologiquement apparentées à des peptides de Vertébrés. *Gen. comp. Endocrinol.*, **65**, 457–468.
27. Malecha, J. (1983) L'osmorégulation chez l'Hirudinée Rhynchobdelle *Theromyzon tessulatum*. Localisation expérimentale de la zone sécrétrice d'un

facteur de régulation de la balance hydrique. *Gen. comp. Endocrinol.*, **49**, 344–351.

28. Malecha, J., Cardon, C. & Tramu, G. (1983) Hormonal control of water balance in the leech *Theromyzon tessulatum* (O.F.M.). 2° Coll. Int. CNRS, Biosynthèse, métabolisme et mode d'action des hormones d'Invertébrés, 29/08–02/09/83 Strasbourg.
29. Malecha, J., Tramu, G., Cardon, C. & Verger–Bocquet, M. (1986) Mise en évidence de peptides apparentés à la vasopressine et à l'ocytocine dans le cerveau de l'Hirudinée Rhynchobdelle *Theromyzon tessulatum* (O.F.M.). *Gen. comp. Endocrinol.*, **64**, 13–20.
30. Marcel, R. & Cardon, C. (1982) Régénération céphalique et clitellogenèse sont-elles gouvernées par la même neurohormone chez *Eisenia foetida? J. Physiol.*, **78**, 987–991.
31. Marcel, R. & Cardon, C. (1983) Purification par H.P.L.C. de la neurohormone cérébrale d'*Eisenia foetida* (Annélide Oligochète). Essai sur la régénération et la clitellogenèse. *Reprod. Nutr. Dévelop.*, **23**, 1003–1009.
32. Marchand, Cl.–R. & Dubois, M.–P. (1986) Immunoreactivity of the hermaphroditic gonad of the snail *Helix aspersa* Müller towards antibodies raised to fragments of pre– pro– opio– melanocortin. *Cell TissuesRes.*, **245**, 337–341.
33. Meijer, L. (1979) Hormonal control of oocyte maturation in *Arenicola marina*. II. Maturation and fertilization. *Develop. Growth Differ.*, **21**, 315–329.
34. Olive, P.J.W. & Bentley, M.G. (1980) Hormonal control of oogenesis ovulation and spawning in the annual reproductive cycle of the polychaete *Nephtys hombergii*. *Int. J. Invert. Reprod.*, **2**, 205–211.
35. Osborne, N.N., Patel, S. & Dockray, G.J. (1982) Immunohistochemical demonstration of peptides, serotonin and dopamine β–hydroxylase-like material in the nervous system of the leech *Hirudo medicinalis*. *Histochem.*, **75**, 573–583.
36. O'Shea, M. & Schaffer, M. (1985) Neuropeptide function: the invertebrate contribution. *Ann. Rev. Neurosci.*, **8**, 171–198.
37. Porchet, M. (1984) Biochemistry of oocyte differentiation in nereids. *Fortsch. Zool.*, **29**, 207–225.
38. Porchet, M., Dhainaut–Courtois, N., Cardon, C. & Bataille, M. (1985) Structure and functions of neuropeptides of polychaete annelids. In: *Neurosecretion and the biology of neuropeptides*, Ed. Kobayashi, Springer, pp. 377–385.
39. Rémy, C. & Dubois, M.–P. (1979) Localisation par immunofluorescence de peptides à α–endorphine dans les ganglions infra-oesophagiens du Lombricidé *Dendrobaena subrubicunda* Eisen. *Experientia*, **35**, 137–138.
40. Rémy, C., Tramu, G. & Dubois, M.–P. (1982) Immunohistological demonstration of a CRF-like material in the central nervous system of the annelid *Dendrobaena*. *Cell Tissue Res.*, **227**, 569–575.
41. Rosca, D.I. (1972) Cercetari asupra reglarii hormonale a schimburilor osmotice la *Hirudo medicinalis*. *Stud. Univ. Babes Bolyai Ser. Biol.*, **1**, 115–132.
42. Rzasa, P.J., Kaloustian, K.V. & Prokop, E.K. (1982) Immunochemical evidence for a gastrin-like peptide in the intestinal tissues of the earthworm *Lumbricus terrestris*. *Comp. Biochem. Physiol.*, **71A**, 631–634.
43. Rzasa, P.J., Kaloustian, K.V. & Prokop, E.K. (1984) Immunochemical evidence for met–enkephalin-like and leu–enkephalin-like peptides in tissues of the earthworm *Lumbricus terrestris*. *Comp. Biochem. Physiol.*, **77C**, 2, 345–350.

44. Schaller, H.C., Hoffmeister, S. & Bodenmuller, H. (1984) Neuropeptide in *Hydra* In: *Biosynthesis, Metabolism and Mode of action of Invertebrate Hormones,* pp. 5–9, Springer–Verlag, Berlin.
45. Schot, L.P.C. & Boer, H.H. (1982) Immunocytochemical demonstration of peptidergic cells in the pond snail *Lymnaea stagnalis* with an antiserum to the molluscan cardioactive tetrapeptide FMRF–amide. *Cell Tissue Res.,* **225,** 347–354.
46. Stent, G.S. & Kristan, W.B. (1981) Neural circuits generating rhythmic movements. In: *Neurobiology of the leech.* eds. K.J. Muller, J.G. Nicholls, & G.S. Trent, pp. 113–146, New York: Cold Spring Harbor Lab.
47. Stent, G.S., Weisblat, D.A., Blair, S.S. & Zackson, S.L. (1982) Cell lineage in the development of the leech nervous system. In: *Neuronal development,* ed. N. Spitzer, pp. 1–44, New York Plenum Press.
48. Sundler, F., Håkanson, R., Alumets, J. & Walles, B. (1977) Neuronal localization of pancreatic polypeptide (PP), and vasoactive intestinal peptide (VIP) immunoreactivity in the earthworm (*Lumbricus terrestris*), *Brain Res.,* **2,** 61–65.
49. Tanaka, K.R. & Weeb, R.A. (1983) Octopamine action on the spontaneous contractions of the isolated nerve cord of *Lumbricus terrestris. Comp. Biochem. Physiol.,* **76C**, 113–120.
50. Zipser, B. (1980) Identification of specific leech neurones immunoreative to enkephalin. *Nature,* **283,** 857–858.

M.C. THORNDYKE & D. GEORGES

Functional aspects of peptide neurohormones in protochordates

Introduction

The collective title protochordates is essentially a convenient 'umbrella' term for a diverse assembly of animals sharing relatively few, although important, common features. The protochordate group is usually thought to comprise three subphyla, the hemichordata, urochordata (tunicata) and cephalochordata, although some authorities consider the hemichordates a separate phylum. There are, however, no published reports on the occurrence of peptides or amines in hemichordates and they will not be considered further here.

The tunicates and cephalochordates are highly specialized marine organisms whose close relationship is perhaps rather superficial and, in reality, based more upon similar, ciliary powered, particle feeding mechanisms which have developed around the possession of a perforated pharynx, rather than any genuine and common phylogenetic background. Notwithstanding such reservations the protochordates are probably the only available extant representatives of ancient groups which "bridged the gap" between invertebrates and vertebrates, and as such could provide useful clues to the origins of certain vertebrate features.

The features which have attracted most attention and which have provided the most useful information are the central nervous system (CNS) and the gastrointestinal tract. The tunicate CNS has a relatively simple organisation which is in many respects rather more comparable with simple invertebrate neurosecretory centres than any part of the vertebrate CNS. Similarly the cephalochordate CNS is a simple anterior elaboration of the dorsal nerve cord (characteristic of these animals), and its structure and function reflects the highly specialized nature of the adult organism as well as its unusual and asymmetric development. The gastrointestinal tract, while being equally specialised, probably provides rather more useful information, particularly in ascidian tunicates where, as in the other protochordates, the suspension feeding process centres around a highly ciliated perforated pharynx. Passage of the food mass through the gut similarly depends on ciliary action and the digestive process includes both intracellular and extracellular phases (Barrington 1965). It is here that the origins of the complex vertebrate plan can be seen: thus the enzyme-secreting zymogen cells

Table 1. *Distribution of Neurohormonal Peptides in Ascidians*

Peptide/Amine	Species	Digestive Tract	Neural Ganglion/ Nervous System	Neural gland
ACTH	*Ciona intestinalis*	*a*	*a,b*	
	Styela plicata			c
Bombesin	*C. intestinalis*	d	e	
Calcitonin/cGRP	*C. intestinalis*	f	g,h,i	
	Styela clava	j	i	
β Endorphin	*C. intestinalis*		e	
	S. plicata		k	k
met Enkephalin	*C. intestinalis*	u (ovary)	b	
Gastrin/ cholecystokinin	*Branchiostoma lanceolatum*	l,m	n	
	C. intestinalis	o	p	
	S. clava		q	
	A. aspersa		q	
Glucagon	*B. lanceolatum*	l,m		
Insulin	*C. intestinalis*	l,m		
	C. intestinalis		r	
	S. clava	s		
LHRH	*C. intestinalis*		b,t*	b,t*
Motilin	*C. intestinalis*		e	
α MSH	*S. plicata*	x		
Neurotensin	*B. lanceolatum*	m		
	C. intestinalis	d,v	e	
Pancreatic polypeptide	*B. lanceolatum*	m		
	C. intestinalis		e	
Prolactin	*C. intestinalis*	e		
	S. plicata	w	x	
Secretin	*B. lanceolatum*	m		
	C. intestinalis		e	
	S. plicata	y		
	S. clava	z		
Somatostatin	*B. lanceolatum*	m		
	C. intestinalis	o	g	
	S. plicata	a2	a2	
Substance P	*C. intestinalis*	d	g,b2	
VIP	*B. lanceolatum*	m		
	C. intestinalis		e	
Serotonin and Biogenic Amines	*C. intestinalis*	e,c2	c2 (young animals)	c2 (young animals)

typically form part of the digestive mucosal epithelium, and scattered amongst them are characteristic, basally granulated, endocrine cells (Thorndyke 1977). These represent the basic elements of the vertebrate digestive plan from which, by regional concentration and specialization, the complex and sophisticated mammalian system can be derived (see Thorndyke & Falkmer 1985 for recent review).

Distribution and occurrence of neurohormonal peptides and amines

The distribution of neurohormonal peptides in protochordates has been studied extensively in recent years, yet relatively few species have been investigated; most of them ascidians including *Ascidiella aspersa, Ciona intestinalis, Styela plicata,* and *Styela clava* (Table 1). In a similar way only the cephalochordate *Branchiostoma lanceolatum* has been investigated for neurohormonal peptides (Table 1). These studies have been the subject of two recent reviews and a more detailed account may be found therein (Thorndyke 1986; Thorndyke & Falkmer 1985). Full details of the regional anatomical distribution of the peptides may be found in Table 1. In the majority of these reports the evidence has relied entirely upon immunocytochemical methods, and the limitations of this must be taken into account (Thorndyke 1986). In only a few instances has immunocytochemistry been supported by either radioimmunoassay of tissue extracts or more rigorous chromatographic analysis and biological assay. Four of these type of studies will be considered briefly here.

Gastrin/cholecystokinin

Gastrin/cholecystokinin-like immunoreactivity in both the gut and CNS of several protochordates has been described (Table 1). While several investigations utilized well-characterised antisera, only quite recently was any attempt made to determine the precise relationship of the endogenous immunoreactive material to gastrin and/or cholecystokinin. Aqueous extracts from the neural complex of *Ciona* (Thorndyke & Dockray 1986) have been studied chromatographically using a panel of antisera to enable accurate discrimination between cholecystokinin and gastrin in

Key to Table 1: a. Georges & Dubois, 1979; b. Georges & Dubois 1985; c. Pestarino 1985b; d. Fritsch *et al.* 1980b; e. Fritsch *et al.* 1982; f. Fritsch *et al.* 1980a; g. Fristch *et al.* 1979; h. Girgis *et al.* 1980; i. Thorndyke & Falkmer 1985; j. Thorndyke & Probert 1979; k. Pestarino 1985a; l. Van Noorden & Pearse 1976; m. Reinecke 1981; n. Thorndyke 1986; o. Fritsch *et al.* 1978; p. Thorndyke & Dockray 1986; q. Thorndyke 1982; r. O'Neil *et al.* 1986a; s. Bevis & Thorndyke 1978; t. Georges & Dubois 1980; u. Georges & Dubois 1984; v. Reinecke *et al.* 1980; w. Pestarino 1983a; x. Pestarino 1984; y. Pestarino & Taglioferro 1983; z. Bevis & Thorndyke 1979; a^2, Pestarino 1983b; b^2, O'Neil *et al.* 1986b; c^2, Georges 1985.
*Found at the junction between neural ganglion and gland.

the separated fractions. Based on its behaviour following fractionation on Sephadex gels, reversed phase HPLC and its ability to dilute in parallel with a Gastrin 17 standard (in a radioimmunoassay using antisera favouring G17 rather than CCK[8]), the *Ciona* peptide is apparently rather more gastrin-like than CCK-like.

Secretin/VIP

Peptides of the Secretin–VIP family have been described in both CNS and alimentary tract of several species (Table 1). In *Styela clava*, preliminary biological and chromatographic analyses of purified stomach extracts (Thorndyke & Bevis 1983) support the immunocytochemical evidence and suggest the existence of a potent secretin/VIP-like peptide when pancreatic secretion in the anaesthetised cat is used to assess biological activity.

Insulin-like peptides

The importance and significance of the presence of insulin-like peptides in the CNS of protochordates has been the subject of much debate, particularly in view of the apparently widespread distribution of similar peptides in invertebrates (Thorndyke & Falkmer 1985; O'Neil *et al.* 1986a). For many years it was held that insulin-like peptides were restricted to the gastro-entero pancreatic (GEP) system of chordates including protochordates. However recent studies using an antiserum raised against purified salmon insulin have shown that insulin-like peptides are present in the neural ganglion of *Ciona* (O'Neil *et al.* 1986a). Furthermore, extracts of the ganglion of *Ciona*, when subject to radioimmunoassay produce dilution curves in parallel with salmon insulin, and indicate approximately 650 $pg.g^{-1}$ of insulin-like material in the *Ciona* neural complex (M.C. Thorndyke & E.M. Plisetskaya, unpublished observations).

Tachykinins

New evidence based on a combination of immunocytochemistry, radioimmunoassay and HPLC has confirmed the presence of at least two types of tackykinin in the gut and neural ganglion of *Ciona* (O'Neil *et al.* 1986b). There is an extensive array of neurones containing Substance P-like immunoreactive material both centrally and in the periphery. In addition a separate population of larger cells in the neural ganglion show neurokinin A-like immunoreactivity. These findings are supported by convincing evidence from radioimmunoassay and HPLC reverse-phase hydrophobic separation of extracts from both neural ganglion and pharynx; where at least 2 peaks of material with C–terminal Substance P-like immunoreactivity together with a quite separate peak of NKA-like material are found.

The ascidian neural complex and the vertebrate pituitary

The relationship of the ascidian neural complex to the vertebrate pituitary gland has been the subject of considerable debate since a possible homology was first suggested over 100 years ago by Julin (1881). With the description of a number of neurohormonal peptides in the complex, interest in the possibility of the neural ganglion/gland as a pituitary homologue has re-emerged. Detailed embryological evidence (Elwyn 1937; reviewed by Dodd & Dodd 1966) confirms that if there is an homology it must be between the neural complex and the vertebrate neurohypophysis, because all parts of the complex derive from the embryonic neural tube. However, all attempts at localization and/or extraction of neurohypophyseal-like peptides have met with little success (Dodd & Dodd 1966; Goossens 1977). Indeed, current evidence from the laboratory of one of the authors suggests that the biologically active principles in the neural complex are more likely members of the tachykinin family of peptides (see above). Because the embryological evidence rules out any homology with the adenohypophysis, the description of Prolactin–, β–endorphin-, αMSH- and ACTH-like immunoreactive material in the neural gland (Pestarino 1985b) cannot alone evidence homology with the vertebrate adenohypophysis. Furthermore, large proteins within the glandular cells could possess amino acid sequences sufficiently similar to those present in all of the peptides noted above, so as to allow cross reactivity with the peptide antisera. Only the careful use of panels of region-specific antisera accompanied by fractionation and HPLC can confirm or deny chemical identities (Rehfeld 1984; Thorndyke 1986). Finally, it is now well established that neurohormonal peptides are of widespread distribution in both gastrointestinal and neuronal sites throughout the animal kingdom (Thorndyke 1986). Therefore the mere localization in the neural gland of a peptide that is also present in the vertebrate pituitary does not establish it as a pituitary homologue. There can be no doubt, however, that neurohormonal peptides are present in significant quantities in the Ascidian neural ganglion (Table 1), but despite the evidence for a relationship between some of these peptides and glandular cycles in the neural complex (see below), present evidence is more in keeping with the suggestion of Dodd and Dodd (1966) that the neurosecretory cells of the ganglion are "more comparable with neurosecretory neurons in the ganglion of invertebrates than anything in the pituitary complex of vertebrates".

Functional aspects of neurohormonal peptides in protochordates

Although much information is now available on the occurrence and distribution of neurohormonal peptides, relatively little is known of their likely role or roles in protochordates. Ascidians are encased within a tough, often thick tunic and do not lend themselves readily to surgical manipulation. This, together with their rudimentary circulatory system, severely limits the degree of in vivo physiological intrusion possible making experimental studies with protochordates notoriously difficult. More progress may be possible by utilizing suitable bioassays for assessment of neurohormonal peptide and/or amine activities *in vitro* (see later). In spite of these difficulties and shortcomings, there have been three major studies which have made some progress in determining the functions of neurohormonal peptides in protochordates and these are considered below.

Neurohormonal regulation of reproductive cycles and spawning in *Ciona*

Georges (Georges 1973, 1977, 1978; Georges & Dubois 1985) has carried out a detailed study of the relationship between reproductive events, tidal period and cyclical changes in the neural complex of *Ciona intestinalis*.

In the English Channel, *Ciona* spawn between June and September, and typically spawning takes place just before dawn with a maximum number of animals releasing eggs twice a month at neap tides. The spawning sequence itself involves the passage of eggs from the ovary into the oviduct and the relaxation of the muscular ring at the oviduct lip to release eggs.

A unique cycle of cellular activity in both neural ganglion and gland is closely correlated with the tidal cycle. Correlated with the period from ebb to low tide the large peripheral neurons in the ganglion are characterised by an active Golgi complex with an accompanying high concentration of Golgi vesicles and vacuoles (Fig. 1). During the succeeding phase of flood to high tide this arrangement is replaced by one showing neurons with large dense bodies. In parallel with these changes, the neural gland also undergoes a rhythm of activity. The epithelial phase occurs during the period flood to high tide and here the gland comprises 2 layers of prismatic glandular cells surrounded by a thin sinuous lamella. With the approach of ebb tide and culminating at low tide this is succeeded by the mesenchymal phase in which, following proliferation, the cells become stellate, accumulate centrally and are finally released into the blood sinus (Fig. 1). In this way, the rhythmic cycles of activity in both neural gland and ganglion are closely linked to tidal events. In a series of experiments in which the neural gland was excised and various tissue extracts injected (Georges 1978), it became apparent that the glandular cycle reflects the production and release, during the mesenchymal phase, of a spawning inhibitory substance (SIS). The interaction of the ganglion, gland and ovary in these cyclical phenomena are complex and not yet fully understood. Experiments where the ganglion only is excised (Georges 1977, 1978) indicate that the presence of neural ganglion is essential for maintenance of the gland cycles. If the ganglion is removed at flood tide (during the epithelial phase) the cycle halts immediately. When it is removed at ebb tide (during the mesenchymal stage) the cycle continues with the release of cells (and SIS) and the gland returns to the epithelial phase at which point the cycle stops.

Removal of the neural gland has no effect on the pattern of events in the neural ganglion. Rather, it is the ovary which appears to play a part in the neuronal cycle (Georges 1978). Removal of the ovary *in vivo*, or culture of the neural complex *in vitro* in the absence of ovarian tissue results in an increase in the dense body phase of the ganglion cycle (Georges 1978). The normal ganglion cycle returns on the introduction of ovarian tissue or extracts, or addition of steroids such as dehydroepiandrosterone and cortisol. The interrelationships here are far from clear and there are a number of unexpected observations in this work which are difficult to explain. Thus, in the experiments described above, the glandular cycle remains *in vivo* or *in vitro* provided the ganglion is present, and is *not* affected by changes in the neuronal cycle induced by ovary ablation or removal. Furthermore, in spite of there being 2 cycles of ganglion and gland with concomitant release of SIS in any 24 h period, spawning only occurs once in 24 h. Interestingly, in those species found in the Mediterranean which has no tides, only 1 cycle per day is seen. Clearly other internal and/or external factors are involved in the synchronization of egg release. Light is obviously crucial since maximum spawning only occurs some time after dark, just before dawn and generally at the time of neap tides (Georges 1978). In

Fig. 1. Diagrammatic representation of the relationship between neural ganglion and neural gland during a single tidal cycle. GCS, Gland Controlling Substance; SIS, Spawning Inhibitory Substance.

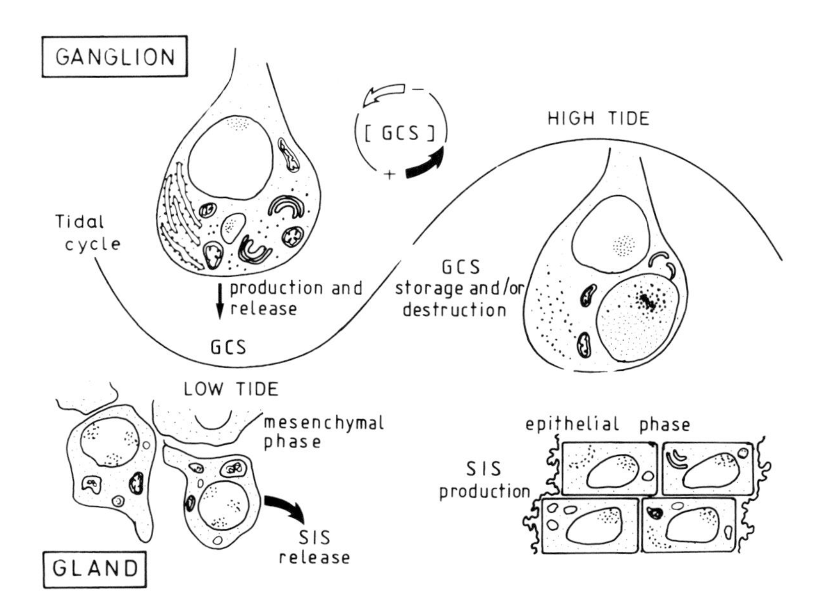

conditions of constant light, although the ganglion and gland cycles continue and eggs are passed into the oviduct, release of eggs from the oviduct by relaxation of the distal sphincter does not take place. One possible explanation of the maximum spawning at neap tide and minimum at spring tide may be that in the latter case there are two complete mesenchymal stages and accompanying SIS release immediately prior to the appropriate spawning period. Conversely in the former situation there is only a single mesenchymal phase with SIS release before the high tide which is here synchronised with the hours before dawn.

It is difficult to interpret the ganglion cycle in terms of changes in peptide profile. Of those neurohormonal peptides described, only changes in ACTH-like immunoreactivity correlate exactly with the cycling phenomena. Met–enkephalin- and LHRH-like immunoreactive material does not exhibit a 24 h cycle but nevertheless declines after spawning (Georges & Dubois 1985). Monosodium glutamate, an agent known to destroy Met–enkephalin cells in mammals, severely depleted the Met–enkephalin-like cells of *Ciona*, and subsequently the neural gland cycle was arrested. It is tempting to speculate that the relationship between ovary, neural ganglion and neural gland somehow involves neurohormones of the ACTH–, Met–enkephalin family, particularly since these are known to be closely linked in vertebrates with the regulation of steroid hormone production and allied reproductive events.

Neurohormonal peptides and reproduction in *Branchiostoma*

The larval biology of cephalochordates has attracted attention largely due to the phylogenetic position of this group and its importance as a primitive chordate (Barrington 1965). Ecological studies indicated a remarkable co-ordination of gamete release within single populations of *Branchiostoma* (Amphioxus) which until recently had no supporting physiological or biochemical data. However a detailed study has now been carried out on the Chinese form of amphioxus, *Branchiostoma belcheri,* the results of which suggest a remarkable co-relation between likely control mechanisms in this species and those known to be present in both agnathans and gnathostomes (Chang *et al.* 1984).

Preliminary investigation of the steroid content in both male and female gonads showed that while high levels of progesterone and testosterone were present in both sexes, oestradiol 17β and oestrone were only elevated in females (Chang *et al.* 1985). Subsequent experiments explored the effect of LH, Prolactin (PRL), LHRH and hCG on both steroid levels and production in the gonads of both male and female individuals. In females a single dose (10ng.animal^{-1}) of LHRH results in a sevenfold increase in progesterone and a fourfold increase in oestradiol (Fig. 2); hCG also enhances progesterone production but oestradiol does not. In males, testosterone levels were increased, with LHRH showing a greater potency than hCG.

The effects of ovine LH and prolactin (PRL) have also been studied in males and females; both hormones elevate progesterone levels with LH being more potent than

PRL and males in particular showing a significant increase in progesterone levels. Testosterone production is stimulated in both sexes but with LH being more potent than prolactin in males, and the reverse being so in females. It is only with the oestrogens that the increase in steroid levels is restricted to females. In an attempt to confirm the specificity of the response, binding studies using radioiodinated LHRH and hCG were carried out. Scatchard plots of ^{125}I–Lh–RH and hCG binding to gonad homogenates indicated genuine specificity with a single class of binding sites for hCG and saturable binding of LH–RH. Immunocytochemical studies demonstrated the presence of LH-, hCG- and FSH-like material in Hatschek's pit,

Fig. 2. Effect of LHRH and hCG on progesterone and estradiol–17β production in female *Amphioxus*. (adapted from Chang *et al.* 1985).

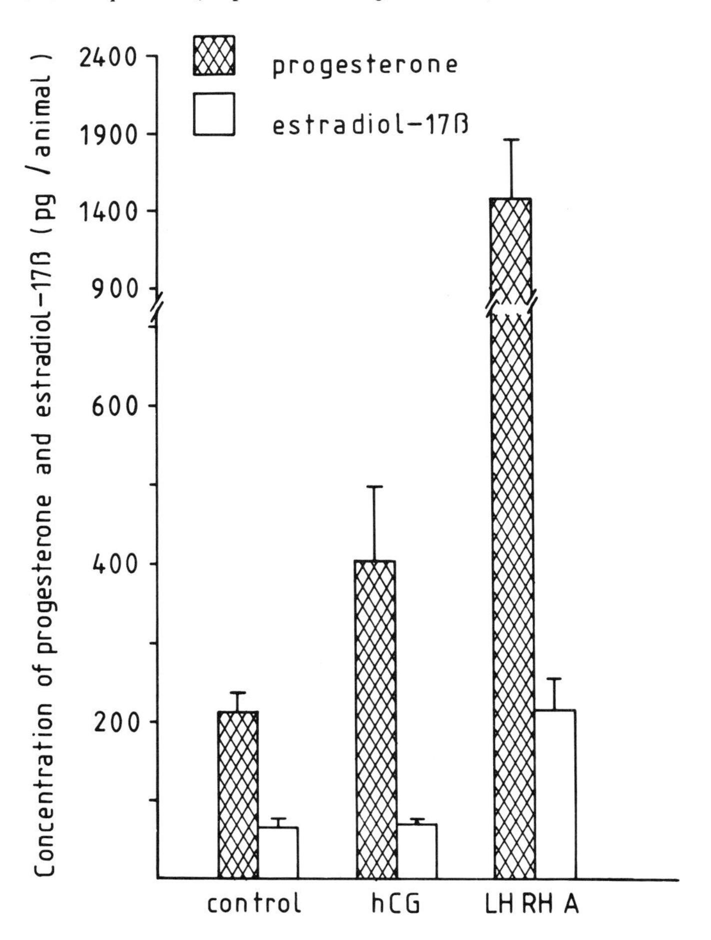

while LHRH and TRH-like molecules appear to be associated with the epithelial layers of this organ.

These findings are of particular interest because for some time Hatschek's pit has been suggested as a likely homologue of the vertebrate pituitary gland (Tjoa & Welsch 1974). Furthermore, the results of Chang and his colleagues described above clearly suggest the presence in *Branchiostoma belcheri* of a regulatory mechanism which might well serve as a model for the primitive vertebrate pattern. However such a proposal should be viewed cautiously because as we have said the cephalochordates are a small and very specialized group of animals, probably as far removed from invertebrate protochordates like ascidians as they are from vertebrate chordates. Furthermore, the experiments described above have yet to be confirmed in other laboratories.

Hormonal control of gut secretion in ascidians

The alimentary canal in Ascidians is characterized by a large number of typical, basally granulated endocrine cells of a type known from studies on other groups, to be involved in the regulation of gut secretory activity (Table 1). Furthermore, immunocytochemical studies suggest that the products of these cells are immunochemically similar to polypeptide hormones such as secretin, bombesin, VIP and others, which have established roles in vertebrate gastrointestinal physiology (Bevis & Thorndyke 1979; Fritsch *et al.* 1982; Thorndyke & Bevis 1983; Thorndyke 1986).

In addition, ultrastructural evidence indicates that the enzyme secreting cells present in the Ascidian gut in many ways represent presumptive pre-pancreatic vertebrate zymogen cells (Thorndyke 1977). It is of considerable interest therefore, to determine what control mechanisms might exist for the regulation of gastrointestinal secretion in Ascidians and whether endogenous polypeptide messengers are involved. Using a novel perfusion technique in which the effect of various peptides on gastrointestinal secretions of the solitary ascidian *Styela clava* could be assessed by the measurement of enzyme and protein levels in the gut perfusate, it was established that certain 'vertebrate' gastroenteropancreatic (GEP) peptides have a significant stimulatory effect on stomach secretory activity. All forms of CCK tested are able to stimulate enzyme secretion *in vivo* (measured as perfusate protein levels) in the *Styela* bioassay (Fig. 3). Interestingly, the large form, CCK33, is in molar terms rather more potent than the shorter form, CCK_8; suggesting perhaps that more than the C–terminal octapeptide is necessary for full receptor activation. Moreover the response to CCK_{33} and CCK_8 appears to be genuine since trials using the specific CCK inhibitor dibutyryl cyclic GMP (Bt_2 cGMP) infused together with either CCK_8 or CCK_{33}, significantly reduced the stimulation of protein release when compared to the effect of CCK_8 or CCK_{33} alone (Fig. 3). In vertebrates the sulphated tyrosine at position 7 from the C–terminus of CCK is of crucial importance for biological activity (Jensen *et al.* 1981). Thus, in mammals if either the tyrosine is transposed to position 6 from the C–terminus (as in gastrin) or is desulphated, then its secretagogue effect on the pancreatic acinar cells is seriously impaired. These changes also reduce its efficacy as a gall bladder contracting agent, the other major role for CCK in vertebrates. Studies on coho salmon (Vigna & Gorbman 1977) however, suggest that in lower vertebrates this requirement is not so rigid, because although sulphation of the tyrosine is important for full activity, change in position from 7 to 6 from the C–terminus has little effect on biological activity. It is, therefore, of considerable interest than in *Styela* the receptors were unable to discriminate between sulphated and non-sulphated CCK. Clearly, the implication here is one of a receptor system which lacks the sophistication seen in vertebrates, and as such perhaps represents a potential phylogenetic forerunner of the chordate plan.

Additional support for such a concept comes from experiments with bombesin,

Fig. 3. Effect of peptides and tissue extracts on protein secretion in the stomach of *Styela clava*. (a) Sulphated and non-sulphated CCK; (b)CCK_{8s} with and without dibutyrylcyclic GMP; (c) bombesin and physalaemin; (d) *Styela* stomach extract and CCK_{33}. Methods according to Thorndyke & Bevis 1984. Arrows show time of injection of test material.

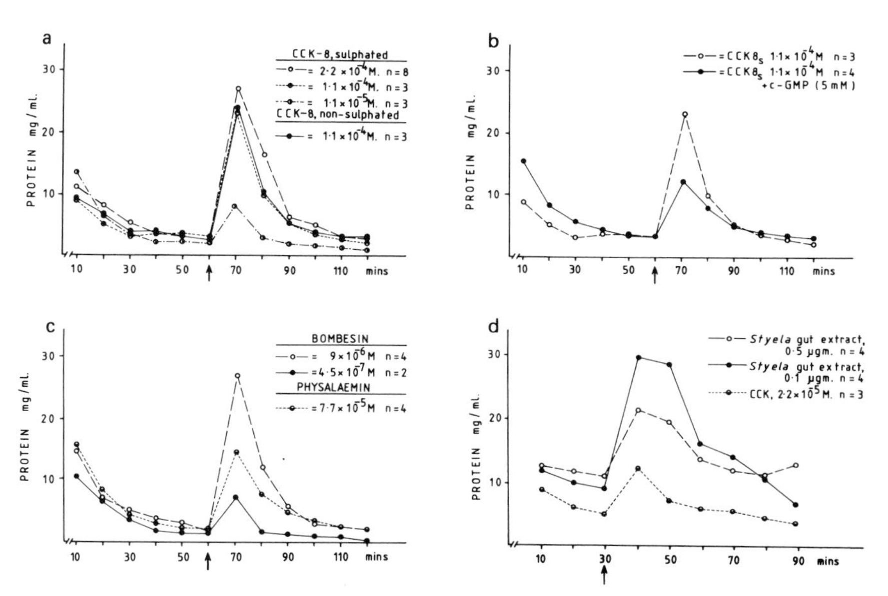

physalaemin and caerulein which are also effective secretagogues, although both physalaemin and caerulein are only weak agonists while bombesin is the most active of all peptides (Fig. 3). Interestingly, secretin, one of the peptides suggested by immunocytochemistry to be most likely present in the gastric epithelium of *Styela*, is totally without effect in the bioassay, as is glucagon a hormone from the same peptide family. This finding is perhaps not so surprising because although secretin is known as a pancreatic enzyme secretagogue in vertebrates, its primary role is the stimulation of fluid and bicarbonate secretion, with its action on enzyme secretion being comparatively weak (Dockray 1979). It has been suggested (Thorndyke 1977; Thorndyke & Bevis 1984) that secretin-like peptides could be more concerned with the regulation of the extensive mucus secreting capacity of the *Styela* gut, although this has not been investigated.

On the basis of current results it is difficult to determine whether the receptors in *Styela* are of a variety of types or if there is a single receptor class with the ability to bind a range of peptides. Notably, all the peptides which are active in *Styela* are thought to operate in vertebrates through a calcium-dependent second messenger system. Secretin which is inactive in *Styela* is thought in vertebrates to utilize cAMP as its second messenger. Thus it may be that in ascidians (represented by *Styela clava*) the primitive receptors involved are calcium-dependent, and only later were cAMP-dependent receptors developed. This aspect of the pharmacology of the receptors has, however, not been investigated.

Rather more problematic is the paradoxical finding that by immunocytochemistry, at least, only secretin-like material is detected in the gut of *Styela clava* (Bevis & Thorndyke 1979) while other experimental approaches, notably those involving the extraction and assay of the endogenous *Styela* material by both immunochemical and biological means, are indicative of a CCK-like moeity. Radioimmunoassay of *Styela* gut extracts (Dockray & Thorndyke, unpublished observations; Thorndyke & Bevis 1983) suggest the presence of a small CCK-like molecule, as does assessment of the same extracts in the Guinea Pig gall bladder bioassay (Thorndyke & Bevis 1983). These findings suggest, therefore, a lack of sensitivity in the immunocytochemical tests, rather than the absence of a CCK-like factor. It is likely that the apparently simple requirements of the receptor in *Styela clava* merely reflect the nature of the endogenous peptide and the natural molecule may well be a molecule (or molecules) with an amino acid sequence with some similarity to CCK yet with other, additional, biologically important sites. In this way full receptor activation may require a number of discrete amino acid groupings. In this respect it is of interest that CCK_{33} is more potent than the shorter CCK_8, suggesting that more than the C–terminal octapeptide is necessary for full activation. Indeed, in the *Styela* bioassay (reported in Thorndyke & Falkmer 1985) the secretory response induced by infusion of homologous extract is considerably larger than that seen with CCK (Fig. 3d). Despite the ambiguities, however, current evidence can be used as the basis for a model describing some of the likely mechanisms involved in the control of gastric secretion in *Styela*. This model is shown in Fig. 4. Until the native peptide or peptides are isolated and sequenced the nature of the relationship between endogenous agonist and receptor will remain open to speculation.

Functions with the potential for aminergic or peptidergic regulation

Although there are relatively few established roles for peptides and amines as regulatory agents in protochordates (see above), there are a number of physiological processes for which it seems highly likely that peptides and/or amines may have a putative modulatory role.

Control of branchial cilia in ascidians

The most dominant feature of ascidians is their large branchial basket and prominent siphonal openings, both of which subserve the filter-feeding process characteristic of this group (Fig. 5). As with the majority of invertebrate particle feeders, the motive force for the feeding water current is provided by an array of

Fig. 4. Diagram to show proposed functional relationship between ascidian gut endocrine cells and digestive processes. Endocrine cells found at the tips of gastric folds in *Styela* (or more scattered in other species) detect food material in the gut lumen and subsequently release a peptide messenger basally. This acts in a local endocrine or paracrine fashion on adjacent zymogen cells.

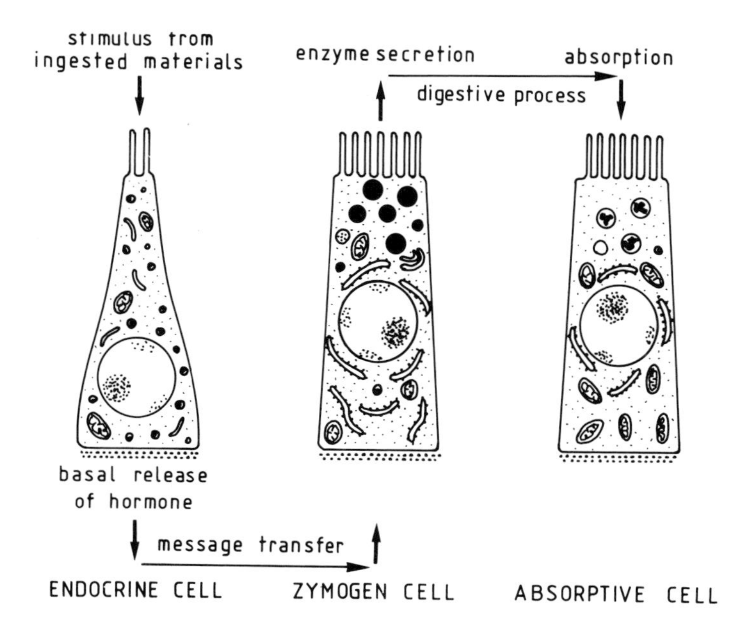

heavily ciliated gill slits or stigmata. These draw the water current through the oral or inhalant siphon, and the suspended food material is trapped in a mucous sheet driven dorsally across the branchial basket wall by ciliary power. This food and mucus mass is finally compacted in the dorsal groove and drawn into the oesophagus. Branchial ciliary movement is intermittent and the stationary or arrested phase is co-ordinated with muscle-driven siphonal closures. This is essential since it allows the stigmata to remain open at the time of body contraction and thus ensure that there is no pressure build up as the animal contracts with a resulting reduction in branchial basket volume (Goodbody 1974).

In an elegant series of experiments Mackie and his colleagues (1974) showed that ciliary control in the branchial sac of *Corella* was largely mediated by way of a network of fine nerve fibres originating as branches of the visceral nerve which arises directly from the neural ganglion. Since no cell bodies have yet been found in the branchial or visceral nerves it is thought that the cell bodies lie in the CNS. Actual contacts between the branchial nerve branches and the ciliated cells are rather few and restricted to the atrial side of the pharynx. Even here, although some nerve endings have been seen lying in indentations at the bases of ciliated cells, no evidence of

Fig. 5. Schematic representation of *Ciona* to show general organisation.

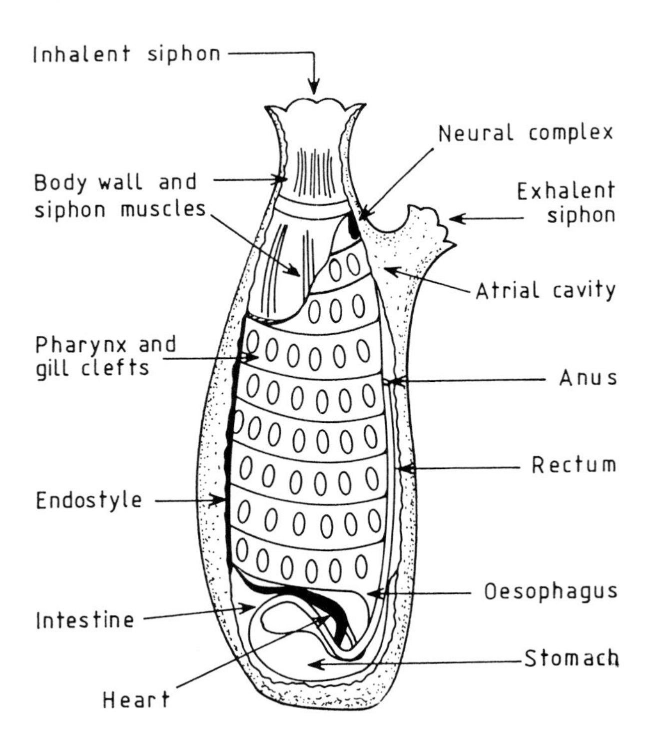

functional specializations have been observed (Mackie *et al.* 1974). In view of the rather sparse direct innervation of the ciliated cells Mackie *et al.* (1974) suggest that cell–cell conduction by way of 'close-junctions' between ciliated cells spreads the wave of exictation or inhibition from the innervated cells to those without innervation. Indeed, stimulation in only one remote region of the basket results in a wave of stimulation spreading over the whole basket. The ciliary arrest response is associated with neuronal electrical activity comprising trains of regular pulses (Mackie *et al.* 1974). This can be achieved by direct electrical stimulation or by tactile stimulation to the branchial basket. The potentials associated with the phenomenon of ciliary arrest have been termed ciliary arrest potentials (CAPS), and the burst of CAPS which accompany the cessation of ciliary beats is also the signal for contraction of body wall muscles (Mackie *et al.* 1974). Clearly this implies that the branchial innervation is not a separate unit but a part of a more widespread peripheral system including the body wall and siphons. Attempts to unravel the likely chemical mediators of the ciliary control processes have yielded somewhat equivocal results. Thus under some conditions curare produces a rapid decay in the electrical response while under others, CAPS and associated ciliary arrests are possible by transmission through a curarized region. This implies that control must be mediated, at least in part, by a non-cholinergic system (Mackie *et al.* 1974). It is possible that amines may be involved since, although serotonin was ineffective in preparations *in vitro*, when applied to intact animals it produced almost continuous ciliary arrest (Mackie *et al.* 1974). Additional support for this possibility has recently emerged with the finding of serotonergic elements in the peripharyngeal bands and dorsal languets in *Ciona*, although it should be pointed out that the serotonin-like material appears to be present in epithelial cells rather than neurons (Georges 1985).

The control mechanisms involved, therefore, are still unclear; the experiments on *Corella* imply a non-cholinergic system, or at least partly so, with the added possibility of aminergic involvement. Moreover, it is still uncertain to what extent control of ciliary activity is totally under neuronal control, or whether other mechanisms need be invoked. The latter seems a strong possibility, because neuronal activity produces only momentary reversal followed by arrest, and it is difficult to see how control over rate of ciliary beating is achieved. Because of the prolonged arrest response induced by nerve stimulation and because direct innervation of ciliary cells is sparse, Mackie and his colleagues (1974) speculate that some blood-borne hormonal mechanism is likely. However, the recent finding of serotonin-like material in the pharynx of *Ciona*, together with the earlier finding of substance P–, neurotensin- and bombesin-like immunoreactive peptides in the gill epithelium of the same species (Fritsch *et al.* 1980), suggests that local control mechanisms may be important.

One of the most important ionic events thought to underly ciliary arrest is an increase in intracellular calcium ion concentration in the ciliated cells (Mackie *et al.*

1974). Many peptides, including bombesin are notable for utilizing intracellular calcium as their second messenger (Jensen *et al.* 1981); this provides additional circumstantial evidence for a peptidergic involvement in ciliary control.

Contractile activity of body wall and siphons in ascidians and pelagic tunicates

The only motor responses shown by ascidians involve contraction of the body wall and siphons. In salps and doliolids, similar activities are utilized as part of locomotory behaviour. Both activities are based upon the contraction of muscle bands which lie in the body wall.

Ascidians. The rhythmic feeding and defensive movements seen in this group fall into two categories: direct responses, where siphons and body wall (or mantle) contract following external stimulation; crossed responses where stimulation of the inside of one siphon results first in contraction of the opposite siphon, followed by contraction of the body wall and finally, closing of the stimulated siphon. Presumably, internal stimulation simulates the presence of a foreign particle, and such a contraction sequence produces a water current directed out of the stimulated siphon followed by its closure. In addition to this squirting response, spontaneous and rhythmic squirting is seen in some species; the latter can have a periodicity of 6–9 min in *Phallusia* (Hoyle 1953).

Physiological studies of these movements show that the system is frequency-dependent, with a single stimulus producing only a small response while 2 or more stimuli produce a strong contraction which when above threshold is independent of stimulus strength (Hoyle 1953). Repetitive stimulation leads to exhaustion and results in only small reflex contractions (Hoyle 1953; Goodbody 1974). The neural ganglion is important in co-ordinating the responses, particularly the crossed response, which is virtually removed if the ganglion is ablated, although both siphons remain responsive to electrical stimulation in the absence of the ganglion (see Goodbody 1974). The most favoured neurotransmitter for this activity is acetylcholine (ACh), again Goodbody (1974) has reviewed the earlier work and discussed the difficulties encountered in the initial identification of acetylcholine in ascidians. A more recent study of the muscle receptors in the ascidian *Halocynthia aurantum* (Kobzar & Shelkovnikov 1985) confirms the existence of both nicotinic and muscarinic cholinergic receptors. An additional number of interesting points emerge from this study: first, the muscarinic response is reduced considerably in the winter months; and second, catecholamines, as well as cGMP and papaverine decrease the response to acetylcholine.

Thus, the basic cholinergic response can be modulated. Because the major part of the somatic innervation is by neurons arising from the neural ganglion (Goodbody 1974), there are perhaps two or three possible explanations which can be put forward. Simple enhancement of the signal may take place centrally in the ganglion, but this is perhaps least likely because seasonal variation in the response can be detected with isolated muscle strips lacking all contact with the ganglion. It is more likely that local modulation of the cholinergic response occurs directly at the receptor site. Regulatory peptides are powerful modulatory agents in lower vertebrates and in particular are recognised as interacting with established neurotransmitters (Thorndyke & Falkmer 1985).

In view of its established role as a smooth muscle stimulant (Erspamer *et al.* 1977) perhaps one of the most likely modulatory candidates in ascidians is substance P (SP); almost 20 years ago small amounts of SP were noted in whole body extracts from *Ciona* (Dahlstedt *et al.* 1959). Beginning earlier and continuing until Dodd and Dodd's (1966) extensive study, numerous attempts were made to identify homologues of the pituitary octapeptides oxytocin and vasopressin in the ascidian neural complex. Even one of the earliest stated there was a substance present in the complex of *Ciona*, which contracted smooth muscle yet was not oxytocin (Pérès 1943).

Dodd and Dodd (1966) concluded from their studies of biologically active molecules in extracts from *Ciona* that "there is at present no satisfactory evidence that any part of the neural complex has pituitary affinities". Despite this, subsequent careful examination of their results reveals some interesting points. One of the tests for oxytocic activity concerns the ability of oxytocin and/or extracts to contract rat

Fig. 6. Depressor effect of extracts of *Ciona* neural complex and branchial region on control of blood pressure in the anaesthetized rat. (1) 4.0 ImU Pitressin (Parke–Davis); (2) 1.0 mg neural complex extract; (3) 1.0 mg Branchial region extract. Adapted from Dodd & Dodd (1966).

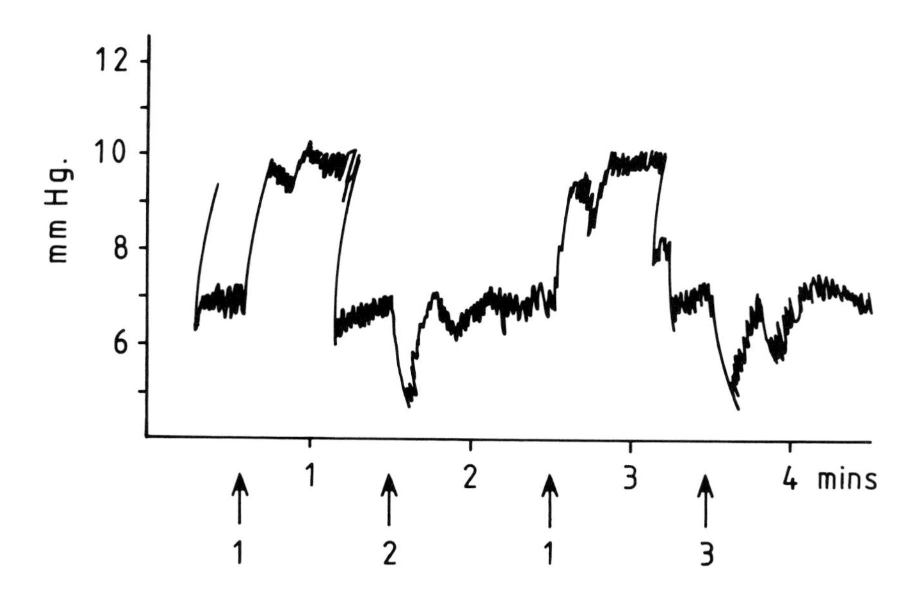

uterus *in vitro*. Dodd and Dodd (1966) showed that extracts from both neural complex and branchial tissues produced a contraction of uterine smooth muscle, but the response was more rapid than that to a synthetic pituitary octapeptide ('Pitocin'). The active molecule was not ACh or serotonin. With the benefit of hindsight, it seems most likely that the active factor was substance P (SP) although this was not suggested by the authors at the time. The activity associated with the branchial tissues is especially interesting in view of the later finding of SP and other peptides in these tissues. In other experiments, in response to extracts from the neural complex Dodd and Dodd (1966) found an unexpected but marked fall in blood pressure in the rat. Although its significance was not appreciated at the time (Fig. 6), this finding is particularly illuminating since SP has potent vasodilatory and vasodepressor effects (Burcher *et al.* 1977).

Finally, recent work (O'Neil *et al.* 1986b; G.S. O'Neil & M.C. Thorndyke, unpublished observations) has identified both SP and the closely related tachykinin, neurokinin A, in both neural complex and body wall. Biological assays on the effects of SP on isolated muscle strips prepared from the body wall of *Ciona* show that while ineffective alone, SP at physiological doses will markedly enhance the normal response to acetylcholine (Fig. 7). Further studies on this particular problem are currently in progress, but present evidence strongly supports the idea of a peripheral, modulatory, role of SP-like tachykinins in the regulation of ascidian body wall contraction.

Doliolids and Salps. These closely-related groups of planktonic tunicates utilize the contraction of muscle bands in the bodywall to provide a water current which is the basis of their 'jet-propulsive' motion (Bone 1982). The essentials of the innervation are similar in both salps and doliolids where the muscle bands are innervated by fine branches of nerves derived from the brain (Bone 1959). Rhythmic contractions of the muscle bands are produced by a complex conducting network involving epithelioneural and neuroepithelial synapses, although the majority of the cell bodies of the motor fibres are located at the periphery of the brain (Bone 1959). The synapses contain numbers of electron-lucent vesicles and, while the full details are not clear, it is likely that this represents an essentially cholinergic system (Bone 1982). It may be that here too peptides and/or amines are involved in the fine modulation of motor control, but, there is as yet no direct evidence for either the occurrence of peptides or a physiological effect for these regulators.

Control of cardiac activity in ascidians

The ascidian heart has attracted attention largely due to its unique ability to reverse its direction of beating (Goodbody 1974). Despite early suggestions that the ascidian heart was under neuronal control (Alexandrowicz 1913), it now seems certain that the heart itself is not innervated, and contractions are generated by a myogenic mechanism involving several pacemaker sites with at least one pacemaker centre at either end of the cardiac tube (Goodbody 1974). There is however, a fine plexus of fibres on the surrounding pericardium, but because these fibres do not penetrate the band of smooth muscle around the pericardium, it was assumed that these neurones must be sensory (Bone & Whitear 1958). In this way these fibres may respond to pressure changes in the heart and form part of a sensory-motor reflex arc to the pericardial muscle and cardiac myoepithelial cells, which is consistent with the

Fig. 7. Effect of Substance P and/or Acetylcholine on a muscle strip preparation from *Ciona intestinalis*. Chart speed 0.24 mm sec^{-1}. Isometric tension recorded. Arrow indicates time of addition of test material. W, wash. (Data kindly provided by Greg O'Neil from unpublished work.)

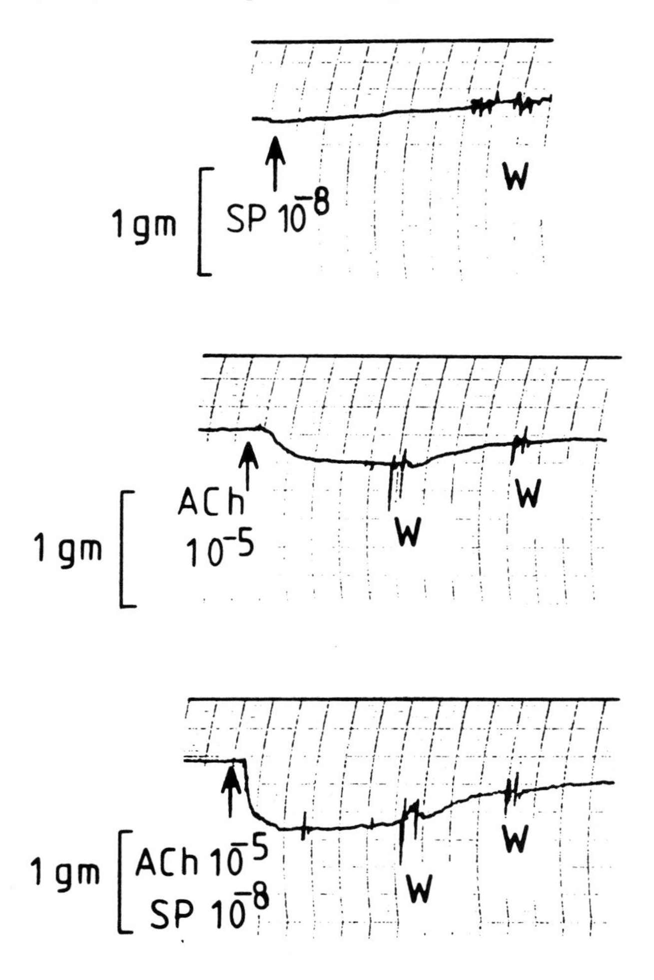

contention that an increase in pressure plays an important part in the phenomenon of heart beat reversal (Kriebel 1968; Goodbody 1974). But what regulates the myogenic pacemaker centres at each end of the heart? The effects of a variety of neurotransmitters and related drugs have been tested on the ascidian heart, with contradictory results. Thus, acetylcholine has been found by various authors to enhance heartbeat, slow down and stop contractions (atropine-sensitive effect) or to have no effect. Similarly, epinephrine can both reduce and enhance heart rate, while serotonin increases beating when applied to the visceral end of the heart and decrease when applied to the branchial end (Goodbody 1974). It is possible that regulatory agents released into the blood at peripheral sites could influence the myoepithelial cells 'en passant' through the heart, but there is no evidence to support this hypothesis. Kalk (1970), along with others, has reported the presence of apical projections from the myoepithelial cells and this author goes on to consider that the material contained therein might be released into the heart lumen to act as a local humoral modulator. Kalk (1970) suggests that the released substance may be acetylcholine, but there is no direct evidence for this.

In view of the contradictory results described above, it remains possible that other chemical regulators such as peptides are involved in the control of heart beat. Unfortunately there is, as yet, little evidence available either for or against this proposal. Preliminary investigations on the occurrence of peptides in the heart of *Ciona* (O'Neil, unpublished observations) indicate the presence of calcitonin gene-related peptide (CGRP) in localized areas of the myoepithelium, and this may well represent the material present in the apical regions of the myoepithelial cells observed by Kalk (1970).

Conclusions

Definitive rôles for many of the peptides and amines present in protochordates remain elusive. Nevertheless, and in spite of the difficulties encountered in the experimental manipulation of protochordates, there are at least 3 areas in which there can be little doubt of a major peptide and/or amine involvement. These are the rhythmic cycles in reproductive activity seen in *Ciona*, and the likely control of reproductive maturity in amphioxus, together with the control of secretory activity in the digestive tract in the ascidian *Styela clava*.

In addition we have identified 3 key areas where it seems highly likely that careful and thorough investigations will be rewarded. In each of these, ciliary control in the branchial basket, muscular activity in the siphons and body wall as well as the regulation of cardiac rhythms, the basic control mechanisms require elucidation. We hope, however, that we have shown that there is sufficient evidence to form a firm foundation for further work by suitably motivated and inquisitive investigators.

Acknowledgements

We thank Greg O'Neil for permission to quote from his unpublished work as well as Ann Edwards and Zyg Podhorodecki for help with manuscript preparation. Much of the experimental work described in this review was carried out with the help of SERC Grants to MCT.

References

Alexandrowicz, J.S. (1913) Zur kenntnis des sympathetischen Nerven-systems einiger Wirbellosen. *Zeitschrift für allgemeine Physiologie,***14**, 358–376.

Anderson, P.A.V. & Bone, Q. (1980) Communication between individuals in Salp chains II. Physiology. *Proc. Roy. Soc. Lond. B.*, **210**, 559–574.

Barrington, E.J.W. (1965) *The biology of hemichordata and protochordata.* Oliver & Boyd, Edinburgh, London.

Bevis, P.J.R. & Thorndyke, M.C. (1978) Endocrine cells in the oesophagus of the Ascidian *Styela clava,* a cytochemical and immunofluorescence study. *Cell Tissue Res.*, **187**, 153–158.

Bevis, P.J.R. & Thorndyke, M.C. (1979) A cytochemical and immunofluorescence study of endocrine cells in the gut of the Ascidian *Styela clava. Cell Tissue Res.*, **199**, 139–144.

Bevis, P.J.R. & Thorndyke, M.C. (1981) Stimulation of gastric enzyme secretion by porcine cholecystokinin in the ascidian, *Styela clava. Gen. Comp. Endocrinol.* **45**, 458–464.

Bone, Q. (1959) Observations upon the nervous systems of pelagic tunicates. *Quarterly J. of Microscopical Sci.*, **100**, 167–181.

Bone, Q. (1982) The role of the outer conducting epithelium in the behaviour of salp oozoids. *J. Mar. Biol. Assoc. UK.* **62**, 125–132.

Bone, Q. & Whitear, M. (1958) A note on the innervation of the pericardium in *Ciona. Publicazioni della Stazione Zoologica di Napoli.* **30**, 337–341.

Bone, Q., Anderson, P.A.V. & Pulsford, A. (1980) The communication between individuals in salp chains I. Morphology of the system. *Proc. Roy. Soc. Lond. B.*, **210**, 549–558.

Burcher, E., Nilsson, G., Anggard, A. & Rosell, S. (1977) An attempt to demonstrate release of Substance P from the Tongue, Skin and Nose of Dog and Cat. In: Substance P. Eds. U.S. Von Euler & B. Pernow. New York, pp. 183–186.

Chang, C.Y., Yi–xun Liu, Ti–tao Zhu & Hong–hua Zhu (1984) The Reproductive Endocrinology of Amphioxus. *Frontiers in Physiological Research.* Ed. D. Carlick & P.I. Korner, Canberra: Australian Acad. Sci.

Chang, C.Y., Liu, Y.H. & Zhu, H.H. (1985) Steroid hormones and their functional regulation in Amphioxus. In: Current Trents in Comparative Endocrinology. Eds. B. Lofts & W.N. Holmes, Hong Kong Univ. Press, Hong Kong. pp. 205–207.

Dockray, G.J. (1979) Comparative biochemistry and physiology of gut hormones. *Ann. Rev. Physiol.* **41**, 83–95.

Dodd, J.M. & Dodd, M.H.I. (1966) An experimental investigation of the supposed pituitary affinities of the Ascidian neural complex. *Some contemporary studies in Marine Science.* Ed. H. Barnes. George Allen & Unwin Ltd., London, 233–252.

Elwyn, A. (1937) Some stages in the development of the neural complex in *Ecteinascidia turbinata. Bull. Neurol. Inst. NY.* **6**, 163–177.

Fritsch, H.A.R., Van Noorden, S. & Pearse, A.G.E. (1978) Localization of somatostatin- and gastrin-like immunoreactivity in the gastrointestinal tract of *Ciona intestinalis* L. *Cell Tiss. Res.*, **186**, 181–185.

Fritsch, H.A.R., Van Noorden, S. & Pearse, A.G.E. (1979) Localization of somatostatin-, Substance P- and calcitonin-like immunoreactivity in the neural ganglion of *Ciona intestinalis* L. *(Ascidiaceae). Cell. Tiss. Res.* **202**, 263–274.

Fritsch, H.A.R., Van Noorden, S. & Pearse, A.G.E. *et al.* (1980a) Calcitonin-like immunocytochemical staining in the alimentary tract of *Ciona intestinalis. Cell Tiss. Res.* **205**, 439–444.

Fritsch, H.A.R., Van Noorden, S. & Pearse, A.G.E. (1980b) Substance P-, Neurotensin- and Bombesin-like Immunoreactivities in the Gill Epithelium of *Ciona intestinalis* L. *Cell Tiss. Res.* **208**, 467–473.

Fritsch, H.A.R., Van Noorden, S. & Pearse, A.G.E. (1982) Gastro-intestinal and neurohormonal peptides in the alimentary tract and cerebral complex of *Ciona intestinalis* (Ascidiaceae) *Cell Tiss. Res.*, **223**, 369–402.

Georges, D. (1973) Functionnement rhythmique du complexe neural et controle neuro-endocrinien de la ponte chez *Ciona intestinalis.* (Tuncier). C.R. 96°*Congres Nat. Soc. Sav. Toulouse*, **3**, 27–39.

Georges, D. (1977) Analyse Fonctionelle du Complexe Neural chez *Ciona intestinalais* (Tuncier, Ascidiace). *Gen. Comp. Endocrinol.*, **32**, 454–473.

Georges, D. (1978) Circadian rhythms in the neural gland of *Ciona intestinalis.* In: *Comparative Endocrinology*, eds. P.J. Gaillard & H.H. Boer. Elsevier/North Holland Biomedical Press, Amsterdam. pp. 161–164.

Georges, D. (1985) Presence of cells resembling serotonergic elements in four species of tunicates. *Cell Tiss. Res.*, **242**, 341–348.

Georges, D. & Dubois, M.P. (1979) Immunological evidence for an ACTH-like antigen in the nervous ganglion of a protochordate *Ciona intestinalis* (Ascidiacea). *Arch. Anat. Micros. Morphol. Exp.* **68**, 121–125.

Georges, D. & Dubois, M.P. (1980) Mise en evidence par les techniques d'immunofluorescence d'un antigene de type LHRH dans le systeme nerveux de *Ciona intestinalis* (Turnicies Ascidiace). *C.R. Adac. Sci. D.* **290**, 29–31.

Georges, D. & Dubois, M.P. (1984) Methionine–enkephalin-like immunoreactivity in the nervous ganglion and the ovary of a protochordate, *Ciona intestinalis, Cell. Tiss. Res.*, **236**, 165–170.

Georges, D. & Dubois, M.P. (1985) Presence of vertebrate-like hormones in the nervous system of a tunicate, *Ciona intestinalis.* In: Current Trends in Comparative Physiology, eds. B. Lofts & W.N. Holmes. Hong Kong Univ. Press, Hong Kong, pp. 55–57.

Girgis, S.I., Galan, F., Arnett, T.R., Rogers, R.M., Bone, Q., Ravazzola, M. & MacIntyre, I. (1980) Immunoreactive human calcitonin-like molecule in the nervous systems of protochordates and a cyclostome, *Myxine. J. Endocrinol.*, **87**, 375–382.

Goodbody, I. (1974) The Physiology of Ascidians. *Advances in Marine Biology* **12**, eds. F.S. Russell & G.M. Yonge. Academic Press, London. pp. 1–149.

Goosens, N. (1977) Immunohistochemistry of the neural complex of tunicates. *Biol. Jb Dodonaea*, **45**, 138–140.

Hoyle, G. (1953) Spontaneous squirting of an ascidian. *J. Mar. Biol. Assoc. UK.*, **31**, 541–562.

Jensen, R.T., Collins, S.M., Pandol, S.J. & Gardner, J.D. (1981) Lessons from studies of secretagogue receptors on pancreatic acinar cells. In: *Gut Hormones.* Eds. S.R. Bloom & J.M. Polak, pp. 133–136. Churchill Livingstone, London.

Julin, C. (1881) Recherches sur l'organisation des Ascidies simples. *Archives de Biologie, Paris*, **2**, 211–232.

Kalk, M. (1970) The organization of a tunicate heart. *Tissue & Cell*, **2**, 99–118.

Krebel, M.E. (1968) Electrical characteristics of tunicate heart cell membranes and nexuses. *J. Gen. Physiol.* **52**, 46–59.

Mackie, G., Paul, D.H., Singla, C.M., Sleigh, M.A. & Williams, D.E. (1974) Branchial innervation and ciliary control in the ascidian *Corella. Proc. Roy. Soc. B.* **187**, 1–35.

O'Neil, G.S., Falkmer, S. & Thorndyke, M.C. (1986a) Insulin-like immunoreactivity in the neural ganglion of the ascidian *Ciona intestinalis. Acta Zoologica.* **67**, 147–153.

O'Neil, G.S., Thorndyke, M.C., Deacon, C.F. & Conlon, J.M. (1986b) Tachykinins in the nervous system of the protochordate *Ciona intestinalis. Can. J. Physiol. Pharmacol.* Suppl. July 1986, p.

Pestarino, M. (1983a) Prolactinergic neurons in a protochordate. *Cell Tiss. Res.* **233**, 471–474.

Pestarino, M. (1983b) Somatostatin-like immunoreactive neurons in a Protochordate. *Experientia* 39, 1156–1158.

Pestarino, M. (1984) Immunocytochemical Demonstration of Prolactin-like Activity in the Neural Gland of the Ascidian *Styela plicata. Gen. & Comp. Endocrinol.* **54**, 444–449.

Pestarino, M. (1985a) Occurrence of β–endorphin-like immunoreactive cells in the neural complex of a protochordate. *Cell Mol. Biol.,* **37**, 27–31.

Pestarino, M. (1985b) A pituitary like Role of the Neural Gland of an Ascidian. *Gen. & Comp. Endocrinol.* **60**, 293–297.

Pestarino, M. & Taglioferro, G. (1983) Occurrence of secretin-like cells in the digestive tract of the ascidian *Styela plicata. Bas. Appl. Histochem.* **27**, 144–148.

Rehfeld, J.F. (1984) Some biochemical and semantic issues in the study of hormone families. In: Evolution and Tumour Pathology of the neuroendocrine system. Eds. S. Falkmer, R. Hakanson & F. Sundler. Elsevier, Amsterdam, pp. 225–230.

Reinecke, M. (1981) Immunohistochemical localization of polypeptide hormones in endocrine cells of the digestive tract of *Branchiostoma lanceolatum. Cell Tiss. Res.* **219**, 445–456.

Reinecke, M., Carraway, R.E., Falkmer, S., Feurle, G.E. & Forssmann, W.G. (1980) Occurrence of Neurotensin-immunoreactive cells in the digestive tract of Lower Vertebrates and Deuterostomian Invertebrates. *Cell Tiss. Res.,* **212**, 173–183.

Thorndyke, M.C. (1977) Observations on the gastric epithelium of ascidians with special reference to *Styela clava. Cell Tiss. Res.,* **184**, 539–550.

Thorndyke, M.C. (1982) Cholecystokinin (CCK)/gastrin-like immunoreactive neurones in the cerebral ganglion of the protochordate ascidians *Styela clava* and *Ascidiella aspersa. Regulatory Peptides,* **3**, 281–288.

Thorndyke, M.C. (1986) Immunocytochemistry and evolutionary studies, with particular reference to regulatory peptides. In: Immunocytochemistry: Modern Methods and applications. Eds. J.M. Polak & S. Van Noorden. John Wright & Sons, Bristol, U.K., pp. 308–327.

Thorndyke, M.C. & Probert, L. (1979) Calcitonin-like cells in the pharynx of the ascidian *Styela clava. Cell Tiss. Res.* **203**, 301–309.

Thorndyke, M.C. & Bevis, P.J.R. (1983) CCK-like and secretin/VIP-like activity in protochordate gut extracts. *Regul.atory Peptdes Suppl.* 2. A 147–148.

Thorndyke, M.C. & Bevis, P.J.R. (1984) Comparative studies on the effects of cholecystokinins, caerulein, bombesin 6–14 nonapeptide, and physalaemin on gastric secretion in the ascidian *Styela clava. Gen. Comp. Endocrinol.* **55**, 251–259.

Thorndyke, M.C. & Falkmer, S. (1985) Evolution of Gastro–Entero–Pancreatic Endocrine Systems in Lower Vertebrates. In: Evolutionary Biology of Primitive Fishes. Ed. R.E. Foreman, A. Gorbman, J.M. Dodd & R. Olsson. Plenum Press, New York, pp. 379–400.

Thorndyke, M.C. & Dockray, G.J. (1986) Identification and localization of material with gastrin-like immunoreactivity in the neural ganglion of a protochordate *Ciona. Regulatory Peptides.* **16**, 269–279.

Tjoa, L.T. & Welsch, U. (1974) Electron microscopical observations on Kollikers & Hatschek's pit and on the wheel organ in the head region of Amphioxus. *Cell Tiss. Res.*, **153**, 175–187.

Van Noorden, S. & Pearse, A.G.E. (1976) The localization of immunoreactivity to insulin, glucagon and gastrin in the gut of *Amphioxus (Branchiostoma) lanceolatus.* Ed. Grillo, T.A.I., L. Liebson, & A. Epple. *The evolution of the Pancreatic Islets.* Pergamon Press, Oxford, 163–178.

Vigna, & Gorbman, A. (1977) Effects of cholecystokinin, gastrin and related peptides on coho salmon gall bladder contraction *in vitro. Am. J. Physiol.* **232**, E485–491.

PART IV

Neurohormones in Molluscs

W.P.M. GERAERTS, E. VREUGDENHIL &
R.H.M. EBBERINK

Bioactive peptides in molluscs

Introduction

In molluscs, bioactive peptides are produced by peptidergic neurons, endocrine glands and other tissues, such as cells of the intestinal tract. These peptides function as neurotransmitters/neuromodulators and (neuro)hormones, and control a wide range of events concerned with behaviour, reproduction, and metabolism. Particular attention has been paid to the peptidergic model systems in *Lymnaea* and *Aplysia,* to FMRFamide and related peptides which also exhibit an extra-molluscan distribution, and to the presence and function of vertebrate peptides in molluscs.

Our knowledge of biologically active peptides in molluscs is expanding rapidly due to the introduction, among other things, of sophisticated chromatographic and sequence techniques, and the methods of molecular biology. A review of this length must, of necessity, be selective. We have attempted to give a critical account of the data concerning the physiological role and the nature of (presumed) bioactive peptides, and avoided discussions of non-relevant details. Recent reviews present a wealth of complementary data (e.g. Joosse & Geraerts 1983; Geraerts & Joosse 1984; Roubos 1984; Rothman *et al.* 1985; Geraerts *et al.* 1987. See also Chapters 2 and 14 this volume).

The FMRFamide family, and opioid peptides*

The FMRFamide family

The neuropeptide FMRFamide was isolated originally from the clam *Macrocallista nimbosa* (Price & Greenberg 1977). In addition to FMRFamide, various related peptides have subsequently been isolated from the brain of a number of species belonging to different classes of the molluscs (Table 1). Immunocytochemical, pharmacological, and neurophysiological evidence suggests that FMRFamide and related peptides act as transmitters/neuromodulators in molluscs (see Chapters 12 and 14 this volume; Price 1986; Joosse & Geraerts 1983; Greenberg *et al.* 1985). For example, in *L. stagnalis* and the stylommatophoran *Helix aspersa,*

---

* Amino acid sequences in standard one letter code; F: phenylalanine; L: leucine; M: methionine; R: arginine.

Table 1. *Molluscan distribution of FMRFamide and related peptides**

– Class Subclass Species	Peptide	Reference
– Bivalves		
Macrocallista nimbosa }		Price & Greenberg (1977)
}	FMRFamide	
Geukensia demissa }		Price (1986)
– Gastropods		
Prosobranchs		
Pomacea paludosa }	FMRFamide	Price (1986)
}	+	
Busycon contrarium }	FLRFamide	Price (1986)
Opisthobranchs		
Aplysia californica }		Schaefer *et al.* (1985)
}	FMRFamide	
Aplysia brasiliana		Lehman *et al.* (1984)
Pulmonates		
Helix aspersa }		Price *et al.* (1985)
Cepea nemoralis }	+	Price (1986)
Succinea campestris }	pQDPFLRFamide	Price (1986)
Lymnaea stagnalis {	FMRFamide	Ebberink & Joosse (1985)
{	+	
{	SDPFLRFamide	
{	GDPFLRFamide	
– Cephalopods		
Octopus vulgaris	YGGFMRFamide	Voight *et al.* (1986)

*Amino acid sequences are in standard one letter code; D=Asp, F=Phe, G=Gly, L=Leu, M=Met, P=Pro, pQ=pyroglutamic acid, R=Arg, S=Ser, Y=Tyr.

immunocytochemical methods have identified various types of FMRFamide containing neurons with axons terminating on other neurons, on muscles as well as synapses containing FMRFamide-positive material. In addition, FMRFamide has been found to affect several aspects of the electrophysiological characteristics of identified neurons and muscles of these species and other species. Recently, evidence has been obtained that FMRFamide functions as an endogenous modulator of opioid activity in molluscs, in a manner analogous to its modulation of vertebrate opioid systems (see below) (Kavaliers *et al.* 1985a). It is possible also that FMRFamide and related peptides act as neurohormones, e.g. in *Helix* where they may arouse the animal from estivation (Greenberg *et al.* 1985).

Table 2. *Amino acid sequences for FMRFamide and some related molluscan and vertebrate peptides*

Origin	Peptide	Reference
molluscs	Phe–Met–Arg–Phe–NH_2 FMRFamide	Price & Greenberg (1977)
molluscs	Phe–Leu–Arg–Phe–NH_2 FLRFamide	Price (1986)
bovine	Tyr–Gly–Gly–Phe–Met–Arg–Phe Heptapeptide	Stern *et al.* (1979)
rat	Tyr–Val–Met–Gly–His–Phe–Arg–Trp–Asp–Arg–Phe–NH_2 γ–MSH	Nakanishi *et al.* (1979)
chicken brain	Leu–Pro–Leu–Arg–Phe–NH_2	Dockray *et al.* (1983)
bovine brain	Phe–Leu–Phe–Gln–Pro–Gln–Arg–Phe–NH_2	Yang *et al.* (1985)

In gastropods, the occurrence of FMRFamide and related peptides has been investigated in a relatively large number of species (Table 1). In prosobranchs, FMRFamide and FLRFamide have been detected, wherease in the opisthobranchs only FMRFamide could be demonstrated. The latter results have been confirmed by molecular cloning indicating that in *Aplysia californica* genes encoding numerous copies of FMRFamide and only one copy of FLRFamide are present. In the pulmonates, several extended FLRFamides are present in addition to FMRFamide. In this group perhaps two types of gene are present, one encoding FMRFamide and the other the extended FLRFamides. These speculations, of course, need to be confirmed by isolation of the genes.

FMRFamide immunoreactivity has been reported in representatives of all major phyla, e.g. the coelenterate *Hydra*, the CNS of crustaceans and insects, and the CNS, pituitary, gastrointestinal tract, and pancreas of various vertebrates (see Chapters 3, 4, 6 and 10 this volume; Joosse & Geraerts 1983). Unfortunately, until now very few studies have been undertaken to isolate and sequence these peptides (see Chapter 10). In chicken and bovine brain, novel peptides have been discovered using antisera against FMRFamide. These novel peptides are listed in Table 2, along with examples of known vertebrate peptides to which FMRFamide is related. FMRFamide and the novel peptides as well as γ–MSH share only the amidated C–terminal sequence, –Arg–Phe–NH_2. On the contrary, the heptapeptide Met–enkephalin–arg^6–Phe^7 shows a remarkable similarity with FMRFamide suggesting possible ancestral relationships. However, the great differences in the structures of their precursors preclude any ancestral relationship. The presence and function of FMRFamide in vertebrates are still under investigation, as are the functions of the novel neuropeptides. Exogenously applied FMRFamide has been reported however, to increase blood pressure (Barnard & Dockray 1984), to inhibit insulin and somatostatin release (Sorensen *et al.* 1984), to excite brain medullary neurons (Gayton 1982), and to modulate opiate-induced analgesis (Yang *et al.* 1985).

Opioid peptides

Both Met– and Leu–enkephalin and the heptapeptide Met–enkephalin–Arg^6–Phe^7 have been isolated from the CNS of *Mytilus* (Leung & Stefano 1984; Stefano &

Fig. 1. Schematic diagram of the CNS of *L. stagnalis* and *A. californica.* (A) Transverse section through the cerebral ganglia of *Lymnaea,* showing the location of the caudodorsal cells (CDC). Further indicated are the peptidergic light green cells (LGC), the lateral lobes (LL) and the endocrine medio- and laterodorsal bodies (MDB and LDB). The peripheries of the intercerebral commissure and the median lip nerves serve as the neurohaemal areas of the CDCs and LGCs, respectively. (B) The dorsal surface of the abdominal ganglion of *Aplysia.* The bag cells are grouped into two large clusters, each on the rostral side of the ganglion. Further indicated are crossing-over axons for electrotonic contacts, and the extent of the neurohaemal area.

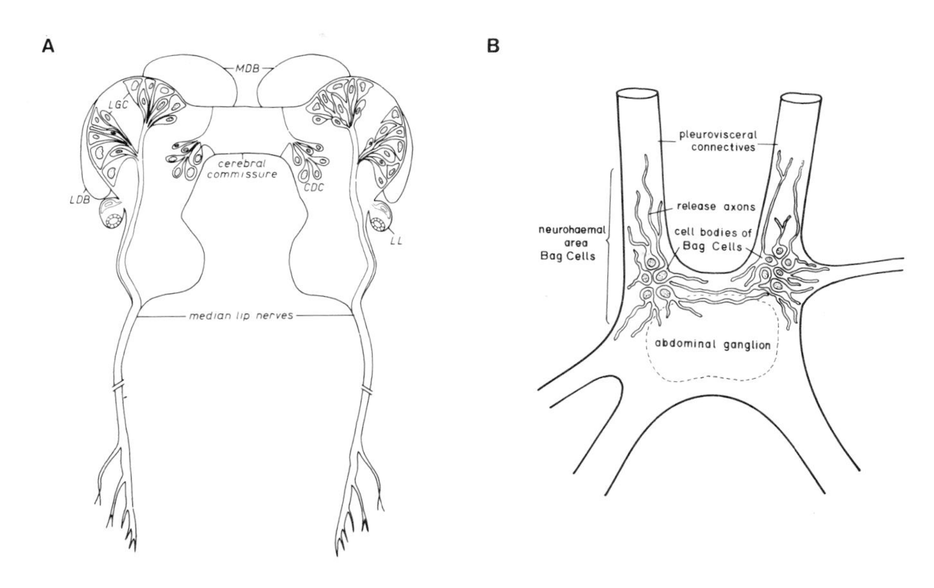

Leung 1984). Receptors for the enkephalins have also been demonstrated suggesting that these opioid peptides have a physiological function in this animal. Recent data indicate that enkephalins and their receptors are also present in the CNS of the freshwater pulmonate *Lymnaea stagnalis* (G.B. Stefano & H.H. Boer, personal communication) and it will now be of interest to determine whether they have a general distribution in molluscs. Studies including several molluscan species indicate that as in vertebrates' opioid systems are implicated in the regulation and modulation of behavioural and physiological functions, such as stress and feeding (Stefano 1982; Kavaliers *et al.* 1985b).

Peptides controlling reproduction

Neuropeptides and egg laying

Neuropeptides which induce egg laying have been demonstrated in all subclasses of the gastropods: prosobranchs (Ram 1977), opisthobranchs (Kupfermann 1967) and pulmonates (Geraerts & Bohiken 1976). The peptidergic systems governing egg laying and associated behaviours have been investigated in great detail in the freshwater pulmonate *L. stagnalis* (CDCs, caudodorsal cells) and the marine opisthobranch *Aplysia californica* (bag cells) (see reviews in Joosse & Geraerts 1983; Geraerts & Joosse 1984; Roubos 1984; Mayeri & Rothman 1985; Rothman *et al.* 1985; Geraerts *et al.* 1987). CDCs and bag cells consist of homogeneous clusters of large neurons (diameter 40–90 μm) that are easily identified. These properties make them particularly convenient for combined ultrastructural, molecular biological, neurophysiological, endocrinological, and behavioural studies. Hence, CDCs and bag cells offer excellent model systems for the study of the basic characteristics of peptidergic neurons. Egg laying in *Lymnaea* and *Aplysia* is characterized by a stereotyped pattern of behaviours involving ovulation of numerous oocytes, packaging of fertilized oocytes, oviposition, and alterations of locomotion and feeding movements. The egg-laying behavioural repertoire differs clearly between the two species. For instance, *A. californica* produces a long egg string that is coiled and fastened to the substrate with a series of specific head movements while feeding is inhibited. In contrast, *L. stagnalis* does not show head movements and deposits the compact egg mass on a substrate that has been cleaned in advance during a period of increased eating movements. The differences in egg-laying behaviour undoubtedly reflect the different life histories of the species, and the adaptations to different environments.

Approximately 100 CDCs are located in two symmetrically arranged clusters in the cerebral ganglia (Fig. 1). The cells use the periphery of the cerebral commissure as their neurohaemal area. Axonal branches of the CDCs project into the interior of the commissure where they release their products, apparently in a nonsynaptic manner. This system is called the collateral system (Schmidt & Roubos 1986). Each cluster of CDCs contains a small subset of specialized cells receiving various inputs, and relaying this input electrotonically to the other cells of the CDC network. In the much larger *A. californica*, the bag cell neurons are far more numerous (about 800), and are located in two distinct clusters of electrotonically coupled cells on the pleural-abdominal connectives adjacent to the abdominal ganglion (Fig. 1). Their neuro-haemal area is located in the connective tissue sheath surrounding the connective nerves and the abdominal ganglion.

CDCs and bag cells are normally silent, but prolonged afterdischarges (bursts of synchronous spiking activity) can be evoked in all cells of the network by the electrical stimulation of a single cell *in vitro*. Discharges with a pattern of action potentials similar to the afterdischarges recorded in isolated brains can be recorded from CDCs (ter Maat *et al.* 1986) and bag cells (Pinsker & Dudek 1977) in freely moving animals, indicating that afterdischarge-like firing patterns are a normal property of these systems. Discharges always precede egg laying. During electrical activity of the CDCs exocytosis of the contents of numerous elementary granules occurs and is accompanied by a massive release of the egg-laying peptide into the blood (Buma & Roubos 1983; Geraerts *et al.* 1984). Studies *in vitro* have shown that electrically active CDCs and bag cells release multiple peptides in a co-ordinated fashion (Fig. 2) (Stuart *et al.* 1980; Geraerts *et al.* 1985). The most extensively studied peptides are the ovulation hormone (CDCH) of *Lymnaea* and egg-laying hormone (ELH) of *Aplysia*. The primary sequence of these peptides has been elucidated (Chiu *et al.* 1979; Ebberink *et al.* 1985). They initiate and regulate egg laying, though do not control all of the events associated with egg laying. These peptides have a direct action on the gonad and affect also identified neurons in the neural feeding and locomotory circuits (Joosse & Geraerts 1983; Mayeri & Rothman 1985). A second extensively studied peptide is the autotransmitter, called CDCA (for CDC autotransmitter) in *Lymnaea,* and α–bag cell peptide (α–BCP) in *Aplysia* (Rothman *et al.* 1985; ter Maat *et al.* 1987). It is supposed that upon depolarization, CDCA is released by a few specialized CDCs that receive the sensory input. After release, it excites other cells of the network leading to discharges of all cells. Thus, it serves not only to spread the excitation over the network, but also to produce a hormonal output that is independent of the suprathreshold input. α–BCP has been shown to affect the activity of neurons in the abdominal ganglion, in addition to the bag cells.

The co-ordinated release of multiple peptides by the CDCs and bag cells raises the question of their mode of synthesis and transport. Bioactive peptides are synthesized in the form of larger precursor molecules which subsequently undergo limited proteolysis to yield smaller active peptides. Indeed, pulse-chase studies show this to be so for CDCH and ELH (Berry 1981; Geraerts *et al.* 1985) (Fig. 2). Recently, the primary structure and organization of the egg-laying peptide precursors has been elucidated with recombinant DNA techniques. The ELH protein precursors of *A. californica* and of a second *Aplysia* species, *A. parvula,* have been fully characterized

(Scheller *et al.* 1982, 1983; Nambu & Scheller 1986). We recently isolated the CDCH precursor. Because the screening of the full-length clones is still in progress, only the protein structure deduced from a partially sequenced stretch of a full length clone is presented along with the structural organization of the ELH precursors of *A. californica* and *A. parvula* (Fig. 3). Comparison of the precursors and peptides is of interest with regard to the evolutionary history of these species. The structural organizations of the ELH precursors of both *Aplysia* species are quite similar, with conservation of all but one of the potential proteolytic cleavage sites characterized by the occurrence of single, dibasic or multiple basic amino acid residues). The overall amino acid homology between the ELH precursors is 66%. The regions containing ELH, α–BCP, and the β–BCPs are even more highly conserved, indicating that these peptides are of importance in egg laying.

Three regions of the CDCH precursor identified so far are of interest: CDCH has a high degree of homology with ELH, α–CDCP with α–BCP, and a region containing 3 neuropeptides, the β–CDCPs, have a remarkable sequence homology with β– and

Fig. 2. Schematic representation of biosynthesis, transport and release of the ovulation hormone (CDCH), the autotransmitter (CDCA), and other peptides. The data are based on *in vitro* pulse-chase experiments (Geraerts *et al.* 1985). Separations of newly synthesized peptides in the CDC system were performed with high performance gel permeation chromatography (HPGPC). Released peptides were separated with HPGPC and further resolved by reverse phase high performance liquid chromatography (boxed within broken lines). At least 9 peptides representing 5 molecular weight classes are released. See text for further details.

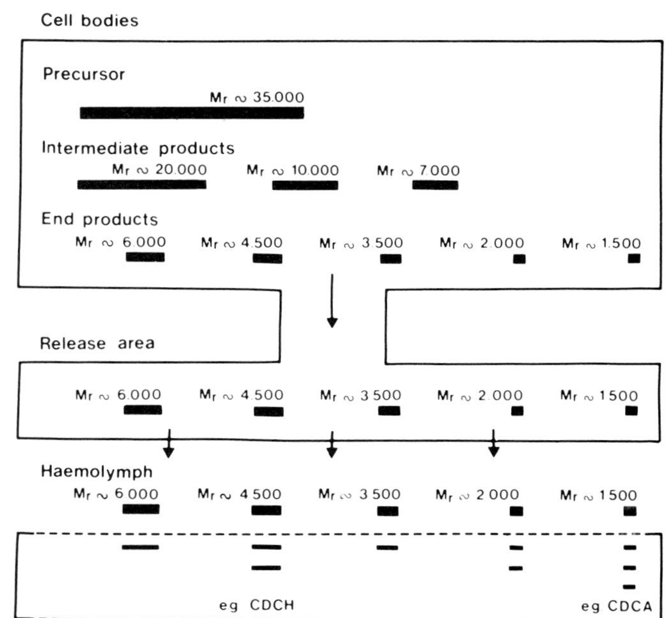

γ–BCP, and with α–CDCP and α–BCP. In comparison with ELH precursors, one extra pentapeptide (β2–CDCP) has been generated in the CDCH precursor. Compared with *A. parvula* and *L. stagnalis* precursors, the δ–BCP region in the *A. californica* precursor is completely divergent. The region between the β–CDCPs and α–CDCP is much shorter than the counterpart on the ELH precursor. This difference suggests the occurrence of one or more deletions in the *Lymnaea* gene during the course of evolution. Another difference between the CDCH and ELH precursors is that, apart from the homologies described above, there are none in the other regions. For example, the sequence of the 27 amino acid acidic peptides (AP, see Fig. 3), thought to play a role as a carrier protein for the basic peptides (α, β, and γBCPs), on the ELH precursor has completely diverged to a smaller *Lymnaea* cognate of 23 amino acids. Also, an extra potential cleavage site occurs between α–CDCP and the β–CDCPs.

The egg-laying peptides, CDCH and the ELHs, consist of 36 amino acids and are basic (Fig. 4). The Gly–Lys–Arg C–terminus, leading to amidation at the C–terminus, is conserved between the three species. The ELHs of both *Aplysia* species exhibit 78% homology, with nearly all of the amino acid alterations in the middle third of the molecule. CDCH shows 44% homology with ELH of *A. californica,* with amino acid alterations predominantly in the middle part of the peptide. α–BCPs share 100% homology, and α–CDCP has 5 uninterrupted amino acids in common with the α–BCPs (Fig. 5), suggesting that α–CDCP exhibits similar functions as α–BCP,

Fig. 3. Comparison of the polyprotein precursor containing ELH from *A. parvula* and *A. californica,* and CDCH from *L. stagnalis.* Of the CDCH precursor only the portion that has been identified so far is given (E. Vreugdenhil, unpublished results). Percentage homologies between regions are indicated. The positions of identified peptides, including ELH and CDCH (stippled), α–BCP and α–CDCP (cross–hatched), β–, γ–BCP and β_1, β_2, β_3–CDCP and acidic peptide (AP) indicated. Known or potential proteolytic cleavage sites are indicated. See text for further details.

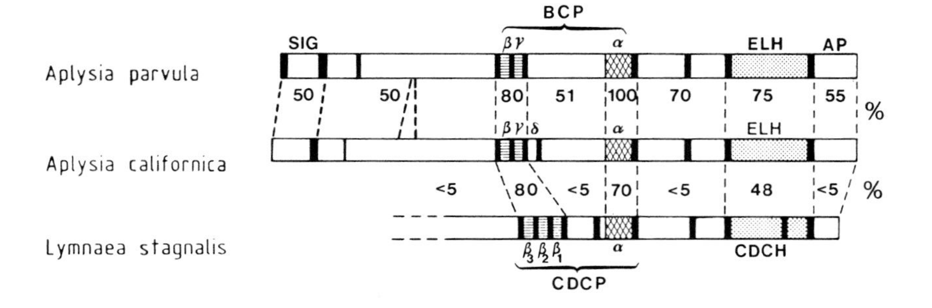

including the autotransmitter function, β_1–CDCP is completely homologous with the *Aplysia* β–BCPs (Fig. 5), and further exhibits strong homology with β_2– and β_3–CDCP, and with the γ–BCPs. Moreover, the pentapeptide sequence Arg–Leu–Arg–Phe–His is also present in α–CDCP and only slightly altered in the α–BCPs. The

Fig. 4. Amino acid sequence homologies among ELH regions of the ELH precursors of *A. parvula, A. californica,* and the CDCH region of the CDCH precursor of *L. stagnalis*. Homologous residues are enclosed within a box. See text for further details.

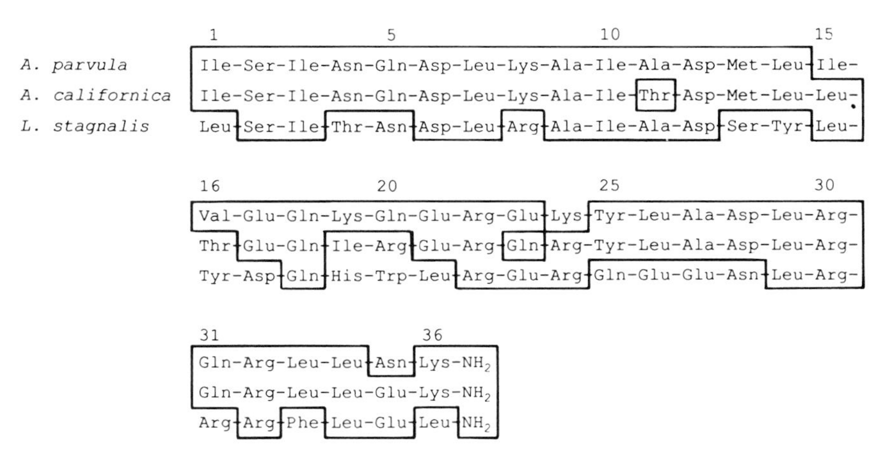

Fig. 5. Amino acid sequence homologies among CDC and bag cell peptides of *A. parvula, A. californica* and *L. stagnalis*. Homologous residues are enclosed within boxes. (Top) α–BCP and α–CDCP. (Bottom) the pentapeptides. γ–BCP–1 and γ–BCP–2 are encoded on two strongly homologous ELH genes in *A. parvula* (Nambu & Scheller 1986). See text for further details.

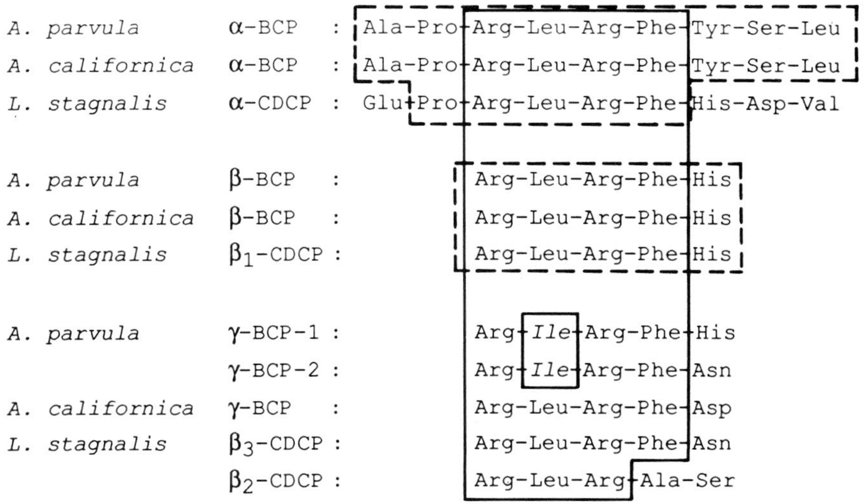

strong homology together with the repetitive character and the generation of an extra pentapeptide (β_2–CDCP) on the CDCH precursor suggest that the pentapeptides may be important in the control of egg laying. Their precise role however is unknown.

The regions on the precursors that exhibit strong homology are peptides that control the initiation of egg laying. Apparently, these peptides have been under intense selective pressure. Other regions on the precursors have only slightly diverged between *A. californica* and *A. parvula.* These species show a high degree of similarity in egg laying and associated behaviours (Nambu & Scheller 1986). The regions have completely diverged however, between *Lymnaea* and *Aplysia,* which show many dissimilarities in egg laying. It is estimated that speciation between *A. californica* and *A. parvula* took place about 140 million years ago (Namby & Scheller 1986), while pulmonate and opisthobranch gastropod subclasses diverged from each other about 350 million years ago (Moore & Pitrat 1960). The Lymnaeidae are evolved from marine-based molluscs through land-based intermediates, and are only distantly related to the Aplysiids. The changes in environments, from sea to land to fresh water, during the course of evolution are probably reflected in species-specific variants of egg laying. Hence, it is very attractive to suggest that (some of the) diverged regions on the precursors play an important role in the control of the species specific variants of egg-laying behaviour.

In *A. californica,* the gene coding for the ELH precursor is a member of a small multigene family consisting of four to five members and sharing approximately 90% sequence homology (Scheller *et al.* 1982, 1983; Mahon *et al.* 1985). ELH is also expressed in a small network of neurons throughout the CNS. Two other members of the gene family, the A and B genes, are expressed in the atrial gland (an accessory sex gland of *A. californica).* The gene products of these members, the A and B peptides, are capable of activating bag cells *in vitro.* However, the atrial gland is not an endocrine gland, and the A and B peptides therefore do not act as hormones. They perhaps have a pheromone-like function.

ELH gene families are present in other representatives of the genus *Aplysia* other than *A. californica* (Nambu & Scheller 1986). Most of the species examined have fewer genes than found in *A. californica.* In *A. parvula,* where an atrial gland is absent, there are only two ELH genes and none of the genes encode the A and B peptides. Southern genomic blotting revealed the presence of a small multigene family in *L. stagnalis,* comprising of four to six members. Northern blotting in combination with *in situ* hybridization and immunocytochemistry demonstrated that CDCH transcripts or related transcripts are present in neural and nonneural tissues of this species (E. Vreugdenhil, unpublished results).

It is very likely that in gastropods the gene families have been generated by duplications of an ancestral gene coding for the egg-laying peptides (Scheller *et al.* 1983). From the data on the ELH genes of the various *Aplysia* species it can be concluded that duplication of the original ELH gene occurred before species divergence within the genus *Aplysia* (Namby & Scheller 1986). In *A. californica,* further gene duplications have occurred, and include genes encoding the A and B peptides expressed in the atrial gland. Similarly, during phylogeny of *L. stagnalis* also more than two genes encoding precursors for CDCH or related peptides have been generated.

Putative gonadotropic peptide hormones

Various aspects of the control of reproduction in molluscs other than egg laying have been studied in detail, e.g. simultaneous hermaphroditism in pulmonates, sex reversal in prosobranchs, and terminal reproduction in cephalopods. These have been reviewed by Joosse & Geraerts (1983). Many of the hormones involved are of unknown or uncertain chemical nature. Here, a short survey will be given of the functional aspects of those hormones which are tentatively identified as peptides. In pulmonates, the dorsal bodies (Fig. 1) produce the female gonadotropic dorsal body hormone that stimulates vitellogenesis, as well as growth, differentiation, and synthetic activity in the female accessory sex organs. The endocrine optic glands of the gonochoristic cephalopods produce a non-sex-specific hormone that stimulates gonadal processes, and the growth of the male and female ducts. In stylommatophorans, a brain factor (or factors) stimulates the differentiation of male gametes and suppresses the development of female gametes. The role of neurohormones in the control of sex reversal in prosobranchs will be described in the next section.

Morphogenetic neurohormones

In the prosobranchs a peculiar variant of hermaphroditism, protandric sex reversal involving dedifferentiation and morphogenesis of sex organs, is encountered in a number of marine species such as *Crepidula fornicata* (Le Gall 1981). Sex reversal in *Crepidula* has been investigated in detail and can be considered a model system for morphogenetic studies. *Crepidula* lives in colonies, often in the form of a chain, with the younger males on top and the older females at the bottom. Animals in the middle of the chain are in a phase of sex reversal. During sex reversal the testis changes into an ovarium and simultaneously but independently of the gonad, the male accessory sex organs are replaced by female organs. Isolated specimens also show sex reversal, but this process is considerably accelerated by social contacts in a colony. Feminizing and masculinizing factors released by the males and females respectively, play an important pheromone-like role (Le Gall 1980). The feminizing factor is registered by the pallial border and the masculinaizing factor by the tentacles. Both factors affect centres in the cerebral ganglia that exert a neurohormonal and nervous control over the meurosecretory cells in the pedal ganglia which produce the hormones that ultimately control sex reversal (Fig. 6).

At least two neurohormones control directly the differentiation and lysis of the penis (Le Gall 1981). A hormone produced by neurons located in the pedal ganglia stimulates the differentiation of the penis in a morphogenetic territory located near the right tentacle (Fig. 6). This hormone is not species specific, at least amongst *C. fornicata, Calyptraea sinensis, Littorina* and *Buccinum* (Streiff *et al.* 1981). Detailed studies on *Crepidula* demonstrate the unique and somewhat enigmatic phenomenon of accumulation of the hormone in haemal lacunae near the penis (Le Gall 1981). Apparently, the circulating level of the hormone is too low to induce penis differentiation, and the haemal lacunae serve to raise local concentrations of hormone above the threshold for differentiation and outgrowth of the penis. During transition to the female phase, release of the hormone is inhibited by the cerebral ganglia. Simultaneously, lysis of the penis is initiated by another hormone produced in the mediodorsal area of the pleural ganglia (Fig. 6). For the regression of the penis the continuous presence of the hormone is needed. The availability of pure hormone

Fig. 6. Environmental, nervous, and neurohormonal factors involved in (A) the differentiation of the male external genital tract, and (B) the dedifferentiation of the male tract and the differentiation of the female external genital tract in *C. fornicata*. Ce, Cerebral ganglia; Pe, pedal ganglia; Pl, pleural ganglia. (Derived from Le Gall 1980.)

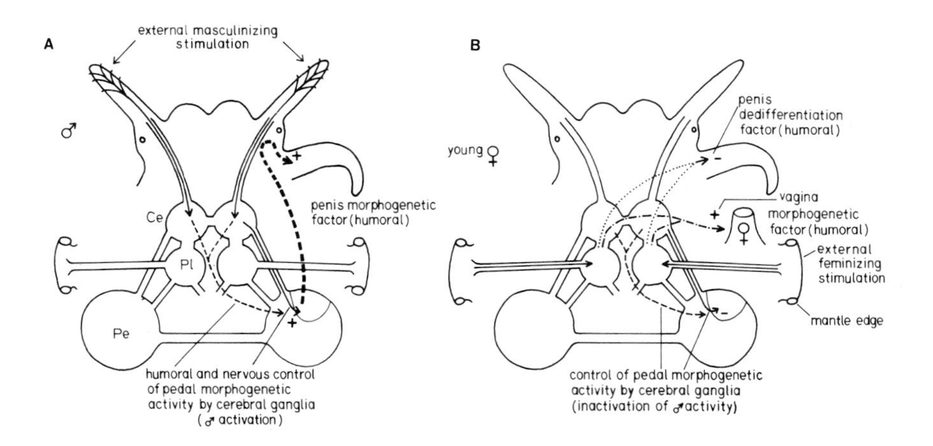

preparations would greatly facilitate in-depth studies of these effects of the neurohormones at the cellular level.

Growth and metabolic hormones

Neurohormones and growth

Extirpation and re-implantation experiments demonstrate that body growth (including the enlargement of the shell) is under neurosecretory control in prosobranchs (Lubet 1971; Le Gall 1980) and pulmonates (Geraerts 1976; Wijdenes & Runham 1977). The neurosecretory centres involved have been identified by Geraerts (1976) in the cerebral ganglia of the freshwater snail *L. stagnalis* (LGCs, light green cells; see Fig. 1) and by Wijdenes & Runham (1977) in the terrestrial slug *Derocerus (Agriolimax) reticulatus* (medial cells). The cells stimulate the proportional growth of all tissues, and co-ordinate the growth of the soft body parts with that of the shell. Cauterization and implantation of the LGCs, as well as injections of homogenates of the median lip nerves (the neurohaemal area of these cells), show that the cells affect various aspects of the metabolism related to growth (reviewed in Joosse & Geraerts 1983). Thus, the LGCs stimulate, probably indirectly, the breakdown of glycogen stores, and stimulate the activity of ornithine decarboxylase, an enzyme which shows high levels of activity in growing animal tissues. Moreover, the LGCs produce a natremic factor, a peptide stimulating the uptake of sodium through the integument (de With & van der Schors 1986). At the moment the significance of sodium uptake for body growth is unclear. An important phenomenon associated with increasing size is the increase in blood volume; for this, an increase in water uptake is necessary, which may upset blood osmolarity. The natremic factor may function therefore to stimulate sodium uptake to regulate blood osmolarity. The LGCs stimulate also various processes of shell growth: formation of the periostracum, the proteinaceous component of the shell; calcium and bicarbonate incorporation into the shell edge; and the maintenance of high concentrations of a calcium-binding protein, important for cellular calcium transport in the mantle edge.

In view of the large number of processes controlled by the LGC, it seems likely that these cells produce a multiplicity of peptides, each controlling one or a few processes involved in growth. Two peptides, the natremic factor and a peptide stimulating calcium incorporation in the shell, have already been purified from the median lip nerve (Ebberink & Joosse 1985). Peptide sequence analyses and molecular genetic experiments are underway to identify the polypeptide precursor and the peptides derived from it.

Putative hormones involved in carbohydrate metabolism

The metabolism of many molluscs (such as bivalves and gastropods) is carbohydrate oriented, and glycogen can be a major energy reserve. The hormones or factors implicated in the control of carbohydrate metabolism are products of the brain and the intestinal tract (Joosse & Geraerts 1983; Plisetskaya & Joosse 1985). Release of the hyperglycaemic factor from the brain of gastropods and bivalves was demonstrated *in vitro* using organ culture techniques and potassium-rich Ringers (Lubet *et al.* 1978; Hemminga *et al.* 1985a and b). In *L. stagnalis*, the major release sites are in the cerebral and buccal ganglia, although other parts of the brain also release substantial amounts of the putative neurohormone. In the bivalve *Mytilus edulis,* the sites of production and release of the factor are entirely confined to the cerebral ganglia. The visceral ganglion of this animal releases a hypoglycaemic factor (Lubet *et al.* 1976). The cells responsible for the production of the factors have not yet been identified although in *Lymnaea,* the peptidergic CDCs and LGCs, and the dorsal bodies (see Fig. 1) can be excluded as possible sources of the hyperglycaemic factor. The peptidergic nature and other molecular characteristics of the factors remain to be elucidated.

Experiments *in vitro* indicate that the major targets for both factors are the cells specialized for glycogen storage which have a general occurrence in molluscs, as well as the adipogranular cells of some bivalve families which store lipid and protein in addition to glycogen (Lubet *et al.* 1976; Hemminga *et al.* 1985b). In *L. stagnalis,* the hyperglycaemic factor simultaneously inhibits glycogen synthesis, stimulates glycogen breakdown, and induces glucose release from isolated glycogen cells (Hemminga *et al.* 1985b).

Various external and internal environmental conditions such as temperature, food quantity and quality, stress, anoxia, and reproductive activities affect carbohydrate metabolism (Joosse & Geraerts 1983), and it is suggested (but not proven) that the hyper- and hypoglycaemic factors are released and function as hormones in response to such stresses.

Insulin-like substance (ILS), demonstrated with antisera against insulin, probably plays a role in the control of carbohydrate metabolism. Cells containing ILS have been identified in various regions of the gut and in the digestive gland of bivalves, prosobranchs and pulmonates, as well as in the brain of *L. stagnalis* (reviewed in Joosee & Geraerts 1983). In a number of bivalves (*Unio pictorina, Anodonta cygnea, Mytilus galloprovencialis, Chlamys glaber ponticus),* injections of ILS-containing gut extracts, or of mammalian insulin stimulate the uptake of glucose by muscles (Plisetskaya & Joosse 1985). Mammalian insulin is similarly effective in a variety of gastropods (Marques & Falkmer 1976; Plisetskaya & Joosse 1985) with the exception of *L. stagnalis,* where no such effects of mammalian insulin or of *Lymnaea* ILS can be demonstrated. Radioimmunoassays indicate that ILS in gut cells of bivalves is released into the blood after a glucose load, pointing to a physiological role of ILS in carbohydrate metabolism. More experimental evidence is needed however, and in view of the crucial function of the glycogen cells in carbohydrate metabolism, studies on the effects of ILS on these cells will be of great interest.

Purification and characterization of ILS have not been reported for any of the mentioned species, but have recently been initiated in *L. stagnalis* (Ebberink & Joosse 1985). ILS, from the digestive system of *Lymnaea* has a molecular weight similar to that of mammalian insulin, but is slightly more hydrophobic. It binds strongly to the insulin receptors of isolated rat fat cells and stimulates their uptake of 2-deoxyglucose. Tests to establish the presence of an α and β chain in ILS have been unsuccessful, perhaps indicating a structure different from vertebrate insulins (Ebberink, unpublished observations).

Kinetic hormones

Peptide hormones and heart beat

In cephalopods, some neurosecretory cells have extensive neurohaemal areas in the anterior vena cava (AVC) (Alexandrowicz 1964) and in the pharyngo-opthalmic vein (POV) (Froesch 1974). Extracts of both AVC and POV, applied to the systemic hear *in vitro,* increase the frequency and amplitude of the heart beat (Blanch *et al.* 1973). The active principle of the AVC of *Octopus vulgaris* has been partially purified and characterized as a peptide with a molecular weight of about 1300 (Blanchi *et al.* 1973). It is not species specific. When tested on the (branchial) hearts of intact freely-moving *O. vulgaris,* AVC and POV extracts each produce a different spectrum of effects which are unlike those of acetylcholine, 5–hydroxytryptamine, adrenaline, histamine or tyramine (Wells & Mangold 1980). There are several reasons to suggest that it is only the product of the AVC that is relevant to the normal performance of the hearts. First, in contrast to the POV material, the AVC material is effective at very low doses. Second, the AVC material is released at a point just upstream of the branchial hearts. Third, release of the AVC material into the blood occurs during physiologically relevant conditions such as stress (Wells & Mangold 1980).

In *Helix aspersa,* two cardioexcitatory neurohormones with identical molecular weight (about 7000), one with a neutral and the other with a basic isoelectric point, have been identified (Lloyd 1978a and b, 1980). Cardioexcitatory material of a similar size has also been demonstrated in *L. stagnalis,* but the presence in this species of two hormones has not yet been investigated (Geraerts *et al.* 1984; Geraerts, unpublished observations). In both species, the cardioexcitatory neurohormones occur in high concentrations in the suboesophageal *(Helix)* or visceral *(Lymnaea)* ganglia and their nerve trunks, as well as in the auricle, but not in the ventricle or in other tissues. Lloyd (1978b) has shown that these peptides are associated with elementary granules (diameter about 122 nm) present in the auricle and in the suboesophageal ganglia of *H. aspersa.* Transport of the peptides via the visceral nerve toward the auricle and release at multiple sites at the peripheries of nerves originating from the suboesophageal *(Helix)* or visceral *(Lymnaea)* ganglia and in the auricle, has been demonstrated *in vitro* (Lloyd 1978b; Geraerts, unpublished observations). Ultrastructural studies provide further evidence that the auriculoventricular junction in particular is a neurohaemal area (Greenberg & Price 1980). The neurons responsible for the production of the neurohormones have not yet been identified.

In gastropods, the heart rate plays a crucial role in a large number of physiological processes. This has been studied in considerable detail in *L. stagnalis* (see Joosse & Geraerts 1983). During starvation the heart rate of *L. stagnalis* drops to about 50% of the values found in fed animals. This decrease in heart rate is the cause of reduced rates of renal filtration, water turnover, and energy metabolism, and is accompanied by changes in the pH and ionic composition of the blood. Re-feeding leads to recovery of normal heart rate within 1–2 h and simultaneously, the blood titers of the cardioexcitatory neurohormone rise to the levels found in fed animals, indicating that the neurohormone is released during feeding (Geraerts *et al.* 1984; Geraerts, unpublished observations).

Neurosecretory factors and hydromineral regulation

Studies on the neurohormonal control of body volume and ionic composition of the blood in molluscs are almost completely confined to the gastropods and in particular to the freshwater pulmonates (Joosse & Geraerts 1983). Like all animals living in freshwater habitats the osmolarity of the blood of freshwater pulmonates is higher than that of the environment. Consequently, there are continuous diffusional (via the integument) and excretory (via the kidney) losses of ions, accompanied by an osmotic influx of water through the integument. To compensate for ionic losses, active uptake mechanisms for active uptake for each of the main ions present in the blood are present in the integument and the intestine. Regulation of the body volume is mainly effected via renal excretion of excess water. Changes in the ionic composition and osmolarity of the environment, or in food availability, considerably affect water and ion fluxes. The various factors which have been implicated in hydromineral regulation are products of neurosecretory systems such as the dark green cells, yellow cells, yellow green cells, and light green cells in *L. stagnalis,* and 3 types of neurosecretory cell in *Helisoma duryi* and *Helisoma trivolvis* (Khan & Saleuddin 1979, 1981; Joosse & Geraerts 1983).

The suggestion that the products of the various neurosecretory systems are involved in hydromineral regulation is based on indirect light and electron microscopical evidence, and there are only a few reports of direct experimental evidence for the presence and functional role of such factors. In *Helisoma,* Khan & Saleuddin (1979, 1981) have suggested from experiments *in vitro* that a putative neurohormone of the visceral ganglia may act directly on the kidney sac epithelium to alter the rate of filtration and pro-urine transport. Similar experiments have demonstrated that the light green cells of *L. stagnalis* produce a peptide, the natremic factor, which has a stimulating effect on the uptake of sodium by the integument. As discussed above in Neurohormones and Growth, the natremic factor probably is one of the peptides from the light green cells involved in the control of body growth. An interesting finding is the possible production and release of a TRH-like substance (TRH = thyroid-stimulating hormone-releasing hormone) by the dark green cells of *L. stagnalis* (Grimm–Jørgensen 1985). TRH-like substance probably alters salt and water fluxes in the skin by changing the secretion of epidermal mucus, which in its turn affects the diffusion of ions and water.

In terrestrial pulmonates, hypercalcaemic factors with a putative neurohormonal function have been demonstrated in *Cryptozona semirugata* (Subramanyam 1973) and *Helix pomatia* (Burton 1977), whereas in the brain of *Arion columbianus* and *Limax maximus* an arginine–vasotocin-like substance has been demonstrated that, upon injection, enhances water permeability of the body wall (Sawyer *et al.* 1985). Kupfermann & Weiss (1976) have shown that the large peptidergic R15 neuron in the abdominal ganglion of the marine opisthobranch *A. californica* produces an antidiuretic neurohormone with a molecular weight of about 1500.

Acknowledgements

We thank Prof. Joosse for critical reading of the manuscript, and Miss Thea Laan for typing of the manuscript.

References

Alexandrowicz, J.S. (1964) The neurosecretory system of the vena cava in Cephalopods. Vol. 1. *Eledone cirrosa. J. Mar. Biol. Ass.UK.*, **44**, 111–32.

Barnard, C.S. & Dockray, G.J. (1984) Increases in arterial blood pressure in the rat in response to a new vertebrate neuropeptide, LPLRFamide, and a related molluscan peptide, FMRFamide. *Regulatory Peptides*, **8**, 209–15.

Berry, R.W. (1981) Proteolytic processing in the biogenesis of the neurosecretory egg-laying hormone in *Aplysia*. 1. Precursors, intermediates and products. *Biochemistry*, **20**, 6200–5.

Blanchi, D., Noviello, L. & Libonati, M. (1973) A neurohormone of cephalopods with cardioexcitatory activity. *Gen. & Comp. Endocrinol.*, **21**, 267–77.

Buma, P. & Roubos, E.W. (1983) Calcium dynamics, exocytosis and membrane turnover in the ovulation hormone-releasing caudodorsal cells of *Lymnaea stagnalis. Cell & Tiss. Res.*, **233**, 143–59.

Burton, R.F. (1977) Haemolymph calcium in *Helix pomatia:* Effects of EGTA, ganglion extracts, ecdysterone, cyclic AMP and ionophore A 23187. *Comp. Biochem. & Physiol.*, **C57**, 135–7.

Chiu, A.Y., Hunkapillar, M.W., Heller, E., Stuart, D.K., Hood, L.E. & Strumwasser, F. (1979) Purification and primary structure of the neuropeptide egg-laying hormone of *Aplysia californica. Proc. Nat. Acad. Sci. USA.*, **76**, 6656–60.

Dockray, G.J., Reeve Jr., J.R., Shively, J., Gayton, R.J. & Barnard, C.S. (1983) A novel active pentapeptide from chicken brain identified by antibodies to FMRFamide. *Nature*, **305**, 328–30.

Ebberink, R.H.M. & Joosse, J. (1985) Molecular properties of various snail peptides from brain and gut. *Peptides*, **6**, suppl. 3, 451–7.

Ebberink, R.H.M., Loenhout, H. van, Geraerts, W.P.M. & Joosse, J. (1985) Purification and amino acid sequence of the ovulation neurohormone of *Lymnaea stagnalis*. *Proc. Nat. Acad. Sci. USA.*, **82**, 7767–71.

Froesch, D. (1974) The subpedunculate lobe of the octopus brain; Evidence for a dual function. *Brain Res.*, **75**, 277–85.

Gayton, R.J. (1982) Mammalian neuronal actions of FMRFamide and the structurally related opioid Met–enkephalin–Arg^6–Phe^7. *Nature*, **298**, 275–6.

Geraerts, W.P.M. (1976) Control of growth by the neurosecretory hormone of the light green cells in the freshwater snail *Lymnaea stagnalis*. *Gen. & Comp. Endocrinol.*, **29**, 61–71.

Geraerts, W.P.M. & Bohliken, S. (1976) The control of ovulation in the hermaphrodite freshwater snail *Lymnaea stagnalis* by the neurohormone of the caudo-dorsal cells. *Cell & Comp. Endocrinol.*, **28**, 350–7.

Geraerts, W.P.M. & Joosse, J. (1984) Freshwater snails (Basommatophora). In: *The Mollusca, Vol. 7 Reproduction*, eds. A.S. Tompa, N.H. Verdonk & J.A.M. van den Biggelaar, pp. 142–207. New York: Academic Press.

Geraerts, W.P.M., Maat, A. ter & Hogenes, Th.M. (1984) Studies on the release activities of the neurosecretory caudo-dorsal cells of *Lymnaea stagnalis*. In: *Biosynthesis, Metabolism and Mode of Action of Invertebrate Hormones*, eds. J.A. Hoffman & M. Porchet, pp. 44–50. Heidelberg: Springer–Verlag.

Geraerts, W.M.P., Vreugdenhil, E., Ebberink, R.H.M. & Hogenes, Th.M. (1985) Synthesis of multiple peptides from a larger precursor in the neuroendocrine caudo-dorsal cells of *Lymnaea stagnalis*. *Neurosci. Lett.*, **56**, 241–6.

Geraerts, W.P.M., With, N. de, Tan, B.T.G., Hartingsveldt, W. & Hogenes, Th.M. (1984) Studies of the characteristics, distribution and physiological role of a large cardioactive peptide in *Lymnaea stagnalis*. *Comp. Biochem. & Physiol.*, **78C**, 339–43.

Geraerts, W.P.M., Vreugdenhil, E. & Maat, A. ter (1987) The peptidergic neuroendocrine control of egg-laying behavior in *Aplysia* and *Lymnaea*. In: *Invertebrate Endocrinology, Vol 2*, eds. H. Laufer & R. Downer, in press. New York: Alan Liss, Inc.

Greenberg, M.J. & Price, D.A. (1980) Cardioregulatory peptides in molluscs. In: *Peptides, integrators of cell and tissue function*, ed. F.E. Bloom, pp. 197–26. New York: Raven Press.

Greenberg, M.J., Price, D.A. & Lehman, H.K. (1985). FMRFamide-like peptides of molluscs and vertebrates: distribution and evidence of function. In: *Neurosecretion and the biology of neuropeptides*, eds. Kobayashi, H.A. Bern & A. Urano, pp. 370–6. Berlin; Springer–Verlag/Tokyo: Japan Sci. Soc. Press.

Grimm–Jørgensen, Y. (1985) Distribution and physiological roles of TRH and somatostatin in gastropods. In: *Current trends in comparative endocrinology*, eds. B. Lofts & W.N. Holmes, pp. 873–7. Hong Kong University Press.

Hemminga, M.A., Maaskant, J.J., Koomen, W. & Jooss, J. (1985a) Neuroendocrine control of glycogen mobilization in the freshwater snail *Lymnaea stagnalis*. *Gen. & Comp. Endocrinol.*, **57**, 117–23.

Hemminga, M.A., Maaskant, J.J. & Joosse, J. (1985b) Direct effects of the hyperglycaemic factor of the freshwater snail *Lymnaea stagnalis* on isolated glycogen cells. *Gen. & Comp. Endocrinol.*, **58**, 131–6.

Kavaliers, M., Hirst, M. & Teskey, G.C. (1985a) Opioid systems and feeding in the slug, *Limax maximus:* similarities to and implications for mammalian feeding. *Brain Res. Bull.*, **14**, 681–5.

Kavaliers, M., Hirst, M. & Mathers, A. (1985b) Inhibitory influences of FMRFamide on morphine- and deprivation-induced feeding. *Neuroendocrinol.*, **40**, 533–5.

Khan, H.R. & Saleuddin, A.S.M. (1979) Osmotic regulation and osmotically induced changes in the neurosecretory cells of the pulmonate snail *Helisoma*. *Can. J. Zool.*, **57**, 1371–83.

Khan, H.R. & Saleuddin, A.S.M. (1981) Cell contacts in the kidney epithelium of *Helisoma* (Mollusca: Gastropoda). Effects of osmotic pressure and brain extracts: a freeze-fracture study. *J. Ultrastruct. Res.*, **75**, 23–40.

Kupfermann, I. (1967) Stimulation of egg laying: possible neuroendocrine function of bag cells on abdominal ganglion of *Aplysia californica*. *Nature*, **216**, 814–5.

Kupfermann, I. & Weiss, K. (1976) Water regulation by a presumptive hormone contained in identified neurosecretory cell R15 of *Aplysia*. *J. Gen. Physiol.*, **67**, 113–23.

Le Gall, P. (1980) Etude expérimentale de l'association en chaine et de son influence sur la croissance et la sexualite chez la crépidule *Crepidula fornicata* (Mollusque mésogastéropode). Thesis, University of Caen, pp. 1–251.

Le Gall (1981) Etude expérimentale du facteur morphogénitique controlant la différenciation du tractus génital male externe chez *Crepidula fornicata* L. (Mollusque hermaphrodite protandre). *Gen. & Comp. Endocrinol.*, **43**, 51–62.

Lehman, H.K., Price, D.A., Greenberg, M.J. (1984) The FMRFamide-like neuropeptide of *Aplysia* is FMRFamide. *Biol. Bull.*, **167**, 460–6.

Leung, M.K. & Stefano, G.B. (1984) Isolation and identification of enkephalins in pedal ganglia of *Mytilus edulis* (Mollusca). *Proc. Nat. Acad. Sci. USA.*, **81**, 955–8.

Lloyd, P.E. (1978a) Distribution and molecular characteristics of cardioactive peptides in the snail, *Helix aspersa*. *J. Comp. Physiol.*, **128**, 269–76.

Lloyd, P.E. (1978b) Neurohormonal control of cardiac activity in the snail *Helix aspersa*. *J. Comp. Physiol.*, **128**, 277–83.

Lloyd, P.E. (1980) Biochemical and pharmacological analysis of endogenous cardioactive peptides in the snail, *Helix aspersa*. *J. Comp. Physiol.*, **138**, 265–70.

Lubet, P. (1971) Influence des ganglions cérébroides sur la croissance de *Crepidula fornicata* Phil. (Mollusque Mesogasteropode). *Comptes Rendues de l'Académie des Sciences, Paris*, **273**, 2309–11.

Lubet, P., Herlin, P., Mathieu, M. & Collin, F. (1976) Tissue de réserve et cycle sexuel chez les lamellibranches. *Haliotis*, **7**, 59–62.

Maat, A. ter., Dijks, F.A. & Bos, N.P.A. (1986) *In vivo* recordings of neuroendocrine cells (caudo-dorsal cells) in the pond snail. *J. Comp. Physiol.*, **158A**, 853–859.

Maat, A. ter., Bos, N.P.A., Geraerts, W.P.M., Jansen, R.F. & Hogenes, Th.M. (1987) Peptidergic positive feedback generates long-lasting discharge in a molluscan neuroendocrine system. In preparation.

Mahon, A.C., Nambu, J.R. Taussig, R., Shyamala, M., Roach, A. & Scheller, R.H. (1985) Structure and expression of the egg-laying hormone gene family in *Aplysia*. *J. Neurosci.*, **5**, 1872–80.

Marques, M. & Falkmer, S. (1976) Effects of mammalian insulin on blood glucose level, glucose tolerance and glycogen content of musculature and hepatopancrease in a gastropod mollusc *Strophocheilus oblongus*. *Gen. & Comp. Endocrinol.*, **29**, 522–30.

Mayeri, E. & Rothman, B.S. (1985) Neuropeptides and the control of egg-laying behaviour in *Aplysia*. In: *Model neural networks and behaviours*, ed. A.I. Selverston, pp. 285–301. New York: Plenum Press.

Moore, R.C. & Pitrat, C.W. (1960) *Treatise on Invertebrate Paleontology, Part 1, Mollusca 1*, Geological Society of America, In, and University of Kansas Press.

Nakanishi, S., Inoue, A., Kita, T., Nakamura, M., Chang, A.C.Y., Cohen, S.N. & Numa, S. (1979) Nucleotide sequence of cloned cDNA for bovine corticotropin–β–liptropin precursor. *Nature,* **278**, 324–7.

Nambu, J.R. & Scheller, R.H. (1986) Egg-laying hormone genes of *Aplysia:* evolution of the ELH gene family. *J. Neurosci.*, **6**, 2026–2036.

Pinsker, H. & Dudek, F.E. (1977) Bag cell control of egg-laying in freely behaving *Aplysia. Science,* **197**, 490–3.

Plisetskaya, E. & Joosse, J. (1985) Hormonal regulation of carbohydrate metabolism in molluscs. In: *Current trends in comparative endocrinology,* eds. B. Lofts & W.N. Holmes, pp. 1077–9. Hong Kong: Hong Kong University Press.

Price, D.A. (1986) Evolution of a molluscan cardioregulatory neuropeptide. *Am. Zool.,* **26**, 1007–15.

Price, D.A. & Greenberg, M.J. (1977) Structure of a molluscan cardioexcitatory neuropeptide. *Science,* **197**, 670–1.

Price, D.A., Cottrell, G.A., Doble, K.E., Greenberg, M.J., Jorenby, W., Lehman, H.K. & Reihm, J.P. (1985) A novel FMRFamide-related peptide in *Helix:* pQDPFLRFamide. *Biol. Bull.*, **169**, 256–66.

Ram, J.L. (1977) Hormonal control of reproduction in *Busicon:* Laying of egg capsules caused by nervous system extracts. *Biol. Bull.*, **152**, 221–32.

Rothman, B.S., Mayeri, E. & Scheller, R.H. (1985) The bag cell neurons of *Aplysia* as a possible peptidergic multitransmitter system: from genes to behavior. In: *Gene expression in brain,* ed. C. Zomzely–Neurath & W.A. Walker, pp. 236–74. New York: John Wiley & Sons.

Roubos, E.W. (1984) Cytobiology of the ovulation-neurohormone producing Caudo-Dorsal Cells of the snail *Lymnaea stagnalis. Int. Rev. Cytol.,* **89**, 295–346.

Sawyer, W.H., Pang, P.K.T., Deyrup–Olsen, I. & Martin, A.W. (1985) Evolution of neurohypophysial hormones: a principle resembling arginine-vasotocin in the gastropod nervous system. In: *Current trends in comparative endocrinology,* eds. B. Lofts & W.N. Holmes, pp. 1153–5. Hong Kong University Press.

Schaefer, M., Picciotto, M.R., Kreiner, T., Kaldany, R.R., Taussig, R. & Scheller, R.H. (1985) *Aplysia* neurons express a gene encoding multiple FMRFamide neuropeptides. *Cell,* **41**, 457–67.

Scheller, R.H., Jackson, J.F., McAllister, L.B., Schwartz, J.H., Kandel, E.R. & Axel, R. (1982) A family of genes that codes for ELH, a neuropeptide eliciting a stereotyped pattern of behaviour in *Aplysia. Cell,* **28**, 707–19.

Scheller, R.H., Jackson, J.F., McAllister, L.B., Rothman, B.S., Mayeri, E. & Axel, R. (1983) A single gene encodes multiple neuropeptides mediating a stereotyped behavior. *Cell,* **35**, 7–22.

Schmidt, E.D. & Roubos, E.W. (1986) Dynamics of neurohaemal and nonsynaptic release of multiple peptides involved in egg-laying behaviour of the pond snail. *Neurosci. Lett.,* Suppl.

Sorensen, R.L., Sasek, C.A. & Elde, R. (1984) Phe–Me–Arg–Phe–amide (FMRFamide) inhibits insulin and somatostatin secretion and anti-FMRF–NH_2 sera detect pancreatic polypeptide cells in the rat islet. *Peptides,* **5**, 777–82.

Stefano, G.B. (1982) Comparative aspects of opioid-dopamine interaction. *Cellular & Molecular Neurobiol.,* **2**, 167–78.

Stefano, G.B. & Leung, M.K. (1984) Presence of Met–enkephalin–Arg^6–Phe^7 in molluscan neural tissues. *Brain Res.,* **298**, 362–5.

Stern, A.S., Lewis, R.V., Kimuru, S., Rossier, J., Geber, L.D., Brink, L., Stein, S. & Udenfried, S. (1979) Isolation of the opioid heptapeptide Met–

enkephalin (Arg^6,Phe^7) from bovine adrenal medullary granules and striatum. *Proc. Nat. Acad. Sci. USA.*, **76**, 6680–3.

Streiff, W., Lebreton, J. & Silberzahn, N. (1970) Non specificité des facteurs hormonaux responsables de la morphogénèse et du cycle du tractus genital male cez les Mollusques Prosobranches. *Annales d'Endocrinologie, Paris*, **31**, 548–56.

Stuart, D.K., Chiu, A.Y. & Sturmwasser, F. (1980) Neurosecretion of egg-laying hormone and other peptides from electrically active bag cell neurons of *Aplysia*. *J. Neurophysiol.*, **43**, 488–98.

Subramanyam, O.V. (1973) Neuroendocrine control of calcium levels in the blood of *Ariophanta semirugata*, a terrestrial pulmonate snail. *Endocrinol. Experientia.*, **7**, 315–7.

Voigt, K.H.C. & Martin, R. (1986) Neuropeptides with cardioexcitatory and opioid activity in *Octopus* nerves. In: *Handbook of comparative aspects of opioid and related neuropeptide mechanism*, ed. G.B. Stefano. Boca Raton, pp. 127–138. Florida: Boca Raton.

Vreugdenhil, E., Geraerts, W.P.M., Jackson, J.F. & Joosse, J. (1985) The molecular basis of the neuroendocrine control of egg-laying behaviour in *Lymnaea*. *Peptides*, **6**, suppl. 3, 465–70.

Wells, M.J. & Mangold, K. (1980) The effects of extracts from neurosecretory cells in the anterior vena cava and pharyngo-ophthalmic vein upon the hearts of intact free-moving Octopuses. *J. Exp. Biol.*, **84**, 319–34.

Wijdenes, J. & Runham, N.W. (1977) Studies on the control of grwoth in *Agriolimax recitulatus* (Mollusca, Pulmonata). *Gen. & Comp. Endocrinol.*, **31**, 154–6.

With, N.D. de & Schors, R.C. van der (1986) Neurohormonal control of Na^+ and Cl^--metabolism in the pulmonate freshwater snail *Lymnaea stagnalis*. *Gen. & Comp. Endocrinol.*, **63**, 344–52.

Yang, H.-Y.T., Fratta, W., Majane, E.A. & Costa, E. (1985) Isolation, sequencing, synthesis and pharmacological characterization of two brain neuropeptides that modulate the action of morphine. *Proc. Nat. Acad. Sci. USA.*, **82**, 7757–61.

G. A. COTTRELL, N. W. DAVIES, J. TURNER & A. OATES

Actions and roles of the FMRFamide peptides in *Helix*

The name 'femerfamide' (FMRFamide) has been given (Greenberg 1982) to the molluscan neuropeptide Phenylalanyl–methionyl–arginyl–phenylalanyl–amide (Price & Greenberg 1977). It is a mnemonic for the contained amino acids and was suggested to avoid prejudicing views about its natural role(s). (During the period of its discovery it was called Cardioexcitatory Neuropeptide because this was one notable action of the peptide. See Price & Greenberg 1977.)

Price and Greenberg and their associates have shown that in molluscs there are several peptides chemically related to FMRFamide (see eg Price 1986): these are FLRFamide,[1] pQDPFLRFamide,[2] GDPFLRFamide[3] and SDPFLRFamide[4]. It is convenient to refer to these peptides with one name, as for instance are the opioid peptides, because like the opioid peptides, they appear to require a particular sequence of amino acids for activity and show a range of effects suggesting that they operate through several receptor types. Consequently we shall refer to the molluscan molecules as the femerfamide peptides.

With the opioid peptides the sequence Tyr–Gly–Gly–Phe (YGGF–) is required for full opioid activity, as with for example Met– and Leu–enkephalin, endorphin and dynorphin. This C–terminal sequence of amino acids has been called the message sequence by Charkin and Goldstein (1981). By contrast, in the femerfamide peptides the C–terminal sequence appears essential for activity, i.e. serves as the 'message-sequence': –Phe (or Tyr)–Met (or Leu)–Arg–Phe–NH_2 (Price 1986).

A number of workers have detected femerfamide-immunoreactivity in central neurones and fine neuronal processes (presumed release-sites), both centrally and in the periphery (Marchand, *et al.* 1982; Boer *et al.* 1980; Schot *et al.* 1984; see also Chapter 2, this volume).

---

[1] Phe–Leu–Arg–Phe–amide

[2] pGlu–Asp–Pro–Phe–Leu–Arg–Phe–NH_2

[3] Gly–Asp–Pro–Phe–Leu–Arg–Phe–NH_2

[4] Ser–Asp–Pro–Phe–Leu–Arg–Phe–NH_2

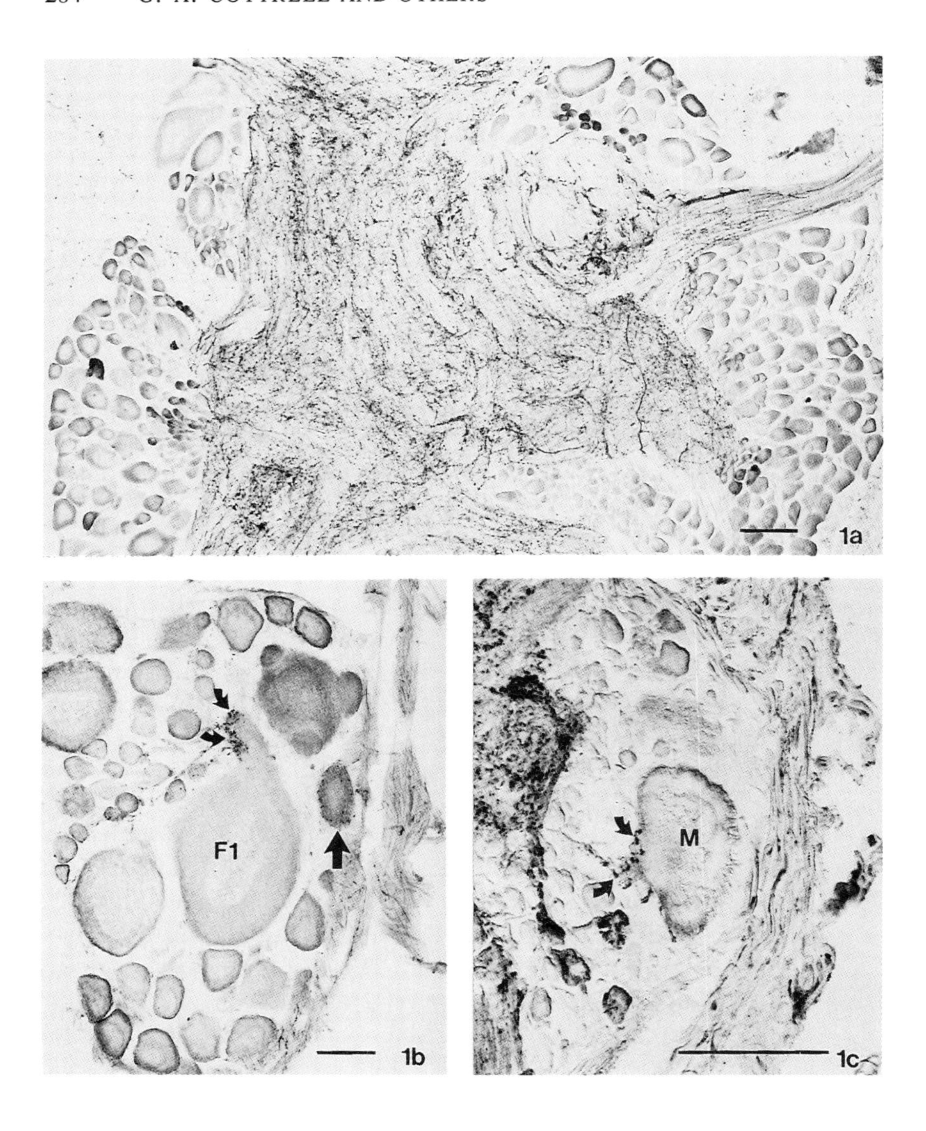

Fig. 1. (a) Right Parietal Ganglion; (b) Right Parietal Ganglion; (c) Buccal ganglion.

In cryostat sections of *Helix* ganglia, peroxidase immunohistochemistry with antiserum directed against FMRFamide shows large numbers of varicose axonal ramifications, many of which are presumed terminal processes with the majority of labelled profiles occurring in the neuropile (a). In addition, some neurones receive a perikaryal innervation of Fa-positive elements (b, large arrow). Some identifiable neurones (b, F1; c, M) are heavily innervated in the axon hillock region (small arrows b, c). Scale bars represent 100 μm.

One immunoreactive neurone in *Helix*, the C3 neurone, sends processes to the tentacle muscle of this snail (Cottrell *et al.* 1983). Stimulation of this single neurone causes contraction of the ipsilateral muscle and FMRFamide itself also induces contractions. This suggests that the femerfamide peptide may be a transmitter at neuro–muscular junctions. Femerfamide peptides may also serve as central transmitters, because they have potent actions on neuronal membranes (Cottrell *et al.* 1984; Boyd & Walker 1985).

In this paper a comparison is made of the actions of different femerfamide peptides on identified central neurones and on the tentacle muscle of *Helix,* and evidence presented for the existence of different receptors for these peptides. We also describe the results of a further histochemical study initiated in part to identify other large femerfamide containing neurones in *Helix,* and we discuss the problems associated with identifying neurones which contain femerfamide peptides.

Immunocytochemistry

Central Nervous System. Immunocytochemical staining with an antiserum generated against FMRFamide (kindly donated by Prof. G.J. Dockray, Liverpool) and another generated against YGGFMRFamide (kindly donated by Dr. D.A. Price, Florida) identified a large number of immunoreactive neuronal cell bodies and processes (Fig. 1.)

Sites of presumed synaptic contact between femerfamide-immunoractive (FaIR) neuronal elements and follower neurones were apparent throughout the ganglia. In addition to the profusion of labelled elements in neuropile regions (Fig. 1), FaIR–axonal, presumed terminal, profiles were seen in intimate association with the axon hillock regions of large and small neurones and investing the perikarya of some smaller cells (Fig. 1).

A search for large identifiable perikarya, which stained with the antiserum to FMRFamide, indicated that the large P or posterior cell of the buccal ganglion might contain the peptide: this neurone stained positively with both antisera but more intensely so with the 'FMRFamide antiserum' in sections fixed in both benzoquinone

Caption for Fig. 1 (contd)

In adsorption studies, the staining reaction was abolished by incubating both antisera for 24 h with 10 μg ml^{-1} FMRFamide, but was not affected by preincubation with equimolar or greater concentrations of metenkephalin–Arg–Phe, metenkephalin, CCK8 oxytocin, neurotensin or substance P. The coupling of FMRFamide to BSA renders the C–terminus unavailable for antibody recognition, and this and the adsorption studies with metenkephalin–Arg–Phe, metenkephalin, and CCK–8, which share parts of the FMRFamide sequence, indicate that the antisera recognised the –Arg–Phe NH_2 sequence of the peptide, and in common with most antisera generated against amidated peptides recognises preferentially, if not exclusively, the amidated form.

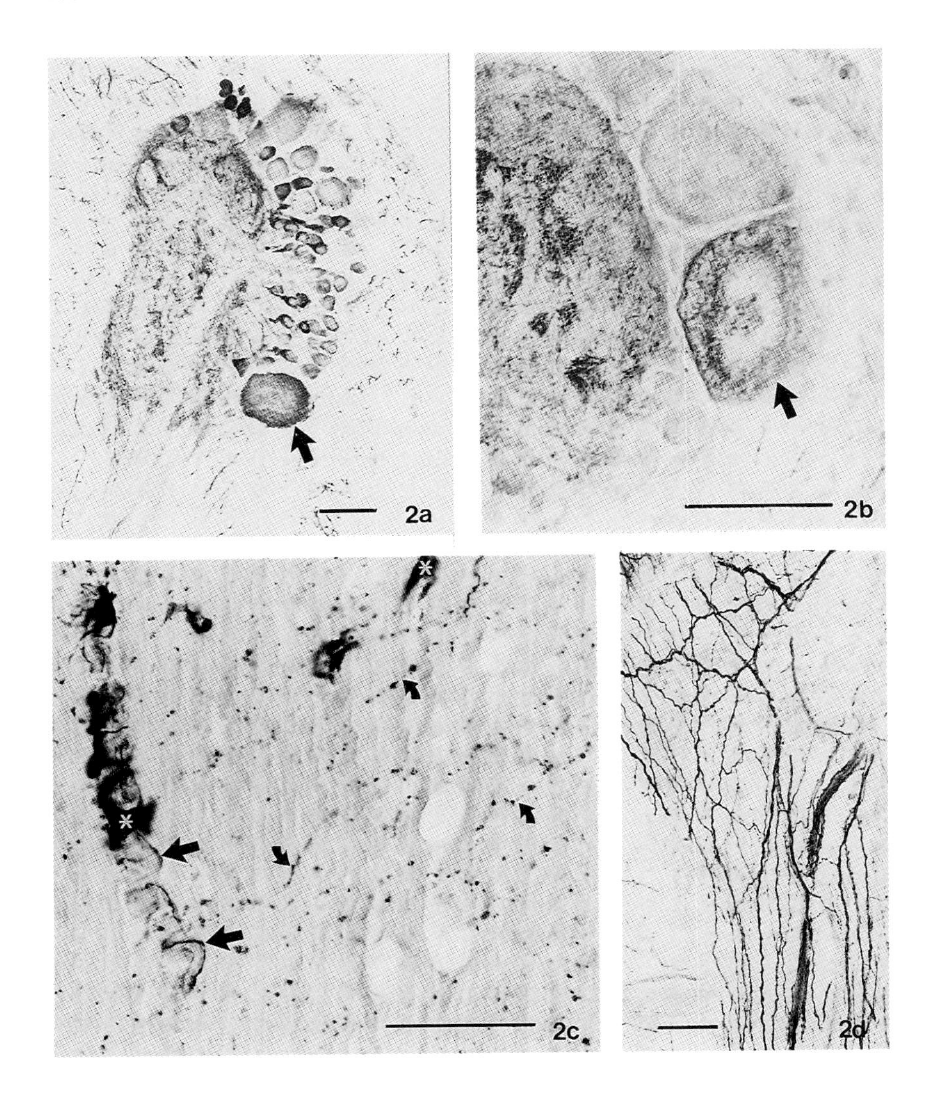

Fig. 2. (a) FMRFamide-like immunoreactivity in P–cells of buccal ganglion of *Helix;* benzoquinone fixed. (b) YGGFMRFamide-like immunoreactivity in P–cells of buccal ganglion of *Helix;* formaldehyde fixed.

Antisera raised against femerfamide peptides also stain numerous large nueronal perikarya in *Helix* CNS. For example, the posterior neurones (P–cells) in the buccal ganglia contained reaction product when stained with antiserum against both FMRFamide (a) and YGGFMRFamide (b). The staining with both antisera was evident after fixation with benzoquinone (e.g. a) or formalin (e.g. b). In (b) note how the adjacent M cell is not stained. (When extracts of dissected P cells were and formaldehyde (Fig. 2a, b). Although the staining was usually less intense than that observed in C3 and some other perikarya, it always exceeded background, unlike the two other large adjacent A and M cells (cf. Cottrell 1978). The P cell was positively identified by its unique somatic serotonergic innervation (Turner & Cottrell 1985).

Because the P neurone is readily identified, more than 15 RIA experiments were made on individually isolated P neurone perikarya. No femerfamide activity was detected in the extracts, nor in extracts prepared from several A or M cells (Price *et al.* 1985). This observation suggested yet again how cautiously judgements on the presence of particular peptides should be made solely on immunocytochemical evidence.

Elsewhere in the ganglia, numerous large cell bodies reacting positively with antiserum were observed, especially in the suboesophageal ganglia. These were, however, less easily identified individually, because other similarly size perikarya were located close to them. In the right parietal and visceral ganglia 2 prominent groups of small FaIR neurones occurred whose processes appeared to project through the neuropile region of the ganglia and out via the major nerve trunks. These neurones may be the source of the particularly rich FaIR innervation of the aorta, and of other blood vessels in the connective capsule surrounding the CNS. The FaIR–processes do not appear to project into the lumen of the aorta (Fig. 2d), but ramify extensively in the muscle layers of the vessel. The function of these processes is not yet known.

Tentacle muscle. Examination of sections of tentacle stained using the FMRFamide antiserum revealed the presence of immunoreactive axonal profiles in both the non-pigmented and the pigmented parts of the tentacle muscle (Fig. 2c). FaIR was also present in the tentacular nerve and ganglion. In the muscle, FaIR-processes were also seen in association with blood vessels (Fig. 2c).

(Caption for Fig. 2 contd)

assayed by RIA, no Femerfamide peptide could be detected (see text a, ×126; b, ×320). (c) In tentacle muscle, two kinds of FaIR processes were apparent. The first contributed to a dense innervation of the muscle itself, by varicose processes (small arrows), in which the inter-varicosity segment were only very lightly stained. The second consisted of thicker, smoother elements, associated with blood vessels penetrating the muscle (large arrows). White asterisks indicate muscle pigment. ×370. (d) Dense networks of FaIR nerve elements were associated with the musculature of blood spaces and major blood vessels associated with the central ganglionic mass. In the walls of the aorta shown here, large incoming fibre bundles give rise to a network of varicose fibres. These fibres are associated with the muscle layers only and do not penetrate into the lumens of the vessels. Scale bar represents 100 μm.

Fig. 3. Diagrammatic representation of the different types of membrane current responses evoked by FMRFamide on different *Helix* neurones. On the left-hand side of the figure, the shape and sign of each response is drawn as it would be recorded under voltage-clamp at the potential indicated (marked X) on the current-response/voltage curve represented on the right.

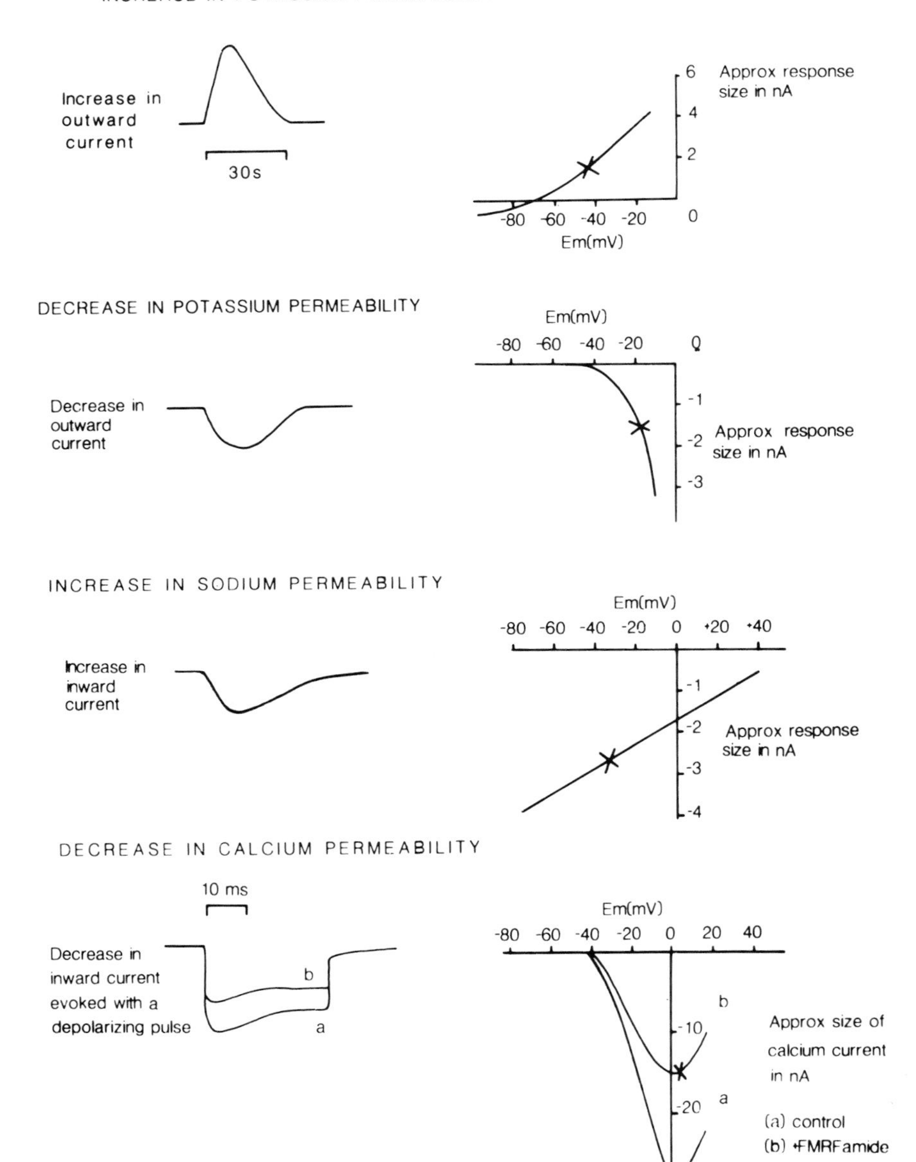

Actions of femerfamide-peptides on central neurones

Early electrophysiological experiments showed that FMRFamide has potent pharmacological actions on molluscan neurones (Cottrell 1978, 1980; Stone & Mayeri 1981; Walker *et al.* 1981). It is now clear that several different responses may be elicited by FMRFamide. These include: an increase in membrane permeability to K^+; a decrease in membrane permeability to K^+; an increase in membrane permeability to Na^+; a decrease in membrane permeability to Ca^{++}; and a change in levels of at least one intracellular messenger, which mediates, amongst other things, the increased permeability to K^+ (Cottrell 1982; Cottrell *et al.* 1984; Colombaioni *et al.* 1985). The different effects on membrane permeability are represented diagrammatically in Figure 3.

It has recently been shown that *Helix* ganglia contain at least two femerfamide peptides, FMRFamide itself and pQDPFLRFamide (Price *et al.* 1985). These were identified using a combination of ion-exchange chromatography HPLC and RIA techniques.

Recent findings suggest that femerfamide peptides mediate the responses by way of several different receptor sites. Evidence which led to this view was obtained by comparing the actions of different naturally occurring and synthetic femerfamide peptides on identified neurones which showed one or more of the responses described above (Cottrell & Davies 1986).

Marked differences have been observed in pharmacological activity between tetrapeptide femerfamides (FMRFamide, FLRFamide etc.) and the heptapeptide femerfamides (pQDPFLRFamide, YGGFMRFamide), but in every case it appears that the sequence –FM(L)RFamide is essential for activity. The differences in pharmacological activity between the tetrapeptides and the heptapeptides on the central neurones are as follows:

1. *Responses resulting from an increase in potassium permeability.* All of the femerfamide peptides tested evoked an increase in conductance to K^+ ions in susceptible neurones. However the response evoked by the heptapeptides was much more rapid in onset and decay than that evoked by the tetrapeptides. Such a difference in time-course of the responses could not be accounted for in terms of proximity of the tip of the micropipette used to apply the peptides because peptide release at different distances from the recorded cell evoked responses with a very similar time-course. On some occasions, when carefully positioning the pipette tip and with larger ejection pressure, it could be seen that the heptapeptides evoked both the fast response and also the slower response (Fig. 4). The fast response evoked by the heptapeptides is thus an extra response not represented separately in Figure 3.

2. *Response resulting from a decrease in potassium conductance.* Both the tetrapeptides and the heptapeptides evoked a decrease in K^+ conductance. This was, however, more easily observed with the heptapeptides, which suggests that they are more potent in this respect than the tetrapeptides (Fig. 5). However, neurones

exhibiting this response also exhibited the response resulting from an increase in potassium conductance and, as described above, the heptapeptides were much less potent in evoking this slow increase in potassium conductance, which normally can

Fig. 4. Application of pQDPFLRFamide (filled triangle) to a neurone in the right parietal ganglion evoked a response which was a combination of both fast and slow increase in conductance to potassium ions. The cell was clamped at −30 mV and the peptide was ejected onto the membrane by pressure. Each record shows the responses observed after placing the tip of the peptide pipette at different positions around the perikaryon of the neurone.

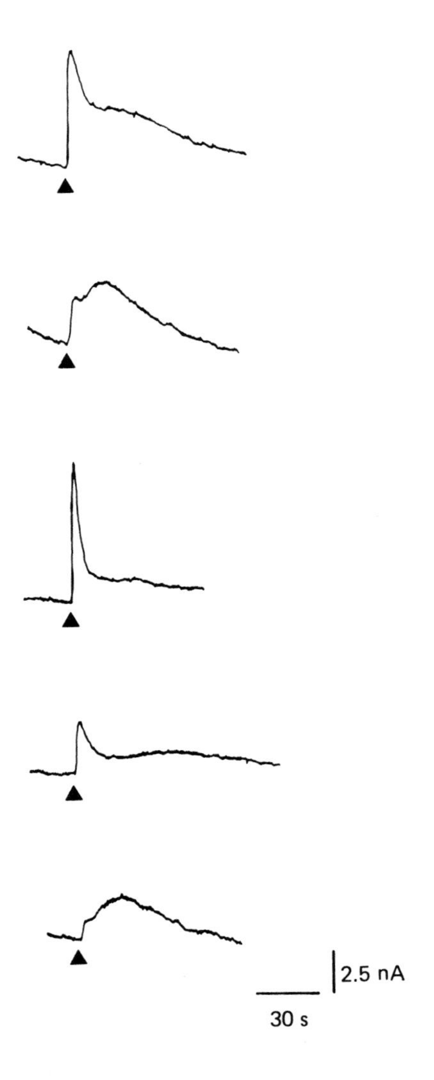

mask to some extent the decrease in conductance. Thus at least part of the apparent difference in potency between the tetrapeptides and the heptapeptides may be due to their marked difference in potency on the *slow increase* in potassium conductance.

3. *Responses resulting from an increase in sodium conductance.* FMRFamide and also FLRFamide and FIRFamide evoked an increase in Na^+ conductance. Repeated applications quite rapidly led to desensitization and, indeed, cross-desensitization occurred (Cottrell & Davies 1986). In contrast, neither pQDPFLRFamide nor YGGFMRFamide evoked this response. Even the pentapeptide PFLRFamide was inactive, but it did not block the effect of FMRFamide.

4. *Response resulting from a suppression of the voltage activated increase in calcium conductance.* There is no experimental information yet available as to the potencies of the different femerfamide compounds on this response (Colombaioni *et al.* 1985).

Different femerfamide receptors on neurones. Comparison of the activity of the different femerfamide agonists led to the view that there are at least 4 types of femerfamide receptors (A, B, C and D) on *Helix* neurones, but that the sequence –Phe(or Tyr)–Met(or Leu)–Arg–Phe–NH_2 is essential for activating each of them (Cottrell & Davies 1986).

According to this hypothesis, the response which involves a slow increase to gK, and which is more sensitive to the tetrapeptides than the heptapeptides, is mediated by

Fig. 5. Different responses induced in the C1 neurone by application of FMRFamide and pQDPFLRFamide. FMRFamide produced mainly a slow increase in conductance to potassium ions, whereas pQDPFLRFamide induced the voltage-dependent suppression of potassium conductance only. V_h, holding potential at which the recordings were made.

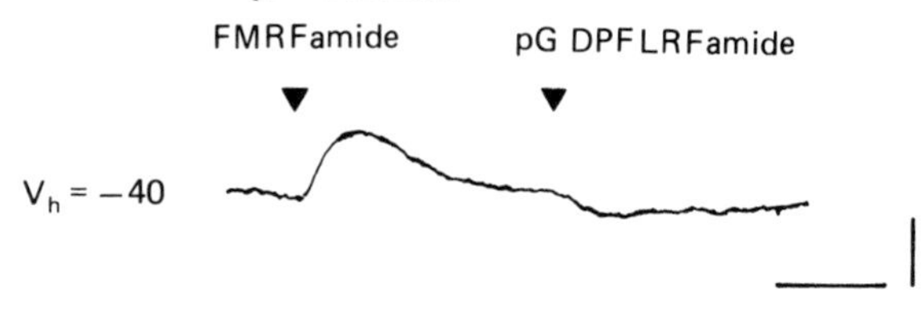

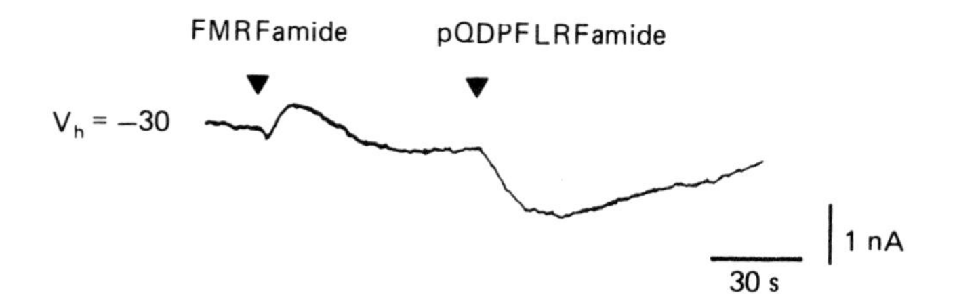

femerfamide receptor A (Fa_A); the fast increase in gK (which is activated only by the heptapeptides) is mediated by Fa_B; Fa_C, on the other hand, is only activated by the tetrapeptides, and is coupled to an increase in membrane conductance, primarily to Na^+; the fourth receptor Fa_D, is coupled to a reduction in gK, involves the formation of an intracellular messenger (Cottrell *et al.* 1984) and appears to be more sensitive to the heptapeptides than the tetrapeptides.

Effects of femerfamide peptides and some related peptides on the tentacle retractor muscle of Helix

Each cerebral ganglion in *Helix* includes a neurone which contains one or more femerfamide peptides. This was established by both immunocytochemistry and analyses of individually dissected neurones (Cottrell *et al.* 1983). The neurone perikaryon contains granulated vesicles similar to those identified by Nagle (1982) as containing FMRFamide in *Macrocallista*. The identified *Helix* neurone, labelled the C3 neurone, sends axonal processes to the tentacle retractor muscle, and intracellular stimulation of each C3 neurone results in contraction of the ipsilateral tentacle retractor muscle. Such observations suggest that one or more of the femerfamide peptides may act as neuromuscular transmitter substances.

Because of the possibility that one or more neuronally released femerfamide peptides may normally exert a physiological action on this muscle, detailed pharmacological experiments tested the effects of different femerfamide peptides and related compounds on the contractile activity of the muscle.

FMRFamide itself contracts the isolated tentacle retractor muscle at concentrations above about 1 nM. The response is, however, rather complex; after adding the pepide, there is invariably a delay of tens of seconds before the onset of the contraction and usually there follows a period during which the muscle exhibits phasic contractions. As the concentration is increased, the most noticeable effect observed under auxotonic recording conditions is an increase in frequency of the phasic contractions, with variable effects on the tension developed. Dose response curves (relating frequency of phasic contractions with log dose) for FnLRFamide, FIRFamide and AcFnLRFamide, are shown in Figure 6. The n–leucine derivative has a similar potency to FMRFamide itself and the acetylated derivative is only slightly less active. The derivative containing iso–leucine is, however, considerably less potent than the other peptides.

It was shown earlier that the amidated form of the vertebrate endogenous opioid YGGFMRF can excite the tentacle muscle at low concentrations and inhibit contractions, but relax the muscle at higher concentrations (Cottrell *et al.* 1983). This particular peptide is interesting in that when amidated it may be considered as a femerfamide peptide (since it incorporates –FMRFamide) or as met–enkephalin–Arg–Phe–amide. Low concentrations of the peptide induce contractions. With increasing

Fig. 6. Log dose response curves for FnLRFamide (filled circles), Acetyl–FnLRFamide (filled triangles) and FIRFamide (open circles). The response is expressed as the percentage of maximum increase in frequency of rhythmic contractions induced by the peptides. At high doses each peptide appears to have some inhibitory effect.

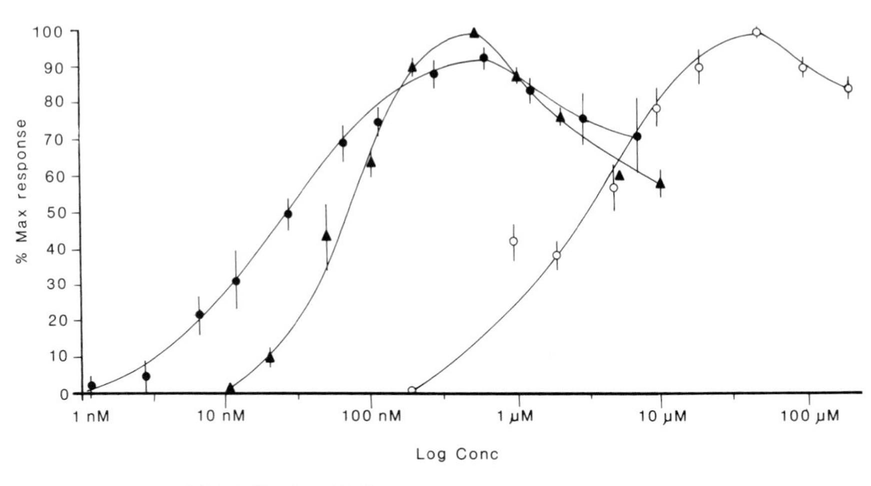

Fig. 7. Log dose response curve for YGGFMRFamide. The response is expressed as the percentage of the maximum interval in frequency of rhythmic contractions. At doses above about 100 μM the peptide exerts a pronounced inhibitory action.

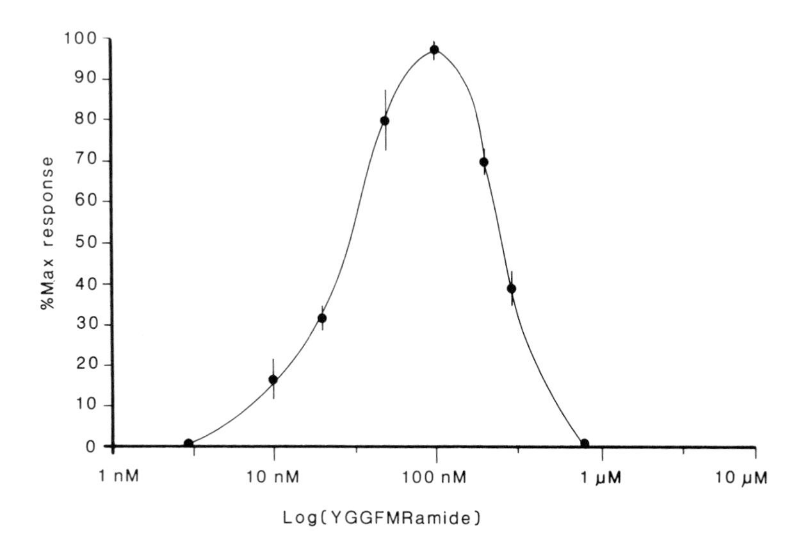

concentrations the frequency of contractions increases to a maximum at about 100 nM above which higher doses result in less frequent contractions and relaxation of the tone of the muscle (Fig. 7). Such higher doses of YGGFMRFamide prevent the stimulatory effect of FMRFamide and the other tetrapeptides (Cottrell *et al.* 1983).

The endogenous heptapeptide pQDPFLRFamide only relaxes the muscle (Fig. 8). The threshold for this effect is less than 1 nM. When tested on the tegument of the tentacle it is even more potent, causing relaxation at doses as low as 0.1 nM. As with YGGFMRFamide, pQDPFLRFamide also prevents contractions by the tetrapeptides. A summary of the relative potencies and overall effects of several femerfamide, and some other, peptides on the *Helix* tentacle muscle is shown in Table 1.

Discussion

Many of the results reported here, and those obtained by other workers, suggest that femerfamide peptides serve as neuronal messengers both centrally and in the periphery of molluscs. Although the immunocytochemical data support this view, the experiments with the 'P–neurone' of the buccal ganglion, which stained positively with antibodies directed against both FMRFamide (Dockray 1985) and against YGGFMRFamide (Price 1982), but which according to radioimmunoassay does not contain a femerfamide-peptide, suggests that caution should be taken in interpreting immunocytochemical data. In many preparations a gradation in intensity of immunoreactivity occurs amongst neurone perikarya. Presumably the most intensely

Fig. 8. Responses of an isolated tentacle muscle to low concentrations of pQDPFLRFamide.

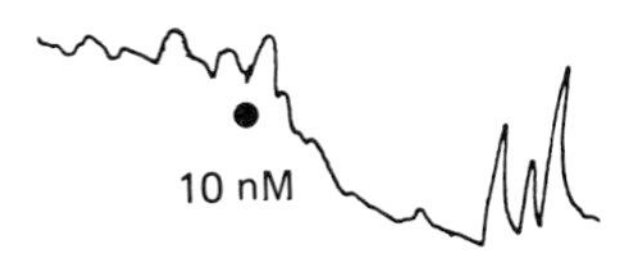

1 min

Table 1. *Comparisons of the potencies of different femerfamide, and related, peptides on the 'excitatory' and 'inhibitory' responses (see text) of the isolated tentacle retractor muscle of* Helix

		Excitatory			Inhibitory		
PEPTIDE	pD_2	Threshold concentration for rhythms (nM)	Concentration for maximum rate of beating (nM)	Maximum rate induced (/min)	Threshold for rhythm inhibition (nM)	Concentration for maximum inhibition (nM)	% inhibition
pQDPFLRFa	–	–	–	–	0.3	*	100
YGGFMRFa	7.54	3	100	10.3	100	700	100
FMRFa	7.02	5	1,000	6.0	1,000	20,000	17
FNLRFa	6.54	1	600	6.5	600	10,000	20
FIRFa	5.47	200	50,000	4.6	50,000	500,000	20
pQDFIRFa	6.05	200	3,000	3.0	5,000	30,000	19
AcFNLRFa	7.15	8	600	6.0	600	20,000	43
tBOC–FNLRFa	6.12	100	10,000	5.5	10,000	500,000	45
tBOC–βAFNLRFa	–	–	–	–	–	–	–
PENTAGASTRIN	–	–	–	–	–	–	–
LEU–ENKEPHALIN	–	–	–	–	–	–	–
YGGFMRF	–	–	–	–	–	–	–

reacting perikarya are the ones most likely to contain one or more of the femerfamide peptides in tissues. However, independent evidence is required to establish with certainty the identity of immunoreactivity in any cell. In tissues where no biochemical or pharmacological evidence exists for the presence of femerfamide peptides even greater care should be taken, because the gradation in staining here may be even more misleading.

It is now established that the femerfamide peptides exert a wide range of actions on central neurones. A similar diversity of actions has already been reported for several transmitters in molluscs – eg ACh (Kehoe 1972), dopamine (Ascher 1972; Berry & Cottrell 1975) and serotonin (Gerschenfeld & Paupardin–Tritsch 1974). The differential effects of the heptapeptides, compared with the tetrapeptides, suggests that there are probably several different central receptors for these peptides. It is particularly interesting that in some cases the action of the extended forms can be completely different from that of FMRFamide itself, for example neuronally released pQDPFLRFamide might be inhibitory on a particular neurone while FMRFamide itself might be excitatory.

Similarly there is a marked difference in effect of the two endogenous *Helix* femerfamide peptides on tentacle muscle. FMRFamide induces a delayed contracture and phasic contractions whereas pQDPFLRFamide potently relaxes the muscle. It is yet to be established which femerfamide peptide is contained in the C3 neurone innervating the muscle. It would also be very interesting to know whether the C3 neurone contains another transmitter. Certainly, it is difficult at present to reconcile the available data on the presence of the femerfamide peptide(s), and the effect of activating the C3–neurone on muscle tension (producing a smooth contraction), with the effects of the *known* endogenous femerfamide peptides described above.

Clearly, there is much still to be learned about these interesting compounds, and it is anticipated that their favourable localization in repeatedly identifiable neurones suitable for detailed electrophysiological, biochemical and molecular biological experiments will enable further direct experiments to be made, which should provide further insight into their actions and roles.

Acknowledgements

We thank the MRC and the SERC for financial support.

References

Ascher, P. (1972) Inhibitory and excitatory effects of dopamine on *Aplysia* neurones. *J. Physiol.*, **225**, 173–209.

Berry, M.S. & Cottrell, G.A. (1975) Excitatory inhibitory and biphasic synaptic potential mediated by an identified dopamine-containing neurone. *J. Physiol.*, **244**, 589–612.

Boer, H.H., Schot, L.P.C., Veenstra, J.A. & Reichelt, D. (1980) Immunocytochemical identification of neural elements in central nervous systems of a snail, some insects, a fish, and a mammal with an antiserum to the molluscan cardio-excitatory tetrapeptide FMRFamide. *Cell Tiss. Res.*, **213**, 21–27.

Boyd, P.J. & Walker, R.J. (1985) Actions of the molluscan neuropeptide FMRF–amide on neurones in the suboesophageal ganglia of the snail *Helix aspersa. Comp. Biochem. & Physiol.*, **81C**, 379–386.

Chavkin, C. & Goldstein, A. (1981) Specific receptor for the opioid peptide dynorphin: Structure activity relations. *Proc. Nat. Acad. Sci. USA.*, **78**, 6543–6547.

Colombaioni, L., Paupardin–Tritsch, D., Vidal, P.P. & Gerschenfeld, H.M. (1985) The neuropeptide FMRF–amide decreases both the Ca^{2+} conductance and a Cyclic 3',5'–adenosine monophosphate-dependent K^+ conductance in identified molluscan neurons. *J. Neurosci.*, **5(9)**, 2533–2538.

Cottrell, G.A. (1978) Actions of a "Molluscan cardio-excitatory neuro-peptide" on identified 5–hydroxytryptamine-containing neurones and their follower neurones in *Helix pomatia. J. Physiol.*, **284**, 130–131P.

Cottrell, G.A. (1980) Voltage dependent and voltage independent actions of the molluscan neuropeptide FMRFamide on snail neurone. *J. Physiol.*, **300**, 41P.

Cottrell, G.A. (1982) FMRFamide neuropeptides simultaneously increase and decrease K^+ currents in an identified neurone. *Nature*, **296**, 87–89.

Cottrell, G.A. & Davies, N.W. (1986) Multiple receptor sites for FMRFamide and related peptides. *J. Physiol*, **382**, 51–68.

Cottrell, G.A., Davies, N.W. & Green, K.A. (1984) Multiple actions of a molluscan cardioexcitatory neuropeptide and related peptides on identified *Helix* neurones. *J. Physiol.*, **356**, 315–333.

Cottrell, G.A., Greenberg, M.J. & Price, D.A. (1983) Differential effects of the molluscan neuropeptide FMRFamide and the related met–enkephalin derivative YGGFMRFamide on the *Helix* tentacle retractor muscle. *Comp. Biochem. & Physiol.*, **75C**, 373–375.

Cottrell, G.A., Schot, L.P.C. & Dockray, G.J. (1983) Identification and probable role of a single neurone containing the neuropeptide *Helix* FMRFamide. *Nature*, **304**, 638–640.

Dockray, G.J. (1985) Characterization of FMRF amide-like immunoreactivity in rat spinal cord by region-specific antibodies in radioimmunoassay and HPLC. *J. Neurochem.*, **45**, 152–158.

Gerschenfeld, H.M. & Paupardin–Tritsch, D. (1974) Ionic mechanisms and receptor properties underlying the response of molluscan neurones to 5–hydroxytryptamine. *J. Physiol.*, **243**, 427–456.

Greenberg, M.J. (1982) Personal communication.

Kehoe, J. (1972) Thre acetylcholine receptors in *Aplysia* neurones. *J. Physiol.*, **225**, 115–146.

Marchand, C.–R., Wijdens, J. & Schot, L.P. (1982) Localisation par la technique cyto–immuno–enzymologique d'un neuropeptide cardio-excitateur (le F.M.R.F.–amide) dans le collier peri-oesophagien d'*Helix aspersa* Muller (Gasteropode, pulmone, stylommatophore). First draft of paper sent in revised to C.R. Acad. Francaise.

Nagle, G.T. (1981) The Molluscan Cardioactive Neuropeptide FMRFamide: Subcellular localization in bivalve ganglia. *J. Neurobiol.*, **12**, 599–611.

Price, D.A. (1982) The FMRFamide–like peptide of *Helix. Comp. Biochem. & Physiol.*, **72**, 325–328.

Price, D.A. (1986) Evolution of a molluscan cardioregulatory neuropeptide. *Am. Zool.*, **26**, 1007–1015.

Price, D.A., Cottrell, G.A., Doble, K.E., Greenberg, M.J., Jorenby, W., Lehman, H.K. & Riehm, J.P. (1985) A novel FMRFamide-related peptide in *Helix:* pQDPFLRFamide. *Biol. Bull.*, **169**, 256–266.

Price, D.A. & Greenberg, M.S. (1977) Structure of a molluscan cardioexcitatory neuropeptide. *Science*, **197**, 670–671.

Price, D.A., Greenberg, M.J. & Cottrell, G.A. (1985) Unpublished.

Schot, L.P.C., Boer, H.H. & Montague–Wajer, C. (1984) Characterisation of multiple immunoreactive neurones in the central nervous system of the pond snail *Lymnaea stagnalis* with different fixatives and antisera adsorbed with the homologous and heterologous antigens. *Histochem.*, **81**, 373–378.

Stone, L.S. & Mayeri, E. (1981) Multiple actions of FMRFamide on identified neurones in the abdominal ganglion of *Aplysia*. *Soc. Neurosci. Abst.*, **7**, 636.

Turner, J. & Cottrell, G.A. (1985) Unpublished.

Walker, R.J., James, V.A. & Roberts, C.J. (1981) The action of FMRFamide and proctolin on *Helix, Hirudo, Limmulus* and *Periplaneta* neurones. In: Adv. Physiol. Sci. **20**, *Advances in Animal and Comparative Physiology*, eds. G. Pethes & V.L. Frenyo. Pergamon Presss, 411–416.

R. TAUSSIG, J.R. NAMBU & R.H. SCHELLER

Evolution of peptide hormones: an *Aplysia* CRF-like peptide

Introduction

The discovery that many central neurones utilize peptides as extracellular chemical messengers has revolutionized our understanding of neuronal signalling. Studies to characterize the structure and functions of neuropeptides have taken various approaches, including purification and biochemical analysis of the peptide products and molecular genetic studies of the genes encoding precursor proteins which give rise to peptide products. These investigations have been greatly aided by the use of non-neuronal tissues, such as epithelial tissue or digestive organs, which are often rich sources of bioactive peptides. Many peptides initially identified in peripheral tissues have been found subsequently in the central nervous system. One preparation, frog skin, has been particularly useful in this regard, and has facilitated the discovery of mammalian peptides related to frog bombesin (Orloff *et al.* 1984).

Invertebrate nervous systems offer unique advantages in the study of neurotransmitter function. Our understanding of the molecular mechanisms underlying neurotransmitter actions have been greatly facilitated by the use of invertebrate systems due to the smaller number of neurons, their simpler organization, and the often large size of their cell soma (see also Chapter 8). In terms of neuropeptide biology and chemistry a number of questions arise: Can neuropeptides related to vertebrate neuropeptides be found in invertebrates? Can neuropeptides characterized in invertebrate systems be used to identify homologous peptides in mammalian systems? Can invertebrate systems be used to gain further insight into the function, regulation, and evolution of neuroendocrine systems?

Several approaches have been employed to address these questions. Initially, investigators used antibodies directed against mammalian neuropeptides to localize immunoreactive material in invertebrate neurons. For example, radioimmunoassays and immunocytochemical tests detect the presence of arginine vasotocin-, oxytocin-, vasopressin-, neurotensin-, and insulin-like material in the nervous systems of coelenterates, molluscs, and/or arthropods (Grimmelikhuizen *et al.* 1982; Carraway *et al.* 1982; Moore *et al.* 1981; Duve & Thorpe 1979).

While the presence of immunoreactive material in these nervous systems suggests the presence of neuropeptides related to mammalian peptides, further structural information is necessary to validate the potential evolutionary relationships between these molecules. Several groups have complemented these data with biochemical techniques to characterize the immunoreactive material. For example it has been shown that arg–vasotocin-like and oxytocin-like material isolated from *Lymnaea* co-elute with synthetic peptide upon separation by high performance liquid chromatography (HPLC) or electrofocusing (Ebberink & Joosse 1986). In addition, an insulin-like substance from *Lymnaea* has a molecular weight similar to mammalian insulin, and competes with ^{125}I–insulin in binding to isolated rat fat cells (Ebberink & Joosse 1986).

Direct amino acid sequencing of peptides isolated from invertebrate species reveals homologies to mammalian peptides. For example Eledoisin, an amidated dodecapeptide purified from the cephalopod, *Eledone,* salivary gland, is related to substance P (Erspamer & Anastasi 1962). These two peptides share 50% overall homology and contain identical carboxy termini (Gly–Leu–Met–amide).

The amidated tetrapeptide Phe–Met–Arg–Phe–amide (FMRFamide), originally isolated from the venus clam, *M. nimbosa* (Price & Greenberg 1977), has recently attracted much interest among neurobiologists. FMRFamide immunoreactivity is detected throughout the animal kingdom, from coelenterates to mammals (Brown *et al.* 1984; Grimmelikhuijzen 1984; Sorenson *et al.* 1984). FMRFamide and FMRFamide-like peptides, for example Phe–<u>Leu</u>–Arg–Phe–amide (F<u>L</u>RFamide), have been subsequently isolated from the nervous sytems of many molluscs including *Helix, Aplysia,* and *Octopus* (Lehmen *et al.* 1982; Voigt *et al.* 1983; Price *et al.* 1985). Anti–FMRFamide antibodies have been used to purify a FMRFamide–like peptide from chicken brain with the sequence Leu–Pro–Leu–Arg–Phe–amide (Dockray *et al.* 1983).

In addition to the presence of immunoreactivity in the brain, FMRFamide has a number of physiological activities in many species. In molluscs, these include cardioactive responses (Painter 1982), effects on muscle contractions (Cottrell *et al.* 1983; Muneoka & Matsuura 1985), and modulation of ionic currents in neurones (Cottrell *et al.* 1984; Boyd & Walker 1985, see also Chapter 14 this volume). In rats, FMRFamide increases arterial blood pressure (Koo *et al.* 1982) and excites brain stem neurons (Gayton 1982). While immunoreactivity and physiological effects of FMRFamide have been demonstrated in mammals, authentic FMRFamide has not been detected in these species to date. Due to the structural similarities between FMRFamide and both the enkephalins and the extended enkephalin-like heptapeptide (Tyr–Gly–Gly–Phe–Met–Arg–Phe) isolated from bovine adrenal medulla, it has been proposed that these peptides may be evolutionarily related (Greenberg *et al.* 1981). Consistent with this idea is the intriguing observation that a computer search of the protein sequence data base reveals only four occurrences of the sequence Phe–Met–

Arg–Phe. While two of these randomly occur in prokaryotic proteins, FMRFamide and Met enkephalin–Arg–Phe are the only eukaryotic sequences which are revealed by this analysis.

The *Aplysia* FMRFamide precursor

An approach of our laboratory has been to clone genes from the marine mollusc *Aplysia californica* which encode the precursors for neuropeptides. The use of differential screening techniques has allowed us to isolate both cDNA and genomic clones specifically expressed in identified *Aplysia* neurons (Scheller *et al.* 1984). One group of these clones was subjected to DNA sequencing and shown to encode the amidated tetrapeptide FMRFamide (Schaefer *et al.* 1985; Taussig & Scheller, unpublished observations). Southern blot analysis demonstrates that the FMRFamide gene is present in a single copy in the *Aplysia* haploid genome. DNA sequencing of one genomic and five cDNA clones revealed that this gene comprises at least two exons: a large exon containing the entire 3' noncoding region and the bulk of the coding sequence, and at least one exon encoding the 5' untranslated region and the amino terminal 36 residues of the precursor. This gene arrangement is similar to those of other *Aplysia* neuropeptide genes which have been characterized (Taussig *et al.* 1984; Mahon *et al.* 1985; Taussig *et al.* 1985).

The predicted precursor protein encoded by this gene is schematized in Figure 1. The precursor is 597 residues in length and contains a hydrophobic signal sequence at the amino terminus, consistent with this being a secreted protein. The carboxy terminal portion of the molecule contains 28 copies of the pentapeptide Phe–Met–Arg–Phe–Gly. These peptides are flanked by basic residues which are recognition sites for proteolytic cleavage (Loh & Gainer 1983), and the glycine residues at the carboxy termini of these cleavage products serve as the substrates for the carboxy terminal amidation, thus producing the amidated tetrapeptide FMRFamide. In addition, spacer peptides (7 and 8 residues in length), resembling the sequence Ser–Val–Asp–Gly–Asp–Val–Asp, separate most of the FMRFamide tetrapeptides. Unlike the repeating FMRFamide residues, these spacer regions are not completely conserved. It is unclear at this time how these spacer peptides function; they may act as neuropeptides or serve to maintain proper spacing between FMRFamide units, assuring proper proteolytic processing and packaging of the precursor.

While we do not yet know whether all of these putative cleavage sites are actually utilized, labelling studies with ^{3}H–phenylalanine *in vivo* demonstrate that most of the isotope which is incorporated into low molecular weight protein is converted into material with co-migrates with synthetic FMRFamide under multiple fractionation schemes (Schaefer *et al.* 1985). This suggests that most of the FMRFamide tetrapeptides are indeed cleaved from the precursor.

In addition, there is a single copy of the related tetrapeptide F$\underline{L}$RFamide present amino terminally to the FMRFamide peptides. The presence of a single leucine-containing peptide with multiple copies of the methionine-containing peptide is reminiscent of the structure of the human enkephalin precursor which contains a single copy of Leu–enkephalin and seven copies of Met–enkephalin (Comb *et al.* 1982). This limited homology is not sufficient to allow one to distinguish an ancestral relationship between these molecules (divergent evolution) from convergent evolution.

The amino terminal portion of the precursor contains several regions flanked by potential proteolytic processing sites (paired basic residues or the putative signal sequence cleavage site). Several of the putative cleavage products share homologies with peptides cleaved from the human pro–opiomelanocortin (POMC) precursor. A predicted cleavage product of 12 amino acids (residues 45–56 of the FMRFamide precursor) is homologous (46%) at 6 positions along the 13 amino acid sequence of the α–melanocyte stimulating hormone (α–MSH). In addition, six positions of the corticotropin like intermediate-lobe peptide (CLIP) are homologous (27%) with a 22 amino acid peptide (residues 21–42) which immediately follows the putative signal

Fig. 1. Schematic representation of the FMRFamide precursor protein. The predicted precursor was deduced from the combined sequence information of the FMRFamide genomic and cDNA clones. A solid region flanked by positively charged amino acids (+) indicates a region of hydrophobic residues which may serve as the signal sequence. A bold arrow following this region denotes the putative cleavage position of the signal sequence. The 28 FMRF tetrapeptides are denoted by parallel horizontal lines. The single FLRF and the carboxy terminal LRF residues of another potential cleavage product are represented by parallel vertical lines. Predicted cleavages at single basic or dibasic residues are labelled with vertical lines or arrows respectively. NH_2 above cleavage sites indicates potential amidation sites. The crosshatched, the slashed lines, and the stippled regions denote the CRF, α–MSH and CLIP homologous sequences respectively.

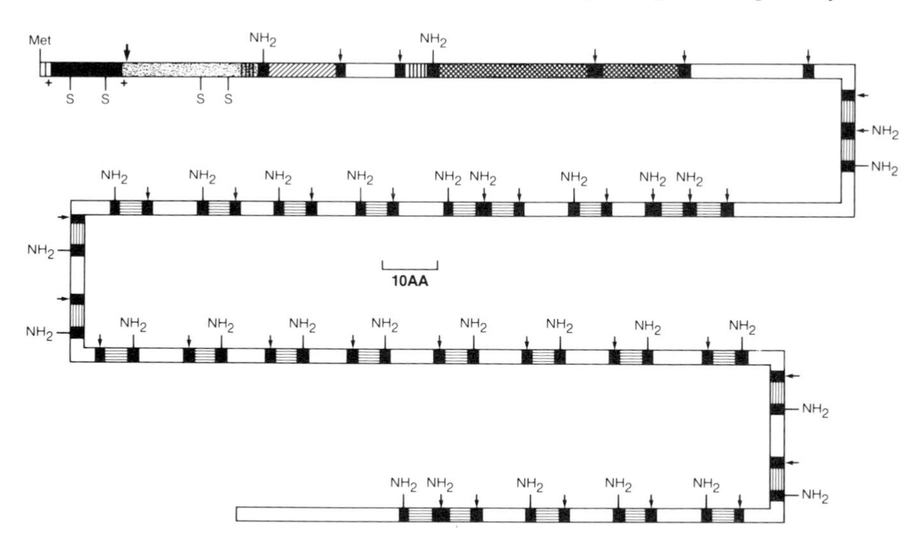

sequence cleavage site (see Figure 1). Whether these peptides are actually cleaved from the FMRFamide precursor is currently under investigation, as are experiments to elucidate the activities of these and other cleavage products.

The CRF-like peptide of *Aplysia*

In addition to POMC homologous peptides, the FMRFamide precursor contains a putative cleavage product which shares homology with ovine corticoptropin releasing factor (CRF). This 42 amino acid peptide contains 8 positional matches with ovine CRF at the amino terminal half of the molecule and 4 positional matches at the carboxyterminal portion resulting in an overall 30% homology. This homology as well as those to the related peptides, sauvagine and urotensin–1 are illustrated in Figure 2.

While a computer search of protein data bases reveals other proteins with homology to the *Aplysia* CRF-like peptide, it must be noted that we have limited our search to peptides and aligned amino and/or carboxy termini, which is equivalent to juxtaposing the cleavage sites of the precursors containing these molecules. By maintaining the alignment of the amino terminal 20 residues of *Aplysia* CRF-like peptide and ovine CRF, and not allowing for random sliding of these sequences, we calculate a probability of finding 8 or more positional matches of approximately 3×10^{-6} (see figure legend 2). Similarly, if we calculate the probability of finding 4 or more positional matches out of 20 (the observed homology between the 20 carboxy terminal residues of ovine CRF and the *Aplysia* CRF-like molecule) we obtain a value of 0.013. The combined probability of finding these observed homologies is approximately 4×10^{-8}. Therefore, it is likely that the observed homology is not fortuitous, and indeed represents an ancestral relationship between these peptides.

Comparison of the sequences of the *Aplysia* CRF-like molecule to the peptides related to ovine CRF are also noted. The *Aplysia* CRF-like peptide is homologous to sauvagine at 6 positions and to urotensin–1 at 8 positions. From these data, we can estimate the evolutionary rates of divergence of CRF-like peptides. Given the divergence times for molluscs (M), fish (F), amphibia (A), and mammals (L) (M to F, A, L 600 million years (MY); F to A, L = 350–400 MY; A to L = 300–350 MY), we calculate an expected divergence rate for CRF of approximately one amino acid substitution every 17.5 MY. This predicts that the *Aplysia* sequence would share 7 amino acids with the mammalian, frog, or fish sequences. The observed homologies to the *Aplysia* sequence are of the order of what we would predict. It is interesting to note that the same divergence rate was determined for another *Aplysia* neuropeptide, the egg-laying hormone (Nambu & Scheller 1986).

It should also be noted that the highest levels of homology are located in the amino terminal 20 residues. For example, ovine CRF and carp urotensin 1 are homologous at 14 positions in the amino terminal half and only 7 positions in the carboxy terminal half of the molecule. Similarly, urotensin–1 and sauvagine contain 13 positional matches in the amino terminal 20 residues of *Aplysia* CRF-like peptide and ovine CRF versus only 4 in the carboxy terminal region is consistent with this observation. This suggests that higher evolutionary constraints have been imposed on the amino terminal portion of this molecule, perhaps reflecting the functional importance of these residues. However, the presence of four consecutive basic residues at positions 24–37 in the *Aplysia* CRF-like peptide leaves open the possibility that this molecule is cleaved *in vivo* into two smaller products; this may in part be responsible for the lower levels of homology between the carboxytermini of the *Aplysia* CRF-like peptide and ovine CRF. If this cleavage does indeed occur, two independently acting products may arise from the region homologous to vertebrate CRF.

Fig. 2. Homologies to vertebrate peptides. Homologies of a 42 amino acid peptide from the *Aplysia* FMRFamide precursor (residues 75–116) to ovine corticotropin releasing factor (CRF) (Furutani *et al.* 1983), frog sauvagine (Montenucchi *et al.* 1983), frog sauvagine (Montenucchi *et al.* 1979), and carp urotensin–1 (Ichikawa *et al.* 1982). The candidate *Aplysia* peptide is flanked by paired basic residues which may serve as processing sites. The standard single letter amino acid code is used to denote the sequences of the peptides. Asterisks above residues in bold denote amino acids in the *Aplysia* peptide which are homologous to at least 2 of the vertebrate peptides. Crosses over bold residues indicate homologous amino acids between the *Aplysia* peptide and one vertebrate peptide. Crosses over bold residues indicate homologous amino acids between the *Aplysia* peptide and one vertebrate peptide. Significance of the observed homologies are strengthened by the fact that amino or carboxy termini were juxtaposed, not allowing for random alignments with known peptide sequences. Alignment of the amino terminal 20 amino acids of the ovine and *Aplysia* CRF peptides result in 8 positional matches. The random chance of getting 8 or more positional matches under the constraint of maintaining alignment of the ends of the peptides is approximated by the expression

$$\gamma = \sum_{i=n\ m} \left(\frac{19}{20}\right)^{m-n} \left(\frac{1}{20}\right)^{n} \binom{m}{n} = 3 \times 10^{-6}$$

(where n = number of positional matches, M = size of the peptide). Further alignment of the carboxy terminal 20 residues reveals 4 matches, = .013. The product of the amino and carboxy terminal probabilities gives a total probability, $= 4 \times 10^{-8}$

```
                     + * + *   *                 *   +     *
Aplysia CRF         -S Q E P D I E D Y A R A I A L I E S E E
Ovine CRF           -S Q E P P I S L D L T F H L L R E V L E
Frog Sauvagine      -  Q G P P I S I D L S L E L L R K M I E
Carp Urotensin I    -N D D P P I S I D L T F H L L R N M I E
                     1                                     20

                           +         +       +       *   *     +
                     P L Y R K R R S A D A D G Q S E K V L H R A
                     M - T K A D Q L A Q Q A H S N R K L L D I A-NH2
                     I - E K Q E K E K Q Q A A N N R L L L D T I-NH2
                     M - A R N E N Q R E Q A G L N R K Y L D E V-NH2
                     21                                         42
```

There are 5 positions where identical amino acids occur in all peptides. Previous structure/function studies on CRF demonstrate that the leucine residues at positions 14 & 15 and 38 & 39 (as numbered in Figure 2) are required for biological activity (Vale *et al.* 1983). Two of these functionally critical residues (positions 15 and 39) are conserved in the *Aplysia* sequence, and the other differ by conservative amino acid substitutions (Alanine and Valine at positions 14 and 38 respectively). Despite all these similarities, it is unlikely that a mammalian CRF antibody or a DNA probe encoding the CRF precursor protein would cross react with the *Aplysia* CRF-like peptide or the FMRFamide gene. Thus, determination of this precursor structure not only allowed the identification of known and/or novel neuropeptides, but also demonstrated evolutionarily related products whose homologies are no longer sufficient for direct immunochemical detection.

The disparate levels of amino acid homology between *Aplysia* CRF-like peptide and the various vertebrate CRF peptides are both perplexing and intriguing. *Aplysia* CRF-like peptide is clearly most homologous to ovine CRF, yet the molluscan phylum is presumably equally divergent from the three different vertebrate classes: osteichthyes, mammalia, and amphibia. Given constant rates of molecular evolution, one would expect similar levels of sequence homology. This discrepancy could be merely the result of stochastic processes which have occurred over many millions of years of evolution; however, the nearly twofold difference in homology between the *Aplysia* CRF-like peptide and those of ovine and frog seems to imply that different selection pressures have acted on this molecule in the different lineages. Thus, the CRF related peptides of fish and amphibians have had different functional constraints imposed on them in the time preceding divergence of the mammalian lineage. In this respect, it will be interesting to determine whether *Aplysia* has CRF-like peptide receptors more similar to mammalian CRF than to those of urotensin or sauvagine.

In mammals, CRF is synthesized and secreted by neuroendocrine cells of the hypothalamus and induces release of POMC peptides from the anterior pituitary gland (for review see Vale *et al.* 1983). CRF has a number of additional actions, including the production of hypotension. Interestingly, sauvagine and urotensin are equally active in eliciting release of POMC peptides (Vale *et al.* 1981) and are actually more potent in inducing hypotension than CRF itself (MacCannell *et al.* 1982). Urotensin I is thought to be important in regulating salt and water balance in fish (Marshall & Bern 1981). In *Aplysia*, the activities of the CRF-like peptide are unknown, but several of the neurons which express the FMRFamide gene innervate cells of the outer body wall epithelium and stimulate the release of mucus (Rayport *et al.* 1983). It is possible that the *Aplysia* CRF-like peptide could serve in an autocrine mode, influencing release of other peptides derived from the FMRFamide precursor, including those with POMC homologies. This situation would be analagous to another *Aplysia* peptide, alpha bag cell peptide, which is thought to produce an autoregulatory effect on the bag cell neurons of the abdominal ganglion (Rothman *et al.* 1983; J. Kauer & L. Kaczmarek, personal communication).

Finally, it is interesting to note that while in mammals, CRF, the POMC peptides, Met–enkephalin, Met–enkephalin–Arg–Phe, and Leu–enkephalin are encoded by as many as 4 different genes (Comb *et al.* 1982; Kakidani *et al.* 1982; Whitefeld *et al.* 1982; Shibahara *et al.* 1983), there is a single gene in *Aplysia* which gives rise to a unique precursor containing peptides with homologies to each of these sequences (Taussig & Scheller, unpublished observations). This finding suggests strongly that the different vertebrate genes arose from a single common ancestor through gene duplication, and that subsequent divergence altered their precise coding capacities and patterns of expression. The ancestral relatedness of the POMC and CRF gene has been suggested previously by Furutani *et al.* (1983).

The different peptide products cleaved from the *Aplysia* FMRFamide precursor may have actions on distinct target tissues, serve to elicit different responses from the same tissue, or act in combination to generate a coordinated response. While the *Aplysia* CRF-like peptide may have a very different function from that of the vertebrate CRF molecules, it is likely that the molecular mechanisms underlying both their biosynthesis and actions are very similar. It is our hope that further investigations of the CRF-like peptide in *Aplysia* may provide insight into these shared mechanisms. Further characterization of the structures and activities of peptides derived from the *Aplysia* FMRFamide precursor will certainly further our understanding of the function and evolution of the neuroendocrine system.

References

Boyd, P.J. & Walker, R.J. (1985) Actions of the Molluscan neuropeptide FMRF–amide on neurones in the suboesophageal ganglia of the snail *Helix aspersa. Comp. Biochem. Physiol.*, **81C**, 379–386.

Brown, R.O., Basbaum, A.I. & Mayeri, E. (1984) Identification of FMRFamide immunoreactivite neurons in the abdominal ganglion of *Aplysia. Soc. Neurosci. Abstr.*, **10**, 691.

Carraway, R., Ruane, S.E. & Kim, H. (1982) Distribution and immunochemical character of neurotensin-like material in representative vertebrates and invertebrates: apparent conservation of the COOH–terminal region during evolution. *Peptides*, **3**, 115–123.

Comb, M., Seeburg, P.H., Adelman, J., Eiden, L. & Herbert, H. (1982) Primary structure of the human Met– and Leu–enkephalin precursor and its mRNA. *Nature*, **295**, 663–666.

Cottrell, G.A., Greenberg, M.J. & Price, D.A. (1983) Differential effects of the molluscan neuropeptide FMRFamide and the related Met–enkephalin derivative YGGFMRFamide on the *Helix* tentacle retractor muscle. *Comp. Biochem. Physiol.*, **75**, 373–375.

Cottrell, G.A., Davies, N.W. & Green, K.A. (1984) Multiple actions of a molluscan cardioexcitatory neuropeptide and related peptides on identified *Helix* neurons. *J. Physiol.*, **365**, 315–333.

Dockray, G.J., Reeve, J.R., Shively, J., Gayton, R.J. & Barnard, C.S. (1983) A novel active pentapeptide from chicken brain identifed by antibodies to FMRFamide. *Nature,* **305**, 328–330.

Duve, H. & Thorpe, A. (1979) Immunofluorescent localization of insulin-like material in the median neurosecretory cells of the blowfly, *Calliphora vomitoria* (Diptera). *Cell Tiss. Res.*, **200**, 187–191.

Ebberink, R.H.M. & Joosse, J. (1986) Molecular properties of various snail peptides from brain and gut. *Peptides,* **6** (suppl. 3), 451–458.

Erspamer, V. & Anastasi, A. (1962) Structure and pharmacological actions of Eledoisin, the active endecapeptide of the posterior salivary glands of *Elodone. Experientia,* **18**, 58–59.

Furutani, Y., Morimoto, Y., Shibahara, S., Noda, M., Takahashi, H., Hirose, T., Asai, M., Inayama, S., Hayashida, H., Miyata, T. & Numa, S. (1983) Cloning and sequence analysis of cDNA for ovine corticotropin-releasing factor precursor. *Nature,* **301**, 536–540.

Gayton, R.J. (1982) Mammalian neuronal actions of FMRFamide and the related opioid Met–enkephalin–Arg6–Phe7. *Nature,* **298**, 275–276.

Greenberg, M.J., Painter, S.D. & Price, D.A. (1981) The amide of the naturally occurring opioid (met)enkephalin–Arg^6–Phe^7 is a potent analog of the molluscan neuropeptide FMRFamide. *Neuropeptides,* **1**, 309–317.

Grimmelikhuijzen, C.J.R. (1984) FMRFamide immunoreactivity is generally occurring in the nervous system of coelenterates. *Histochem.,* **78**, 361–381.

Grimmelikhuijzen, C.J.P., Dierickx, K. & Boer, G.J. (1982) Oxytocin /Vasopressin-like immunoreactivity is present in the nervous system of hydra. *Neurosci.,* **7**, 3191–3199.

Ickikawa, T., McMaster, D., Lederis, K. & KJobayashi, H. (1982) Isolation and amino acid sequence of Urotensin I, a vasoactive and ACTH-releasing neuropeptide, from the carp (*Cyprinus carpio*) urophysis. *Peptides,* **3**, 859–867.

Kakidani, H., Furutani, Y., Takahashi, H., Noda, M., Morimoto, Y., Hirose, T., Asai, M., Inayama, S., Nakanishi, S. & Numa, S. (1982) Cloning and sequence analysis of cDNA for porcine – neo-endorphin/dynorphin precursor. *Nature,* **298**, 245–249.

Koo, A., Chan, W.S., Ng, W.H. & Greenberg, M.J. (1982) Microvascular vasodilatory effect of FMRFamide and Met–enkephalin–Arg^6–Phe^7amide in the rat. *Microcirculation,* **2**, 393–412.

Lehman, H.K., Price, D.A. & Greenberg, M.J. (1982) The FMRFamide-like neuropeptide of *Aplysia* is FMRFamide. *Biol. Bull.,* **167**, 460–466.

Loh, Y.P. & Gainer, H. (1983) Biosynthesis and processing of neuropeptides. In: *Brain Peptides,* eds. D.T. Krieger, M.J. Brownstein & J.B. Martin, pp. 79–116. New York: John Wiley & Sons.

MacCannell, K.L., Lederis, K., Hamilton, P.L. & Rivier, J. (1982) Amunine (ovine CRF), Urotensin I and Sauvagine, three structurally-related peptides, produce selective dilation of the mesenteric circulation. *Pharmacology,* **25**, 116–120.

Mahon, A.C., Nambu, J.R., Taussig, R., Shyamala, M., Roach, A. & Scheller, R.H. (1985) Structure and exprerssion of the egg-laying hormone gene family in *Aplysia. J. Neurosci.* **5**, 1872–1880.

Marshall, W.S. & Bern, H.A. (1981) Active chloride transport by the skin of a marine teleost is stimulated by Urotensin I and inhibited by Urotensin II. *Gen. Comp. Endocrinol.,* **43**, 484–491.

Montecucci, P.C., Henschen, A. & Erspamer, V. (1979) Structure of sauvagine, a vasoactive peptide from the skin of a frog. *Hoppe Zeyler's Z. Physiol. Chem.,* **360**, 1178.

Moore, G.J., Thornhill, J.A., Gill, V., Lederis, K. & Lukowiak, K. (1981) An arginine vasotocin-like neuropeptide is present in the nervous system of the marine mollusc *Aplysia californica*. *Brain Res.*, **206**, 213–218.

Muneoka, Y. & Matsuura, M. (1985) Effects of the molluscan neuropeptide FMRFamide and the related opioid peptide YGGFMRFamide on *Mytilus* muscle. *Comp. Biochem. Physiol.*, **81**, 61–70.

Nambu, J.R. & Scheller, R.H. (1986) Egg-laying hormone genes of *Aplysia* evolution of the ELH gene family. *J. Neurosci.*, in press.

Orloff, M.S., Reeve, J.R., Ben–Avram, C.M., Shively, J.E. & Walsh, J.H. (1984) Isolation and sequence analysis of human bombesin-like peptides. *Peptides*, **5**, 865–870.

Painter, S.D. (1982) FMRFamide inhibition of a molluscan heart is accompanied by increases in cyclic AMP. *Neuropeptides*, **3**, 19–27.

Price, D.A. & Greenberg, M.J. (1977) Purification and characterization of a cardioexcitatory neuropeptide from the central ganglia of a bivalve mollusc. *Prep. Biochem.* **7**, 261–281.

Price, D.A., Cottrell, G.A., Doble, K.E., Greenberg, M.J., Jorenby, W., Lehman, H.K. & Riehm, J.P. (1985) A novel FMRF-related peptide in *Helix:* pQDPFLRFamide. *Bio. Bull.*, **169**, 256–266.

Rayport, S.G., Ambron, R.T. & Babiarz, J. (1983) Identified cholinergic neurons R2 and LP1 control mucus release in *Aplysia*. *J. Neurophysiol.*, **49**, 864–876.

Rothman, B.S., Mayeri, E., Brown, R.O., Yuan, P.M. & Shively, J.E. (1983) Primary structure and neuronal effects of alpha-bag cell peptide, a second canditate transmitter encoded by a single gene in bag cell neurons of *Aplysia*. *Proc. Natl. Acad. Sci. USA*, **80**, 5753–5757.

Schaefer, M., Picciotto, M.R., Kreiner, T., Kaldany, R.R., Taussig, R. & Scheller, R.H. (1985) *Aplysia* neurons express a gene encoding FMRFamide Neuropeptides. *Cell*, **41**, 457–467.

Scheller, R.H., Kaldany, R.R., Kreiner, T., Mahon, A.C., Nambu, J.R., Schaefer, M. & Taussig, R. (1984) Neuropeptides: mediators of behaviour in *Aplysia*. *Science*, **225**, 1300–1308.

Shibahara, S., Morimoto, Y., Furutani, Y., Notake, M., Takahashi, H., Shimizu, S., Horikawa, S. & Numa, S. (1983) Isolation and sequence analysis of the human corticotropin-releasing factor precursor gene. *EMBO J.*, **2**, 775–779.

Sorenson, R.L., Sasek, C.A. & Elde, R.P. (1984) Phe–Met–Arg–Phe–amide (FMRF–NH2) inhibits insulin and somatostatin secretion and anti-FMRF–NH2 sera detects pancreatic polypeptide cells in the rat islets. *Peptides*, **5**, 777–782.

Taussig, R., Kaldany, R.R., Rothbard, J.B., Schoolnik, G. & Scheller, R.H. (1985) Expression of the L11 neuropeptide gene in the *Aplysia* central nervous system. *J. Comp. Neurol.*, **238**, 53–64.

Taussig, R., Picciotto, M.R. & Scheller, R.H. (1984) Two introns define functional domains of a neuropeptide precursor in *Aplysia*. In: *Molecular Biology of Development*, **19**. Eds. E.H. Davidson & R.A. Firtel, pp. 551–560. New York: Alan R. Liss, Inc.

Vale, W.W., Rivier, C., Spiess, J. & Rivier, J. (1983) Corticotropin releasing factor. In: *Brain Peptides*. eds. D.T. Kreiger, M.J. Brownstein & J.B. Martin, pp. 961–974. New York: John Wiley & Sons.

Voigt, K.H., Geis, R., Kickling, C. & Martin, R. (1983) Homologous sequences of bioactive peptides from vertebrate opioid precursor systems in nervous tissues of *Octopus vulgaris*. In: *Handbook of Comparative Aspects of Opioid and Related Neuropeptide Mechanisms*, ed. G.B. Stefano. Boca Raton: CRC Press.

Whitefield, P.L., Seeburg, P.H. & Shine, J. (1982) *The human pro-opiomelanocortin gene: organization, sequence, and interspersion with repetitive DNA. DNA*,**1**, 133–143.

INDEX